536 TYLER

Heat & thermodynamics

T
K

...k is to be returned on or before

HEAT AND THERMODYNAMICS

being a second edition of

INTERMEDIATE HEAT

By

F. TYLER

B.Sc., Ph.D., F.Inst.P.

Second Master, and Senior Science Master,
Queen Elizabeth's Grammar School, Blackburn

Member of the panel of Senior Examiners for
Advanced and Special Physics for the Joint
Matriculation Board of the Northern Universities

LONDON

EDWARD ARNOLD (PUBLISHERS) LTD.

© *F. Tyler, 1964*
First published 1949
Reprinted 1952, 1955, 1957, 1959, 1962
Second Edition 1965

By the same author:

A LABORATORY MANUAL OF PHYSICS
GENERAL PHYSICS:
    HEAT, LIGHT AND SOUND
    MAGNETISM AND ELECTRICITY
    MECHANICS AND HYDROSTATICS
PROBLEMS IN PHYSICS
EXAMPLES IN PHYSICS
MECHANICS AND PROPERTIES OF MATTER
ALTERNATIC CURRENTS AND ELECTRONICS

Made and printed in Great Britain by William Clowes and Sons, Limited
London and Beccles

PREFACE

THE re-planning of the A-Physics syllabuses of the various examining Boards has provided me with the opportunity of revising my " Intermediate Heat " of which this current volume is a second edition. It has not been felt necessary to make radical over-all changes in the textual matter of the first publication. The coverage provided there is still fully adequate to embrace the varied and widespread contents of the different syllabuses, whilst the extra material contained in such sections as the Kinetic Theory of Gases, Radiation, Conduction, and Thermodynamics should continue to meet the demands of open scholarship candidates and others.

Where it was thought necessary sections have been re-written, and a number of minor alterations and corrections have been effected. The main change, however, is in the selection and re-grouping of the chapter questions where almost 200 new questions, selected from recent papers of the Examining Boards and the Oxford and Cambridge Group Scholarship Papers, have been included. Help in providing solutions to these questions has been generously given by two of my senior students, D. J. Ashton and G. H. Redman, and I am most happy here to acknowledge their able assistance.

Finally, I would like to express my appreciation to the many users of " Intermediate Heat " for their support and for the many encouraging letters and valued comments received during the period of its publication: I sincerely trust that in its present form, and under its new title, it will continue to serve their needs.

F. T.

Blackburn,
August, 1964

Key to reference letters of questions from examination papers:

A. Associated Examining Board for G.C.E. examinations.

C. University of Cambridge Local Examinations Syndicate.

L. University of London School Examinations.

N. Northern Universities' Joint Matriculation Board.

O. Oxford Local Examinations Board.

O. and C. Oxford and Cambridge Schools Examinations Board.

S. Southern Universities' Joint Board for School Examinations.

W. Welsh Joint Education Committee.

[ ] Denotes S-paper question.

Camb. Schol. Cambridge University Group Scholarship Papers.

Oxford Schol. Oxford University Group Scholarship Papers.

Other sources—as referenced in text.

CONTENTS

CHAPTER XIV. SOME ASPECTS METEOROLOGICAL PHYSICS   Page 378

CHAPTER I

THE MEASUREMENT OF TEMPERATURE

1.01 Introductory

The sense of touch enables us to assess in an approximate way the relative "hotness" of bodies. Judgments made in this way are very unreliable however, as the sense of touch is a variable criterion and is not capable of distinguishing between the effects due to "hotness" and the conductivity of the body. Thus due to the better conductivity of metals, a piece of brass or iron will seem colder to the touch than the wood of the bench on which it is placed. It is thus manifestly impossible to obtain a precise estimate of the "degree of hotness" of a body by methods relying on sense impressions.

Physical science is largely concerned with the accurate measurement of quantities. Accurate experimental data is an essential pre-requisite for the development of theory, and at a later stage provides the means of checking and assessing the reliability of such theories. In that branch of physics dealing with the study of heat it is clear therefore that something better than the rough estimate of the relative "degree of hotness," or *temperature* as it is called, is required than that provided by physiological sensations. The problem is therefore to construct a scale of temperature which shall be precise and consistent, and to devise a means of assessing the temperature of a body against this scale with a high degree of accuracy.

1.02 Temperature Scales

In the construction of a temperature scale the first necessity is to chose two fixed temperatures, the difference between which defines the *fundamental interval*. This interval is then divided into a number of small divisions, each known as a "degree of temperature." The numerical value of any degree of temperature will depend on the number of subdivisions in the fundamental interval, and the fixing of the arbitrary zero of the temperature scale. The two fixed points chosen should be constant and easily reproducible at all times under specified conditions. The points invariably chosen are the freezing and boiling points of water, and the numerical values of these points on the centigrade scale of temperature are 0° and 100° respectively. The lower fixed point (*ice point*) is the equilibrium temperature between ice and air saturated water at normal atmospheric pressure. The upper fixed point (*steam point*) is the equilibrium temperature between liquid water and

1*

its vapour at the pressure of one standard atmosphere. Standard atmospheric pressure is defined as the pressure due to a column of mercury 760mm. high having a mass of 13·5951gm. per cm.[3] subject to a gravitational acceleration of 980·665cm. per sec.[2], and is equal to 1,013,250 dynes per cm.[2]

For the subdivision of the fundamental interval, that is, for the actual definition of the unit of temperature, it is necessary to chose some property of matter which depends on hotness. This property (for example the volume of a given mass of mercury, the pressure of a gas at constant volume, or the resistance of a piece of platinum wire, &c.) should be capable of precise measurement at the two fixed points, and its value at any other point defines the temperature at that point in terms of the proportion of the change of the property between the two fixed points. Thus if X_{100} is the value of the property at the upper fixed point, X_0 its value at the lower fixed point, and X_t at some other point, then the temperature t in degrees centigrade of this point is defined by the equation:

$$\frac{t}{100} = \frac{X_t - X_0}{X_{100} - X_0} \quad \ldots \ldots \quad [1]$$

On this scale one degree centigrade is defined as the change of temperature which will cause a change of magnitude $\dfrac{X_{100} - X_0}{100}$ in the property of the body.

1.03 Liquid-in-Glass Thermometers

In the choice of a suitable thermometric substance it is essential that the property of the substance in terms of which the temperature scale is to be defined, should be capable of exact evaluation at the two fixed points. In addition, the property selected must vary uniformly with temperature over as wide a range as possible. The expansion of matter when heated, particularly the expansion of liquids, was first used for temperature definition. Thermometers were constructed in which the liquid, contained in a small thin-walled glass bulb, was capable of expansion along the fine bore of a thick-walled hermetically sealed capillary tube. With a tube of uniform bore the change in length of the liquid column is proportional to its expansion. If the length of the column is l_0 at the ice point, l_{100} at the steam point, the temperature t at which the column has length l_t is defined as follows:

$$\frac{t}{100} = \frac{l_t - l_0}{l_{100} - l_0}$$

Mercury is usually chosen for use in liquid-in-glass thermometers for the following reasons:

(i) It is easily seen so that even a very fine thread is readily visible. With such narrow capillary tubes an appreciable increase of length results for a small change of temperature, in other words the thermometer can be made quite sensitive.

(ii) It has a uniform coefficient of expansion over a wide range of temperature within the limits of its freezing point at −39°C., and its boiling point at 357°C.

(iii) It does not wet the glass, which is an important consideration in constructing thermometers of very fine bore.

(iv) It is a good conductor of heat, and hence quickly takes up the temperature of the body.

(v) It has a low specific heat, and thus the thermometer bulb absorbs little heat.

1.04 The Mercury Thermometer: Errors and Corrections

Mercury-in-glass thermometers, though inexpensive and convenient to use, suffer from the following defects which limit their use for accurate scientific work. The errors arising from these defects vary with different thermometers, and will further depend on the type of glass used. An indication of the size of the errors is given in brackets at the end of each paragraph.

(i) *Non-uniformity of the bore of the capillary tube.*—The impossibility of getting an exactly uniform bore results in unequal increments in length for successive equal increments of temperature. Or since the assumption of bore uniformity is invariably made, this means that equal extensions of the thread will not define equal temperature intervals (0·1°C.).

(ii) *Internal pressure change.*—When used in the vertical position (this error does not affect a thermometer used in the horizontal position) the hydrostatic thrust of the column of mercury causes a slight distension of the bulb. In consequence of this, readings registered by the thermometer in this position will be too low (0·001°C. per cm. change of pressure).

(iii) *External pressure changes.*—The bulb suffers contraction due to the inward thrust of the liquid in which it is immersed, resulting in readings which are too high. Atmospheric changes also bring about errors of the same sort as those listed here and in (ii) above (0·001°C. per cm. change of pressure).

(iv) *Exposed stem correction.*—Only when the full stem as well as the bulb of the thermometer are immersed in the heated liquid will the temperature of the mercury be the same throughout. When an appreciable length of the stem is exposed the mercury here will be cooled by an unknown amount, which will cause the thermometer to under-read. The error arising under this heading depends not only on the length of the exposed thread, but also on the temperature difference between the hot liquid and the surrounding air. The required correction to be added to the observed reading t is given by the expression $A n (t-t')$ C.,°

where n is the number of degrees of mercury column exposed and t' its mean temperature. A is a constant, the value of which varies between 0·00015 and 0·00016 according to the nature of the glass used.

(v) *Glass envelope corrections.*—Glass responds slowly when subject to heat treatment. In consequence of this, a sudden cooling, as for example when a thermometer at room temperature is plunged into ice, will produce a depression below the zero mark due to the failure of the bulb to contract quickly. The creep to zero may be a matter of several hours, different qualities of glass responding differently in this respect (0·01 to 0·05°C.).

A similar error, known as the *secular change*, may affect the ice point for a period of several years after the thermometer has been made. The very slow contraction of the glass bulb causes a slight rise at the ice point, which should be checked at periodic intervals to obtain the necessary zero correction (0·01°C. per annum).

1.05 The Mercury Thermometer: Extensions of Range

In practice the range of a mercury-in-glass thermometer is from temperatures somewhat above its freezing point at −39°C. to temperatures up to 300°C. The upper limit of the range can be greatly extended however, by the introduction of some inert gas such as nitrogen in the space above the mercury. As the mercury expands the pressure of the gas is raised with consequent increase in the boiling point of mercury. In this way such a thermometer may be satisfactorily used for temperatures up to 600°C.

The extension of the lower limit necessitates the replacement of mercury by other liquids. Alcohol (B.P. 78°C.) may be used to record temperatures down to its freezing point at −114·9°C., whilst the use of liquid pentane permits of a downward extension of the range to −200°C. It is important to note that alcohol thermometers and others with ranges that do not cover both the ice and steam points, cannot be graduated directly. In such cases it is necessary to standardise the thermometer by comparison with a mercury or other thermometer. Temperatures above 600°C. or below −200°C., are not measured by liquid-in-glass thermometers.

1.06 The Constant Volume Gas Thermometer

It has been seen that on account of the uncertain behaviour of the glassy envelope, and for other reasons, the mercury thermometer is unreliable as an instrument for accurate work. With different specimens of glass having their own peculiar properties, it will be extremely unlikely that any two mercury-in-glass thermometers will record exactly the same reading at any given temperature. For standardising purposes gas thermometers are now invariably used. These thermometers, employing a gas as the thermometric substance, are constructed for use either for measuring the pressure of the gas when contained at constant volume, or with the pressure constant to measure the volume changes. The former method has been found to be much the more convenient in practice, and as the two thermometers give identical temperature scales for a gas obeying Boyle's law, only the constant volume instrument will be described here.

Among the main advantages of the constant volume gas thermometer over the mercury-in-glass thermometer may be mentioned the following:
(i) Gases produce large proportionate increases of volume and pressure when heated, and in consequence gas thermometers are more sensitive, and the correction for the expansion of the envelope is relatively unimportant.
(ii) Gases can be obtained in a high state of purity, and thus the thermometers are reproducible to the same degree of accuracy in whatever place they are required for use.
(iii) They can be used to cover a very large range of temperature (see section 1.08).
(iv) These thermometers give very close agreement with the thermodynamic scale of temperatures (see section 13.07).

Gas thermometers are however large and cumbersome for routine work, and can only be used in one position. Their important use is for the calibration of other thermometers such as the mercury, platinum resistance, or thermo-electric thermometers, which are more convenient to manipulate.

1.07 Jolly's Constant Volume Gas Thermometer

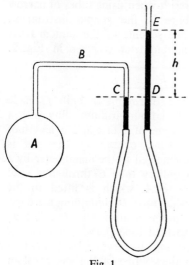

Fig. 1

The simple type of constant volume gas thermometer due to Jolly is shown in Fig. 1. It consists of a bulb A containing a fixed mass of gas which is completely immersed in the body whose temperature is required. The bulb A is connected by narrow tubing B to an open mercury manometer. Since the gas contained in B is not at the temperature of the body, the bulb A should be of sufficiently large volume to reduce errors due to the "dead space" in B. In using the instrument, the movable limb of the manometer is adjusted so as to maintain the level of the mercury at C. The pressure of the gas at constant volume is obtained by adding (or subtracting) the difference of mercury level (h) at C and E to the barometric height H, which should be taken at the beginning and again at the end of the experiment.

The temperatures defined by the constant volume gas thermometer are given by:

$$\frac{t}{100} = \frac{p - p_0}{p_{100} - p_0}$$

p_0, p_{100}, and p being respectively the pressures of the gas at the ice point, steam point, and at the temperature of the body. If the corresponding mercury level differences are h_0, h_{100}, and h, then

$$\frac{t}{100} = \frac{(H + h) - (H + h_0)}{(H + h_{100}) - (H + h_0)}$$

giving

$$t = 100 . \frac{h - h_0}{h_{100} - h_0}$$

if H is constant throughout the experiment.

1.08 The Standard Constant Volume Gas Thermometer

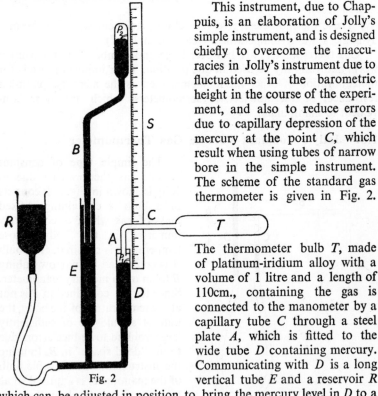

This instrument, due to Chappuis, is an elaboration of Jolly's simple instrument, and is designed chiefly to overcome the inaccuracies in Jolly's instrument due to fluctuations in the barometric height in the course of the experiment, and also to reduce errors due to capillary depression of the mercury at the point C, which result when using tubes of narrow bore in the simple instrument. The scheme of the standard gas thermometer is given in Fig. 2.

The thermometer bulb T, made of platinum-iridium alloy with a volume of 1 litre and a length of 110cm., containing the gas is connected to the manometer by a capillary tube C through a steel plate A, which is fitted to the wide tube D containing mercury. Communicating with D is a long vertical tube E and a reservoir R

Fig. 2

which can be adjusted in position to bring the mercury level in D to a fixed index pointer P_1 thus keeping the volume of the gas constant when taking readings. The pressure of the gas is measured directly from the readings of a barometer against a scale S whose zero is adjusted at P_1. The barometer tube B which moves vertically in the tube E, is bent over at the top so as to bring the widened end of the closed tube in vertical alignment with the tube D. In taking a reading of the pressure, the barometer tube is adjusted until the mercury level is just in contact with a small index pointer P_2 inside the vacuum of the barometer tube. The distance P_1P_2 then gives the pressure of the gas which is read off against the scale by verniers (not shown) at P_1 and P_2.

The International Committee of Weights and Measures adopted in 1887, as a practical standard scale of temperature, the centigrade scale of the hydrogen thermometer between the ice and steam points, the pressure of the hydrogen at the ice point to be 1 metre of mercury. Temperature measurements between the fixed points obtained by the standard hydrogen gas thermometer are accurate to 0·005°C. The range of the instrument is from temperatures just above the absolute zero up to 1500°C. For low temperature work hydrogen or helium gas is used (sometimes neon). At temperatures above 500°C. hydrogen diffuses through the envelope, and for work at high temperatures nitrogen gas is used in a platinum or platinum-iridium bulb.

1.09 The Platinum Resistance Thermometer

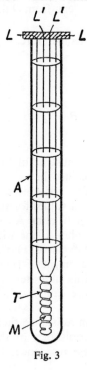

Fig. 3

As indicated in the previous section gas thermometers, being cumbersome in use, are not satisfactory for routine work. Other thermometers are needed for such work which are easier to manipulate, and whose readings can be related to the gas scale. The most satisfactory thermometer of this type is the platinum resistance thermometer. Following up the earlier work of Siemens, Callendar found that the variation in resistance of pure platinum with temperature provided the means to establish a very accurate practical scale of temperature. A form of the platinum resistance thermometer is shown in Fig. 3. The thermometer tube of fused porcelain A contains a coil T of fine platinum wire non-inductively wound on a strip of mica M. The platinum coil is joined to thick leads LL of copper, silver, or platinum, and these, together with an identical pair of dummy leads $L'L'$, pass through an ebonite cap to the Wheatstone bridge arrangement used for measuring the resistance of the platinum coil. A number of mica discs (five only shown) are placed in the tube to insulate the leads from one another, and also to prevent convection currents in the air of the tube.

A special form of Wheatstone bridge, devised by Callendar and Griffiths for use with the thermometer, is shown in Fig. 4. In this arrangement the ratio arms contain equal resistances P and Q. The dummy leads L' with a variable resistance box R', and the leads L attached to the platinum spiral R are connected into the other two arms as indicated. Between these resistance arms there are uniform resistance wires AB and CD of resistance ρ ohm per cm. length, over which moves a sliding contact S connected to the galvanometer. The contact slider and the wires with which it is in contact are all of the same metal to avoid errors due to thermo-electric effects.

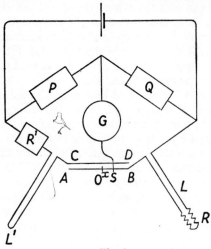

Fig. 4

Suppose a balance is obtained with S a distance x cm. from O the mid point of AB. Then if

R = resistance of the platinum coil at the required temperature,

L = resistance of the leads,

L' = resistance of the dummy leads,

l = length in cm. of AO,

$$\frac{R' + L' + (l + x)\rho}{R + L + (l - x)\rho} = \frac{P}{Q} = 1 \quad \text{Since } P = Q$$

$$\therefore \quad R = R' + 2x\rho$$

Now the platinum chosen for the resistance coil must be of fine wire so that a reasonably high resistance can be obtained in a small space, and should be of such purity that the ratio of its resistances at the steam and ice points is not less than 1·386. Typical values for the coil are $R_0 = 25·6$ ohm with a fundamental interval of 10 ohm. A rise of 1°C. will then cause a change of resistance of 0·1 ohm. If the wire AB has a resistance of 0·005 ohm per cm. length, the movement (x) of S required to register this change is 10cm., that is a movement of 1mm. corresponds to a change in temperature of $\frac{1}{100}$°C. This indicates the order of accuracy with which readings of temperature can be made with the platinum resistance thermometer.

To calculate the temperature given by the platinum resistance thermometer it is necessary to obtain the values of the resistance of the platinum spiral at the ice and steam points. Let these resistances be

R_0 and R_{100}, and let R be the resistance of the platinum spiral at some unknown temperature. This temperature t_p is defined by the equation:

$$\frac{t_p}{100} = \frac{R - R_0}{R_{100} - R_0} \qquad \dots \dots \quad [2]$$

A scale of temperature is thus established which increases uniformly with the resistance, that is equal increments of resistance define equal increments of temperature. In a later section the character of the various temperature scales defined in this way will be discussed more fully. It is however of interest to see how temperatures on the platinum scale correlate with those on the gas scale of temperatures previously discussed. Callendar showed that the resistance of pure platinum could be expressed in terms of the gas scale temperature t by the following formula:

$$R = R_0 (1 + At + Bt^2)$$

Where A and B are constants for the platinum wire and R and R_0 its resistance at the gas temperature t and at the ice point respectively. The resistance R_{100} at the steam point will then be given by:

$$R_{100} = R_0 (1 + 100A + 10,000B)$$

By substituting these values for R and R_{100} in equation 2, an equation relating the platinum temperature t_p to the corresponding gas temperature t will be obtained. Thus:

$$\frac{t_p}{100} = \frac{At + Bt^2}{100A + 10,000B}$$

from which

$$t - t_p = \frac{10^4 B}{A + 100B} \left\{ \left[\frac{t}{100} \right] - \left[\frac{t}{100} \right]^2 \right\}$$

$$\text{or } t - t_p = \delta \left\{ \left[\frac{t}{100} \right]^2 - \left[\frac{t}{100} \right] \right\} \qquad . \quad . \quad [3]$$

$$\text{where } \delta = - \frac{10^4 B}{A + 100B}$$

The value of δ for pure platinum is very nearly $1 \cdot 50$, and the purity of the platinum used in the thermometer should be such that δ is not greater than $1 \cdot 51$. The value of δ is obtained by determining the resistance of the thermometer at the boiling point of sulphur (the corresponding gas scale temperature for which is $444 \cdot 60°C.$), and then calculating the platinum temperature for this point from equation 2. On substituting the two sulphur point temperatures in equation 3 a value for δ can be found. Using this value of δ in equation 3 a table can then be drawn up showing the difference between the gas and platinum temperatures against different gas temperatures, and from this table the gas temperature corresponding to a calculated platinum temperature can be obtained.

Alternatively the value of t for an observed value of t_p can be obtained by the method of successive approximations. For example, if t', t'', t''', &c. represent the first, second, third, &c. approximations for $t_p = 400°C.$, we have (taking $\delta = 1\cdot50$) and starting with $t=t_p = 400°C.$ and then using the calculated value of t' for the second approximation, and so on,

$$t' - t_p = \delta\left\{ \left[\frac{t}{100}\right]^2 - \left[\frac{t}{100}\right] \right\} = 1\cdot5 \times 10^{-4}t(t-100)$$

$$= 1\cdot5 \times 10^{-4} \times 400 \times 300 = 18°C. \text{ or } t' = 418°C.$$

$$t'' - t_p = 1\cdot5 \times 10^{-4} \times 418 \times 318 = 19\cdot9 \text{ or } t'' = 419\cdot9°C.$$

$$t''' - t_p = 1\cdot5 \times 10^{-4} \times 419\cdot9 \times 319\cdot9 = 20\cdot1 \text{ or } t''' = 420\cdot1°C.$$

Continuing this process indefinitely leads to a value of t between $420\cdot1$ and $420\cdot2°C.$ The table given below shows corresponding values of t and t_p calculated in this way for $\delta = 1\cdot50$, and serves to show the divergence between the temperatures defined by the two scales.

TABLE 1

t_p°C.	t°C.	t_p°C.	t°C.
0·0	0·0	400·0	420·2
50·0	49·6	600·0	654·4
100·0	100·0	800·0	910·8
200·0	203·1	1000·0	1197·0

A change of 1 per cent in the value of δ used for calculating the values of t produces a change of less than $0\cdot1$ per cent in the value of t.

With the platinum resistance thermometer determinations of temperature can be taken over a very wide range from $-200°C.$ to $1200°C.$ Accurate readings are obtained within this range, but particularly so between $0°C.$ and the sulphur point. It is convenient to use in practice, and since the leads connecting the thermometer to the measuring bridge can be of any desired length, the temperatures of furnaces can be conveniently and effectively taken at a distance. "Stem" corrections are eliminated by the use of dummy leads, and with pure platinum there is no zero change. It has, however, one disadvantage in that it cannot register rapidly varying temperatures, chiefly on account of the low thermal conductivity of the containing tube and the need for balancing the bridge. The thermo-electric thermometer is well suited for this purpose and is described in the following section.

1.10 The Thermo-Electric Thermometer

This thermometer depends upon the Seebeck effect, namely that if two dissimilar metals such as copper and iron are joined in series to make a complete circuit (Fig. 5), then on heating one of the junctions a current flows round the circuit. Such an arrangement is called a thermo-electric couple, and the e.m.f. established round the circuit depends on the nature of the metals used to form the couple, and also on the temperature difference between the hot and cold junctions. The cold junction is usually maintained at the ice point, and it has been shown experimentally that when the other junction is at some temperature t, the thermo-electric e.m.f. set up depends upon the temperature according to the quadratic law:

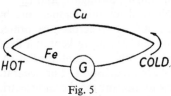

Fig. 5

$$e = A + Bt + Ct^2 \quad . \quad . \quad . \quad . \quad . \quad [4]$$

Where A, B, and C are constants depending on the metals used. This is shown graphically in Fig. 6. The temperature at which the e.m.f. is a maximum (for the two metals) is called the *neutral temperature t_N*.

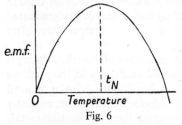

Fig. 6

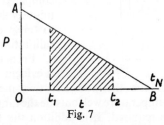

Fig. 7

It is more usual to represent the thermo-electric behaviour of two metals graphically by plotting their *thermo-electric power P* (that is the change in the thermo-electric e.m.f. per degree centigrade in temperature between the hot and cold junctions) against temperature. If t is measured from the ice point, the constant A is zero in equation 4 which becomes:

$$e = Bt + Ct^2$$

Thus $P = \dfrac{de}{dt} = B + 2Ct$

The graph of P against t is thus a straight line (see Fig. 7) cutting the temperature axis at the neutral temperature (for which $\dfrac{de}{dt} = 0$). From this graph the e.m.f., established when the junctions are at two temperatures t_1 and t_2, is $e = \displaystyle\int_{t_1}^{t_2} P\,dt$, and is equal to the area of the shaded portion shown. Standard textbooks on electricity give the thermo-electric line diagrams for a variety of metals against lead, the e.m.f. established for any two metals with junctions at two given temperatures being obtained in the above manner from the area intercepted between the two lines and the ordinates through t_1 and t_2.

It can thus be seen that it is possible to determine temperature by the measurement of the thermo-electric e.m.f. set up across a calibrated

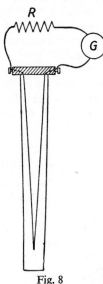

Fig. 8

thermo-couple, one junction being kept at a fixed temperature whilst the other is maintained at the unknown temperature in question. For accurate work it is necessary to measure the thermo-electric e.m.f. using a sensitive potentiometer calibrated against a standard cell,* but for routine work where a high degree of accuracy is not required, it is sufficient to use a milli-voltmeter or a high resistance galvanometer connected directly across the two wires. A ballast resistance R (see Fig. 8) may be included in the circuit if the resistance of the recording instrument is not sufficiently great, so as to make the temperature resistance variation of the metal wires comprising the junction small compared with the total circuit resistance. Used in this way the recording instrument can be graduated to give direct readings of temperature after calibration.

From the previous outline of electrical theory it is evident that to avoid ambiguity in the readings of the thermometer, two metals must be chosen with a neutral temperature well outside the range of readings for which the thermometer is required. If in addition the thermo-electric lines of the two metals are steeply inclined to one another, increased accuracy of measurement will result for small temperature changes. The following couples have been found satisfactory for the temperature ranges stated:

Up to 300°*C.*—Copper (or iron) against constantan (*Cu* 60, *Ni* 40). Such a couple gives a comparatively large e.m.f. (see table 2), but is unreliable above the temperature indicated on account of oxidation troubles.

Up to 1100°*C.*—"Chromel" (an alloy of chromium and nickel) against "Alumel" (an alloy of aluminium and nickel).

Up to 1500°*C.*—Platinum against platinum-rhodium or platinum-iridium. These latter couples have thermo-electric lines which are almost parallel, and so for an extended range give readings of e.m.f. which are approximately proportional to temperature differences. For this range the thermo-couple is calibrated by readings taken at the melting points of antimony, silver, and gold.

---

* See the author's book " A Laboratory Manual of Physics,"—*Arnold.*

Provided there is no risk of contamination a thermo-couple can be inserted in the source without a protecting sheath. Used in this way the "time-lag" is very small. In all other cases sheaths of metal, or of high temperature fused silica or porcelain, must be used to protect the thermo-couple. Such sheaths must be thick enough to withstand the mechanical strain at high temperatures, but not so thick that the "time-lag" is appreciably increased. The electrical insulation of wires and leads must also be effective so as to eliminate any leaks which would introduce serious discrepancy in the readings.

Thermo-electric thermometers, though not as accurate as resistance thermometers, have several advantages which make them suitable for general industrial use. On account of their small thermal capacity and low "time-lag" they can be used for the measurement of varying temperatures. They have a very wide range from $-250°C$. to nearly $1600°C$. (with special precautions up to $1800°C$.), and in addition they are cheap and convenient in use.

TABLE 2

E.m.f. in millivolts of different thermo-couples at the temperatures indicated—one junction being at $0°C$.

Temperature of hot junction in °C.	Copper-Constantan	Chromel-Alumel	Platinum-Platinum-Rhodium (10%)
100	4	4·1	0·64
200	9	8·1	1·44
300	15	12·2	2·32
400	—	16·4	3·25
600	—	24·9	5·22
800	—	33·3	7·33
1000	—	41·3	9·57
1500	—	—	15·50

1.11 Measurement of Very High Temperatures— Radiation Pyrometry

The measurement of temperatures beyond the upper limit of the thermo-electric thermometer is carried out by radiation pyrometers. These are of two types, (a) the optical type, (b) the total radiation type. Accurate readings are possible up to $3000°C$. and upwards by these instruments, which depend upon the laws of radiation. A detailed discussion of radiation pyrometers is deferred to chapter XII.

1.12 Arbitrary Nature of Temperature Scales

Temperatures are analytically defined with respect to some particular property of a substance as the proportionate change in the value of this property compared with the change between the ice and steam points. Then as indicated in section 1.02:

$$\frac{t}{100} = \frac{X_t - X_0}{X_{100} - X_0}$$

It necessarily follows that temperatures defined in this way will vary uniformly with the chosen property of the substance. Clearly, however, the temperature scale thus defined will not agree with a scale obtained in relation to some other property *unless the two properties themselves show identical relative changes when compared with one another*. In practice no such exact relationship has been found to hold good, and consequently temperatures defined by equation 1 are as numerous and distinct as the properties which define them. An instance of the discrepancy between the numerical evaluation of temperatures on the platinum and gas scales has already been given in section 1.09. Examples 22 and 26 at the end of this chapter further illustrate this point.

As no particular property can be singled out as being more fundamental than another it is clear that all scales of temperature defined in relation to specific properties of matter are arbitrary. It is of course possible to calibrate all scales of temperature with reference to one particular scale and to obtain empirical relationships between the other scales. In this way temperatures can be co-ordinated and systematised without however having any fundamental significance. The use of any such arbitrary scale as the standard scale is necessarily unsatisfactory, and it is clear that the ideal scale must be one which is not in any way dependent on any particular property of matter.

In chapter XIII it will be shown that an ideal scale can be conceived on thermodynamical grounds. This scale, due to Kelvin, has the theoretical advantage of specifying exactly the equality of temperature intervals, and defining an absolute zero from which all temperature measurements can be made. Unfortunately this scale cannot be realised in practice, but Kelvin showed that a scale based on the properties of a "perfect" gas would be identical with the thermodynamic scale. All gases show departures in varying degree from the gas laws, and in consequence different gases give temperature scales which show deviations from one another. At low pressures, however, there is close agreement with the standard gas equation, particularly for the more permanent gases such as hydrogen and helium. Thus a scale of temperature defined by a gas thermometer containing hydrogen at low pressure gives the closest practical realisation of the centigrade thermodynamic scale. For this reason the scale of temperature defined by the hydrogen gas thermometer is accepted as the standard scale to which all other scales are ultimately referred.

1.13 International Scale of Temperature

As indicated in section 1.06 the gas thermometer is too cumbersome in practice for the numerical evaluation of temperature in routine work.

Accordingly, to avoid experimental difficulties, the International Committee of Weights and Measures adopted in 1927 a practical scale of temperature known as the *International Temperature Scale*. On this scale a series of freezing points and boiling points of substances in a given state of purity are obtained by the constant volume gas thermometer. These fixed points, which are conveniently and accurately reproducible, together with specified means of interpolation, provide a practical scale of temperature which gives the closest possible realisation of the thermodynamic scale, and at the same time permits of uniformity of temperature statement. The basic fixed points are given below:

(a) Temperature of equilibrium between liquid oxygen and gaseous oxygen at the pressure of one standard atmosphere (Oxygen point) −182·97°C.

(b) Temperature of equilibrium between ice and air-saturated water at normal atmospheric pressure (Ice point) 0·000°C.

(c) Temperature of equilibrium between liquid water and its vapour at the pressure of one standard atmosphere (Steam point) 100·000°C.

(d) Temperature of equilibrium between liquid sulphur and its vapour at the pressure of one standard atmosphere (Sulphur point) 444·60°C.

(e) Temperature of equilibrium between solid silver and liquid silver at normal atmospheric pressure (Silver point) 960·5°C.

(f) Temperature of equilibrium between solid gold and liquid gold at normal atmospheric pressure (Gold point) 1063°C.

The scale is divided into four parts for the purposes of interpolation. The experimental methods specified are:

From −190°C. to 0°C.—Measurement of the resistance of a standard platinum resistance thermometer using the formula:

$$R_t = R_0 [1 + At + Bt^2 + C (t - 100) t^3]$$

The constants R_0, A, B, and C being obtained by calibration at the points (a), (b), (c), and (d).

From 0°C. to 660°C.—Measurement of the resistance of a standard platinum resistance thermometer using the formula:

$$R_t = R_0 (1 + At + Bt^2)$$

The constants R_0, A, and B being obtained by calibration at the points (b), (c), and (d).

From 660°C. to the gold point.—Measurement of the electromotive force of a standard platinum versus platinum-rhodium thermo-couple, one junction being at 0°C., the temperature t of the other junction being defined by the equation:

$$e = A + Bt + Ct^2$$

The constants A, B, and C being determined by calibration at the freezing point of antimony (a secondary point on the international scale with numerical value 630·5°C.), and at points (e) and (f).

Above the gold point.—Comparison of the intensity of monochromatic radiation emitted by a "black body" at the temperature t

in question with the intensity of the same radiation emitted by a black body at the temperature of the gold point (see section 12.16).

1.14   Absolute Zero

An "absolute" zero of temperature is specified in terms of the laws of heat on the thermodynamical scale of temperature (see section 13.07). No temperatures lower than this are possible. The scale of temperatures numerically evaluated with this point as zero is known as the Absolute or Kelvin Scale, and temperatures stated on this scale are designated $°A$ or $°K$. In experimental work, particularly at very low temperatures, it is desirable to express temperatures from the absolute zero of temperature, and this involves the determination of the number of degrees between it and the ice point. This is realised practically using the constant volume gas thermometer and extrapolating to zero pressure. The average value of the thermodynamic temperature of the ice point determined in this way by recent workers using the constant volume helium thermometer is 273·16°.

For approximate purposes the absolute zero is taken as −273°C.; for more accurate work its value is taken as −273·2°C., or as −273·16°C. where there is need for the greatest accuracy. The question of the absolute scale of temperature is more fully discussed in section 5.05.

QUESTIONS.   CHAPTER 1

1.   Discuss the measurement of temperature, and explain how temperature is defined on the scale of a given type of thermometer.

When used to measure temperatures of about 300°C. the readings on the scale of an accurate mercury thermometer are about 2° higher than those of an accurate air thermometer. Why is this, and what reasons, if any, are there for adopting one scale rather than the other?   (O. and C.)

2.   Describe fully, with diagrams, types of thermometer which depend on (a) thermoelectric effect, (b) the change of resistance with temperature, (c) the radiation of heat.

Indicate in the case of *each* of these thermometers its useful working range of temperatures and discuss its uses compared with the ordinary mercury-in-glass thermometer.   (A.)

3.   Describe and explain, with relevant theory, a constant volume gas thermometer.

Give a careful explanation of what is meant by a difference of 1°C. as registered by (a) a mercury-in-glass thermometer, (b) a constant-volume gas thermometer.

Why is temperature measurement by a gas thermometer regarded as less arbitrary than a mercury-in-glass type?   [A.]

4.   Explain the principle of a constant volume gas thermometer and describe a simple instrument suitable for measurements in the range 0°C. to 100°C. What factors determine (a) the sensitivity and (b) the accuracy of the instrument you describe?

A certain gas thermometer has a bulb of volume 50cm.[3] connected by a capillary tube of negligible volume to a pressure gauge of volume 5·0cm.[3]. When the bulb is immersed in a mixture of ice and water at 0°C. with the pressure gauge at room temperature (17°C.), the gas pressure is 700 mm. Hg. What will be the pressure when the bulb is raised to a temperature of 50°C. if the gauge is maintained at room temperature? You may assume that the gas is ideal and that the expansion of the bulb can be neglected. (O. and C.)

5. Explain what is meant by the *temperature* of a body. How may *a scale of temperature* be defined? Give the definition of the absolute gas scale, and explain why it is termed " absolute ".

A constant volume gas thermometer, filled with an ideal gas, is correctly calibrated at the ice- and steam-points. In making a certain measurement, the bulb is placed in an enclosure at 40°C., but the capillary (whose volume is 5 per cent of the whole volume occupied by the gas) protrudes from the enclosure and is at an average temperature of 25°C. What temperature will the thermometer read? [O. and C.]

6. State what you mean by a scale of temperature and describe how you would measure a temperature of about 300°C. in terms of the perfect gas scale when the conditions of the experiment make it impossible to use a perfect gas thermometer directly. (Camb. Schol.)

7. What is a *scale of temperature*? Illustrate your answer by reference to (a) a mercury-in-glass scale, (b) the constant volume hydrogen scale, (c) the international scale.

When using a constant volume air thermometer in the usual way the bulb was placed in melting ice and it was found that the level of mercury in the open tube was 10·0cm. below the level of the mercury in the closed tube. The bulb was then placed successively in steam and warm water and the level in the open tube was 50·0cm. and 3·0cm., respectively, above the other level. Sketch the apparatus in the first position and deduce the temperature of the water. (S.)

8. Explain the principles underlying the establishment of a scale of temperature. Why is hydrogen chosen as the standard thermometric substance?

Describe a simple form of constant-volume gas thermometer and explain how you would determine the melting-point of paraffin wax on the Centigrade Scale of the thermometer. (W.)

9. Distinguish between the " constant pressure " and " constant volume " scales of temperature based on the thermal properties of a gas. Show that, in the case of an ideal gas, the two scales will agree exactly.

Two vessels, A and B, of volumes 1 litre and 3 litres respectively, contain air and are connected by a tube of small cross-section, fitted with a valve which is initially closed. The air in A is at a temperature of 20°C. and a pressure of 65cm. of mercury, that in B at 50°C. and 90cm. of mercury. Calculate the gas pressure when the valve is opened and the temperature is adjusted to 30°C. (W.)

10. Explain what is meant by a *temperature on a Centigrade Scale.*

Describe a platinum resistance thermometer and state how you would use it to find a temperature on the platinum scale. (Experimental details are not required.)

Why does the value thus found not agree with the value obtained using a gas thermometer?

What are the practical advantages of the resistance thermometer compared with the gas thermometer? (L.)

11. Explain how a centigrade temperature scale can be established. Why would you not necessarily expect the temperature of a body as measured by scales established with different types of thermometer to agree?

Describe the thermometer you would use, and the procedure you would employ, for measuring a rapidly changing temperature. Give reasons for your choice. (N.)

12. What is meant by the *fundamental interval* of a temperature scale? How is $t°C$. defined on (a) the mercury-in-glass scale, (b) the constant volume gas scale, (c) the platinum resistance scale? What is the importance of the gas thermometer in the field of temperature measurement?

The resistance of a given wire at various temperatures (measured on the mercury-in-glass scale) is as follows:

$t°C$.	0	10	20	30	40	50	60	70	80	90	100
Resistance in ohms	2·50	2·54	2·58	2·62	2·67	2·71	2·76	2·81	2·87	2·93	3·00

Plot a graph of these values and from it find (i) the temperature on the resistance scale corresponding to 55°C. on the mercury scale, (ii) the temperature on the mercury scale corresponding to 20°C. on the resistance scale. (N.)

13. Discuss the concept and physical significance of *temperature* and describe in detail three methods of measuring it which depend on different physical properties of the thermometric substances.

A scale of temperature is defined by the following equation
$$t = a \log X + b$$
where X is the magnitude of a property of a substance at a temperature t and a and b are constants. Find the temperature in Centigrade degrees when X has the value X_C. [N.]

14. Discuss the concept of temperature. What type of thermometers would you use to measure temperatures of (a) −50°C. and (b) 500°C. and how would you calibrate them?

What meaning can be attached to the phrases " the temperature of interstellar space " and " the temperature in the space between the walls of a vacuum flask filled with liquid air "? (Oxford Schol.)

15. Explain exactly how you would calibrate a platinum resistance thermometer if you were told to use melting ice and steam as your two fixed points.

Would you expect such a thermometer to show the same temperature as a mercury thermometer if the two thermometers were placed in a bath of liquid at about 200°C.?

Give reasons for your answer. (O.)

16. Describe the thermo-electric thermometer, and mention any special advantages it has over the platinum resistance thermometer for general industrial use.

17. The temperature variation of the resistance of platinum on the Centigrade Gas Scale is given by the relation $R_t = R_0(1 + at + bt^2)$. Show that if a gas scale temperature t, measured on the platinum scale, is denoted by t_p, the relation between t and t_p is given by:
$$t - t_p = \delta \left\{ \left(\frac{t}{100} \right)^2 - \frac{t}{100} \right\}$$
where δ is a constant.

If the platinum temperature corresponding to 60° on the Centigrade Gas Scale is 60·36°, what gas scale temperature corresponds to 150° on the platinum scale? (Camb. Schol.)

CHAPTER II

THE MEASUREMENT OF HEAT

2.01 Introductory

We have seen in Chapter I that the temperature of a body can be directly observed and accurately specified according to scales of temperature based on physical properties of substances which are capable of exact measurement. Subsequently we shall investigate certain changes in the physical condition of a body which result when its temperature changes, but we are now concerned with an enquiry into the cause of these changes. If a vessel containing water is placed over a bunsen flame there is a steady rise in temperature. If the vessel is now removed from the burner, the temperature falls steadily. It is natural to assume that something has been communicated to the water to effect the rise in temperature, and that the subsequent fall in temperature results from the water parting with this "something." The agent responsible for these changes we call heat. It is important to realise that the passage of heat to or from a body is only *inferred*, and that these simple experiments provide no evidence as to the nature of heat. This latter question will be dealt with at some length in a later chapter.

2.02 Units of Heat

Before any precise measurements of heat exchanges can be made it is of course necessary to establish an agreed unit of heat in terms of which heat quantities can be expressed. Using a source of constant heat supply it is a simple matter to show that the quantity of heat given to a substance is proportional to its mass and to the temperature rise produced. We can thus define our unit as the quantity of heat required to raise the temperature of a given mass of a specified substance by a given amount. Water is taken as the standard substance, the unit of heat, called the *calorie,* being defined as the heat required to raise the temperature of one gram of water through 1°C.

Accurate investigation (see section 2.04) has shown that the unit of heat so defined varies slightly with the position on the temperature scale at which the measurement is taken. Thus the quantity of heat required to raise the temperature of one gram of water from 0° to 1°C. is not quite the same quantity as would be required in the temperature interval 20° to 21°C. or for the interval 50° to 51°C. It is therefore essential in defining the calorie that the degree interval of temperature involved should be exactly specified. The interval usually chosen is from 14·5° to

15·5°C., and this defines the *15°C. calorie*. Some experimenters have selected the range of temperature from 19·5° to 20·5°C. and so have used the *20°C. calorie,* but for general practical work involving wide ranges of temperature it is customary to use the *mean calorie* which is defined as the hundredth part of the quantity of heat required to raise the temperature of one gram of water from 0°C. to 100°C. The calorie so defined has a value very nearly equal to the generally accepted standard 15°C. calorie. Although it is necessary to define the calorie in terms of some standard specified temperature for accurate work, the quantities of heat required to raise the temperature of one gram of water by 1°C. at different parts of the scale differ only slightly from one another. Some idea of the extent of this fluctuation can be obtained from Table 4. For ordinary practice, and for the purpose of the discussion in this book, it may be assumed that these quantities of heat are equal and that the calorie is the same at whatever temperature it is defined.

The calorie as defined above is a small unit, and if a larger unit of heat is required the *kilo-calorie,* or *large calorie,* is used. This is defined as the heat required to raise the temperature of one kilogram of water through 1°C. In the British system of units the unit of heat is defined as the quantity of heat to raise the temperature of one pound of water through 1°F., this unit is called the **British Thermal Unit** (B.Th.U.). In gas engineering a larger unit, equal to 100,000 B.Th.U.'s, and called the *therm* is used. Another common unit used is the *pound °C.* or *pound calorie* which is the quantity of heat required to raise the temperature of one pound of water through 1°C. This unit is clearly $\frac{9}{5}$ times as large as the B.Th.U.

2.03   Specific Heat

If the same quantity of heat is given to identical masses of different substances, it is found that the resulting temperature rise differs in each case. This is usually summed up by saying that the substances have different *thermal capacities,* the thermal capacity being defined as the quantity of heat required to raise the temperature of a body by one degree. The specific thermal capacity, that is the thermal capacity of one gram of a substance, is clearly a quantity which is characteristic of the substance and is called the *specific heat*. Defined in this way the numerical evaluation of the specific heat of a substance must clearly be accompanied by an expression of the corresponding units of measurement. Thus in the C. G. S. system of units the specific heat will be expressed in calories per gm. per °C. Since the value of the calorie varies slightly with temperature it is as well to realise that the value of the specific heat given in this way will show corresponding slight variation with temperature. The variation resulting in this way is quite distinct from the variation of specific heat with temperature discussed in section 2.26.

The specific heat of a substance is alternatively defined as the ratio of the quantity of heat required to raise the temperature of a given mass of the substance through a given range of temperature, to the quantity of heat required to raise the temperature of an equal mass of water through the same range of temperature. Defined in this way the specific heat of a substance, being the ratio of two similar quantities, has no units. In addition it will be evident that its value is independent of the system of units used.

From a consideration of these definitions and the definition of the calorie it is clear that the value of the specific heat of a substance in the C. G. S. system of units will be the same whichever definition of specific heat is used. There has been considerable discussion as to which definition shall be generally adopted and present opinion seems to be hardening in favour of the first definition given. In this book it will be understood also that this definition is adopted, and in consequence all specific heat values will have their associated units specified.

TABLE 3

Specific heats of common substances in cal. per gm. per °C.

Substance	Temperature °C.	Specific Heat
Solids		
Brass..	0	·088
Eureka	18	·098
Ebonite......................................	20–100	·33
Glass (crown)	10–50	·16
Glass (flint)	10–50	·12
Indiarubber	15–100	·27–·48
Marble	18	·21–·22
Paraffin Wax	0–20	·69
Liquids		
Alcohol (ethyl)...........................	0	·547
Benzene......................................	10	·34
Ether (ethyl)...............................	18	·56
Glycerine	18–50	·58
Olive Oil	7	·47
Paraffin oil	20–60	·51–·54
Turpentine	18	·42

In many heat exchange experiments use is made of a small metal vessel usually made of copper to contain water or other liquid concerned in the heat exchange process. Such a vessel is called a *calorimeter,* and if its mass is *m* grams and *s* is the specific heat of the material from which it is made, then clearly its thermal capacity will be *ms* calories per °C. Thus during the heat exchange process the calorimeter will absorb or give up the same quantity of heat as would *ms* grams of water. This mass of water, having the same thermal capacity as that of the calorimeter, is called the *water equivalent* of the calorimeter, and is frequently used to facilitate calculation in heat exchange problems.

2.04 Specific Heat of Water

To obtain the relative values of the calorie as discussed in section 2.02 it was necessary to obtain the value of the specific heat of water at the various temperatures concerned. The variation of the specific heat of water over a wide range of temperature has been very accurately measured by Callendar and Barnes using their constant flow calorimeter. The details of their method will be discussed in section 6.09, but some of their results for the specific heat of water are given below.

TABLE 4

Specific heat of water at various temperatures

Temperature °C.	Specific Heat	Temperature °C.	Specific Heat
0	1·0094	40	·9982
5	1·0054	45	·9983
10	1·0027	50	·9987
15	1·0011	60	1·0000
20	1·0000	70	1·0016
25	·9992	80	1·0033
30	·9987	90	1·0053
35	·9983	100	1·0074

The specific heat is a minimum at 37·5°C., and the values given above are those deduced by Callendar from the results of the experiments and calculated relative to the specific heat at 20°C.

The results obtained by Callendar and Barnes were later confirmed by an independent method devised by Callendar known as the

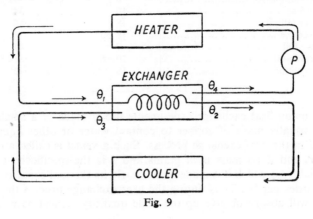

Fig. 9

continuous mixture method. This is an ingeneous combination of the mixture and continuous flow methods of calorimetry. A steady stream of distilled air-free water is circulated by a pump P (Fig. 9) through a *heater* maintained at a constant temperature to a spiral in a part called the *exchanger*. From here it passes to a *cooler* maintained at a constant temperature, and thence past the spiral in the heat exchanger from which it emerges for the process to be continuously repeated. In the exchanger there is an exchange of heat between the two streams of water, the inflow and outflow temperatures of which are taken in the steady state by platinum resistance thermometers. If these temperatures are respectively θ_1 and θ_2 for the water stream from the heater, and θ_3 and θ_4 for the return stream from the cooler, and if the water circulates at the rate of m gm. per second through the system, we have

$$ms(\theta_1 - \theta_2) = ms'(\theta_4 - \theta_3) + h$$

where s and s' are respectively the mean specific heats of water for the temperature ranges θ_1 to θ_2 and θ_3 to θ_4, and h is the amount of heat lost per second by radiation from the exchanger.

The value of h was made small by suitable design of the apparatus, and its value determined by repeating the experiment with different rates of flow but with the same mean temperatures as before. The equation above then enables the value of s' to be obtained in terms of s, and by adjusting the temperatures of the heater and cooler, the relative specific heats of water at different mean temperatures can be obtained over a wide temperature range.

2.05 Calorimetry

The experimental study of heat exchanges, and the measurement of the quantities of heat involved is called calorimetry, the apparatus designed to make such measurements being known as calorimeters. The various methods used are

(*a*) Method of mixtures (see section 2.06).

(*b*) Method of cooling (see section 2.10).

{*c*) Electrical methods (see sections 2.12 and 2.13).

(*d*) Continuous flow methods (see section 6.09).

(*e*) Methods depending on latent heat (see sections 2.22—2.24).

In all the methods except those referred to under (*e*) changes in temperature have to be measured. Latent heat calorimeters possess the special advantage of not requiring the measurement of temperature changes, the quantity of heat being measured in these cases by the latent heat required to effect the change of state of a substance at a fixed temperature. The measurements to be taken here therefore are those of mass, not temperature.

2.06 Method of Mixtures

This is the oldest method of measuring the heat exchanges involved in determining the specific heats of solids and liquids, and it still remains the method most generally adopted in simple laboratory practice. The principle of the method is as follows. A mass m_1 of a substance whose specific heat s_1 is required is heated to a temperature θ_1. It is then quickly transferred to a calorimeter of mass m_2 and specific heat s_2 containing a mass m_3 of water at a temperature θ_2. If the final temperature of the mixture is θ then, taking the specific heat of water as 1, the specific heat s_1 of the solid can be calculated by equating the heat lost by the solid to the heat gained by the calorimeter and contents. That is

$$m_1 s_1 (\theta_1 - \theta) = (m_2 s_2 + m_3)(\theta - \theta_2)$$

$$\text{or } s_1 = \frac{(m_2 s_2 + m_3)(\theta - \theta_2)}{m_1(\theta_1 - \theta)} \quad \cdot \quad \cdot \quad \cdot \quad \cdot \quad [5]$$

The method can be adapted to find the specific heat of a liquid by placing a known mass of it in the calorimeter and introducing a mass of a solid of known specific heat heated to a suitable high temperature as above.

In order to obtain reasonably accurate results by this method it is necessary to take many precautions. The main points requiring attention are the rapid transfer of the heated solid to the calorimeter, the efficient stirring of the liquid in the calorimeter to ensure uniformity of temperature, the avoidance of heat losses by evaporation, and the reduction and correction for heat losses by convection and radiation. Unless these points receive careful attention, the simple heat relation given above cannot strictly be applied. The student will be familiar with the simple precautions taken to obtain reliable results by this method, the most usual arrangement being to support the calorimeter on a cork base inside a double-walled enclosure provided with a lid. Polishing the calorimeter reduces losses due to radiation (although these are small at the temperatures usually involved), and lagging the calorimeter with cotton wool will diminish losses due to convection. Care must be taken however to ensure that the cotton wool remains dry otherwise the losses due to evaporation will be enhanced, and the damp layers of cotton wool will increase the thermal capacity of the calorimeter by an unknown amount. In spite of these precautions however, there is still a leakage of heat from the calorimeter the effect of which is to reduce the final temperature observed to some value lower than that which would otherwise have been attained.

Accordingly it becomes necessary to devise some method of eliminating or correcting for these heat losses. One method used is, after a preliminary experiment to ascertain the range of temperatures involved, to cool the calorimeter and its contents below the temperature of its surroundings by an amount equal to half the rise of temperature estimated. The assumption made is that during the first stage of the experiment the calorimeter will receive an amount of heat from its surroundings equal to that given out by the calorimeter at the later stages when its temperature is higher than that of its surroundings. This method of compensating for the heat losses can only at best be very approximate as the rate of supply of heat to the calorimeter by the hot body is not constant. The rate will rapidly diminish as the equilibrium temperature is attained, and consequently the time during which the temperature of the calorimeter is above that of its surroundings will be greater than the time during which it is below this temperature.

With a good conductor, *e.g.*, copper or other metal, the equilibrium temperature is rapidly attained and the correction for heat losses will in general be very small if the usual precautions are taken. However a bad conductor such as a piece of rubber or glass, gives up its heat slowly to the contents of the calorimeter, and the estimation of the radiation correction* is then a matter of some importance. The method of estimating this correction will be dealt with after a consideration of the laws of cooling.

2.07 Newton's Law of Cooling

A hot body freely suspended in air will cool by the process of radiation, which takes place equally in all directions, and by the convection and conduction of the air surrounding it. Provided the temperature of the body is not greatly in excess of that of its surroundings, the chief factor promoting cooling is the convection of the air surrounding the body, the part played by radiation being relatively small. Convection losses result from the air in contact with the body becoming heated by contact with the hot surface (conduction) with the subsequent formation of an ascending current of hot air and a movement of cooler air towards the body, which in turn carries away some of its heat.

Newton was the first to investigate the cooling of a body freely exposed to the air, but his observations were taken with the body placed in a steady current of air and, consequently, the conditions of his experiments were not quite the same as for the simple case of "natural convection" described above. From the results of his observations Newton put forward an empirical law to express the relation between the temperature of the body and the rate of loss of heat. This law states that *the rate of loss of heat of a body is directly proportional to the excess of the temperature of the body above that of its surroundings*. The law is found to be true for quite large temperature differences, *provided* the body cools under the influence of a strong draught of air. If, however, the cooling takes place as a result of "natural convection" in otherwise

---

* The term is unhappily chosen as for the temperature differences usually met with, the major losses are due to convection and evaporation.

H.T.–2+

still air conditions, the law is only approximately true for small temperature excesses.

In calorimetric experiments the calorimeter is surrounded by an outer jacket which completely excludes any outside draught. Newton's law is thus not strictly applicable in these cases, but since the temperature differences involved are usually small, the law may be applied with sufficient accuracy to calculate the small correction to be applied for heat losses during the experiment.

In mathematical form Newton's law may be written:

$$- \frac{d\theta}{dt} = k (\theta - \theta_r) \quad \ldots \ldots \quad [6]$$

where $- \dfrac{d\theta}{dt}$ is the rate of cooling at the temperature excess $(\theta - \theta_r)$, θ_r

being the temperature of the surroundings. The constant k is a numerical factor which depends on the nature of the body and the extent and emissivity of its surface. Rearranging equation [6] we have:

$$\frac{d\theta}{\theta - \theta_r} = -kdt$$

and this on integration becomes

$$log_e(\theta - \theta_r) = -kt + C$$

where C is a constant of integration.

Now when $t = o$, $\theta = \theta_o$ (θ_o being the initial temperature of the body), and therefore

$$log_e(\theta_o - \theta_r) = C$$

hence we have

$$log_e \left(\frac{\theta - \theta_r}{\theta_o - \theta_r} \right) = -kt$$

$$\text{or } \theta' = \theta_o'e^{-kt}$$

where θ', θ_o' represent the excess temperatures of the body, that is temperatures measured from θ_r.

The cooling curve is thus an exponential curve (Fig. 10) for a body cooling under conditions where Newton's law applies.

2.08   Verification of Newton's Law of Cooling

In verifying Newton's law of cooling it is necessary to reproduce the conditions stipulated by Newton in the enunciation of his law, that is, the observations should be taken with the body cooling in a current of air. This can be done with the body placed in the stream of air from an open window or in the draught caused by an electric fan placed near to the body. A copper calorimeter half filled with water heated to a temperature of about 60°C. is placed on a cork slab and allowed to cool under the above conditions. The readings of a $\frac{1}{10}$°C. thermometer, which is also used as a stirrer, are taken at two minute intervals until the water cools to within a few degrees of the temperature (θ_r) of the air stream. A cooling curve (Fig. 10) is then plotted, and from this curve

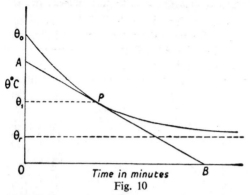

Fig. 10

gradients are taken at given temperatures (say every 5° from the initial temperature θ_o). Thus the gradient $\dfrac{AO}{OB}$ at P corresponding to the temperature θ_1 gives the rate of cooling $\dfrac{d\theta}{dt}$ at this temperature, and since the thermal capacity of the calorimeter is constant throughout the experiment, this rate of cooling is proportional to the rate of loss of heat by the calorimeter. Thus if the rates of cooling $\dfrac{d\theta}{dt}$ are plotted against the corresponding excess temperatures (measured from θ_r) a straight line graph (Fig. 11) should result if Newton's law is obeyed.

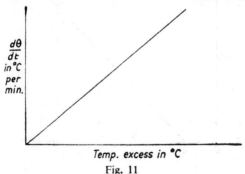

Fig. 11

2.09 Specific Heat of a Bad Conductor: Radiation Correction in Calorimetry

We are now in a position to consider in more detail the method of applying the correction for radiation losses in fixing the final temperature of the mixture in simple calorimetric experiments. In dealing with metals where the transfer of heat takes place quickly, the correction is usually ignored. In the case of a badly conducting solid like rubber or glass however, a considerable time elapses before the hot solid gives up its heat to the water, during which time there is an appreciable loss of heat by radiation, &c. from the calorimeter and the final temperature indicated is well below the value that should be used in calculating the result. The method of obtaining the correction to be applied is illustrated in the following experiment to find the specific heat of a glass stopper.

A stopper of known mass is heated for a considerable time (at least 30 minutes) in a steam heater, after which it is quickly transferred to a weighed calorimeter containing a known amount of water, the calori-

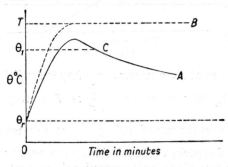

Fig. 12

meter being supported on a cork base inside a double-walled enclosure containing water. Using a $\frac{1}{10}$°C. thermometer readings of the temperature are taken at minute intervals from the moment of transfer, and are

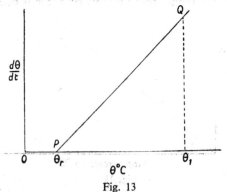

Fig. 13

continued after the mixture has attained its highest temperature as the calorimeter and contents cool through about 5°C. A temperature-time curve (graph A in Fig. 12) is now drawn, and the rate of cooling at some point C corresponding to a temperature θ_1 is obtained by drawing a tangent to the curve at C. Since the rate of cooling at the room temperature θ_r is zero, a straight line graph PQ (Fig. 13) can be drawn showing the rate of cooling in °C. per minute against θ°C. from these two values. By Newton's law of cooling, the rate of cooling at any other temperature is proportional to the excess temperature of the body over the room temperature, and hence from this straight line, the rate of cooling at any other temperature can be read off. From Fig. 12 the mean temperature during the first minute interval is taken, and the rate of cooling in °C. per minute corresponding to this temperature is obtained from Fig. 13. This will give the amount by which the ordinates at the end of the minute interval in graph A must be increased to obtain the corrected temperature. The correction for the next minute interval is obtained in the same way, and the ordinate at the end of this interval is increased by the sum of the corrections for the two intervals. Proceeding in this way the corrected curve B is obtained from which the final temperature T corrected for radiation losses, is read off. The specific heat of the stopper is then calculated using this corrected temperature from equation [5].

The following table of results obtained in an experiment to determine the specific heat of glass in the form of a stopper will serve to illustrate the method of applying the correction for radiation losses described above. In this case the temperature readings were taken at the end of every ¼ minute for the first eight time intervals, and subsequently at minute intervals. A graph was drawn from these observations (similar to curve A of Fig. 12) and from it the rate of cooling at 22 °C. was found to be 0·46 °C. per min. The room temperature was 14·0 °C. and a graph showing the rate of cooling against the excess temperature was drawn (as in Fig. 13) from which the mean rates of cooling and thus the temperature corrections for each time interval were obtained as indicated.

Time interval (¼ min.)	θ °C. observed	Mean temperature	$\dfrac{d\theta}{dt}$ in °C. per min. at mean temperature (from rate of cooling graph)	Cooling in °C. during interval	Correction to temperature at end of interval	Corrected temperatures
0	14·0					14·0
		15·7₅	0·10	0·02₅	0·02₅	
1	17·5					17·5₂
		19·1₅	0·29₅	0·07	0·09	
2	20·8					20·9
		21·4	0·42₅	0·11	0·20	
3	22·0					22·2
		22·3₅	0·48	0·12	0·32	
4	22·7					23·0
		22·9	0·51₅	0·13	0·45	
5	23·1					23·5₅
		23·2	0·53	0·13	0·58	
6	23·3					23·9
		23·3₅	0·54	0·13	0·71	
7	23·4					24·1
		23·4	0·54	0·13	0·84	
8	23·3₅					24·2
		23·2	0·53	0·53	1·37	
12	23·1					24·5
		22·9	0·51	0·51	1·88	
16	22·7					24·6
		22·4₅	0·49	0·49	2·37	
20	22·2					24·6
		22·0	0·46	0·46	2·83	
24	21·7₅					24·6

The further experimental details were as follows:—

Weight of empty copper calorimeter	= 32·32gm.
Weight of calorimeter + water	= 69·25gm.
Weight of water	= 36·93gm.
Weight of glass stopper	= 29·35gm.
Initial temperature of glass stopper	= 99·5°C.
Initial temperature of water	= 14·0°C.
Final temperature (corrected as above)	= 24·6°C.

Then if s = specific heat of the glass stopper $29·35 \times s \times (99·5 - 24·6) =$ $(36·93 + 0·1 \times 32·32)(24·6 - 14·0)$ from which

$$s = \frac{40·16 \times 10·6}{29·35 \times 74·9}$$

$$= ·194$$

Without correcting for radiation losses in this experiment a value for s, which is approximately 13 per cent lower than the value derived above, is obtained.

The method of applying the cooling correction in other cases will be described in sections 6.11(a) and 6.11(b).

2.10   Specific Heats of Liquids by the Method of Cooling

Investigations on the specific heats of liquids by the method of cooling were first made by Dulong and Petit whose work was later extended by Regnault. The method is based upon the assumption that when a body cools in a given enclosure, the rate of loss of heat at a given temperature depends upon the excess temperature (θ) of the body above the enclosure, and on the nature and extent of the surface of the body. Thus if $\dfrac{dH}{dt}$ represents the rate of loss of heat when the excess temperature is θ, we have

$$-\frac{dH}{dt} = Af(\theta) \quad \cdots \cdots \quad [7]$$

where A is a constant depending on the extent and emissivity of the surface of the body and $f(\theta)$ is some unknown function of the excess temperature. In the particular case where Newton's law of cooling applies $f(\theta) = \theta$, but it must be emphasized that the method of cooling as described below is independent of the particular form of the function in equation [7], and is thus not dependent on the validity of any particular law of cooling under the conditions of the experiment.

The usual way of applying the method to the determination of the specific heat of a liquid is as follows: A copper calorimeter (Fig. 14) provided with a copper lid (to reduce losses due to evaporation) through which passes a stirrer of stiff copper wire, is almost filled with water that has been heated to a temperature of about 70°C. The calorimeter and contents are placed on corks inside a large double-walled vessel containing cold water between the walls. This vessel should be sufficiently large to ensure that the temperature of the enclosure does not vary appreciably during the experiment. Data for a cooling curve is now obtained by taking readings of the thermometer every two minutes until

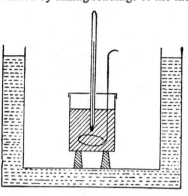

the water has cooled to within a few degrees of the temperature of the enclosure. The water should be kept continuously stirred throughout to ensure that the temperature shown by the thermometer shall also be the temperature of the walls of the calorimeter. The experiment is now repeated using the same volume of some other liquid (say paraffin) in the same calorimeter which is allowed to cool under identical conditions.

Fig. 14

Now if M_1, M_2 be the masses of water (specific heat 1) and paraffin (specific-heat s) in the calorimeter (specific heat s_1) whose mass (with lid and stirrer) is m, the rates of loss of heat in the two cases at some excess temperature θ_1 are respectively,

$$(M_1 + ms_1) \left[\frac{d\theta}{dt} \right]_1 = Af(\theta_1)$$

$$\text{and } (M_2 s + ms_1) \left[\frac{d\theta}{dt} \right]_2 = Af(\theta_1)$$

where $\left[\frac{d\theta}{dt} \right]_1$ and $\left[\frac{d\theta}{dt} \right]_2$ are the corresponding rates of fall of temperature in °C. per min. at the excess temperature θ_1 obtained by taking the gradients $\frac{OA}{OB}$ and $\frac{OA'}{OB'}$ at the points P and Q from the cooling curves (Fig. 15) for the two liquids.

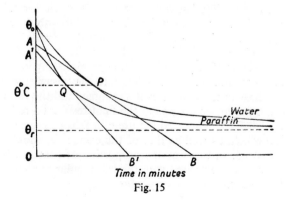

Fig. 15

Now the rates of loss of heat at the temperature excess θ_1 are identical in the two cases, and hence,

$$(M_1 + ms_1)\left[\frac{d\theta}{dt}\right]_1 = (M_2 s + ms_1)\left[\frac{d\theta}{dt}\right]_2$$

from which the specific heat s of the paraffin may be calculated. Performed in this way the method gives the specific heat of a liquid at a particular temperature. For a mean value over a given temperature range equation [7] can be integrated assuming the specific heat constant. Thus

$$-dH = A f(\theta)dt$$

$$\text{or} \quad dt = -\frac{Cd\theta}{A f(\theta)} \quad \cdots \cdots \quad [8]$$

where C is the thermal capacity of the calorimeter and its contents. On integrating equation·[8] the time of cooling from a difference of temperature θ_1 to a difference of temperature θ_2 will be

$$t = -\frac{C}{A}\int_{\theta_1}^{\theta_2}\frac{d\theta}{f(\theta)} = \frac{C}{A}\left[F(\theta_1) - F(\theta_2)\right]$$

where $F(\theta) = \int\frac{d\theta}{f(\theta)}$

If t_1, t_2 represent the times taken for the calorimeter containing first water and then the paraffin to cool between the temperatures θ_1 and θ_2 in each case, we have, since the values of $F(\theta_1)$, $F(\theta_2)$ and A are the same in both cases,

$$\frac{t_1}{t_2} = \frac{C_1}{C_2} = \frac{M_1 + ms_1}{M_2 s + ms_1}$$

from which the average value of s between the temperature range θ_1 to θ_2 can be calculated.

The method of cooling has not been found suitable for solids, but with liquids consistent results can be obtained if the observations are taken carefully and the liquid is kept continuously stirred. There is no transference or mixture with this method which has the special advantage of yielding values of the specific heat at specified temperatures. Any variation in the conditions however invalidates the assumption that the rate of loss of heat in the two cases is the same, and the full effect of any such variation will be present in the final result. To reduce these uncertainties Callendar advocated the use of a fairly large calorimeter, the surface of which, as well as that of the enclosure, should be permanently blackened. In this way the proportion of heat lost by direct radiation is increased compared with the less regular methods of conduction and convection.

2.11 Metallic Block Calorimeters

The use of water as a calorimetric substance is restricted to a modest range of temperatures above 0°C. In determining the specific heat of a body at high temperatures the rise in temperature resulting on its transfer to a water calorimeter may be such as to cause serious loss of heat due to evaporation of the water. A calorimeter in which this objection is overcome is shown in Fig. 16. It was designed by Nernst and used by him in conjunction with Koref and Lindemann to determine specific heats of substances at both high and low temperatures.

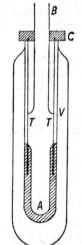

Fig. 16

A large copper block A weighing about 400gm. takes the place of water as the calorimetric substance and is placed inside a vacuum vessel V, where it is made to fit perfectly by the use of Wood's fusible metal. A weighed piece of the substance whose specific heat is required is heated to a known high temperature and introduced into the calorimeter through the glass tube B which passes through a second copper block C, closing the vacuum vessel at the top. The whole apparatus is enclosed in a water-tight sheath of thin copper and is immersed in a constant temperature bath during the experiment, the temperature of the copper block C being maintained at the temperature of the bath. The rise in temperature of the copper block A is determined by means of constantan-iron thermo-couples TT whose junctions are fused into glass tubes and sunk into the copper blocks A and C, good thermal contact being obtained again by the use of Wood's fusible metal.

Heat losses through the sides of the apparatus are reduced to a minimum by the use of the vacuum vessel, and to reduce convection losses through the neck of the apparatus the glass tube B is closed by a wad of cotton wool after introducing the body. On account of the high thermal conductivity of copper, the equilibrium temperature is quickly obtained and there is none of the troublesome stirring required in water calorimetry. In addition the lower specific heat of copper results in a greater rise in temperature than with a water calorimeter containing the same mass of water. In making determinations of specific heat with this calorimeter it is first calibrated using a substance of known specific heat, lead being used for low temperatures, and water for high temperatures.

2.12 Drew's Electrical Method

The principle of the method of electrical heating in calorimetry is the supplying of heat to a known mass of a substance by means of a conductor through which passes a measured current. Using Joule's equivalent the electrical energy supplied is then converted into heat units, enabling the necessary calculations to be made for the determination of the specific heat of the substance. Electrical methods have the advantage that the amount of heat supplied is under complete control and is capable of being measured to a high degree of accuracy. The heat can be supplied at the exact point it is required, almost all the heat being absorbed by the body. The method is suitable for the determination of specific heats at both high and low temperatures, and since the conditions of the experiment are capable of such careful adjustment it is a convenient method of finding specific heats over small ranges of temperature.

A very simple way of applying the method of electrical heating has been suggested by Drew. A measured mass of the solid of cylindrical shape is wrapped in a very thin mica sheet round which is wound several turns of resistance wire through which a known current is passed for a measured interval of time. The resulting rise in temperature of the solid is measured by means of a thermo-couple, one junction of which is inserted into a small hole bored in the specimen.

If the potential difference across the heating coil is E volts and the current supplied is I amperes, then the amount of electrical energy supplied in t seconds is EIt joules. Thus the heat produced is $\dfrac{EIt}{J}$ calories where J is Joule's equivalent in joules per calorie. This heat is supplied to M gm. of the substance of specific heat s and to m gm. of the resistance wire of known specific heat s_1. Hence if the temperature rises from $\theta_1°$ to $\theta_2°$C. in t seconds we have

$$(Ms + ms_1)(\theta_2 - \theta_1) = \frac{EIt}{J}$$

from which s can be calculated. Determinations of specific heat can be made at very high temperatures in this way, a correction for heat losses to the surroundings being necessary to obtain an accurate result.

2.13 Nernst and Lindemann's Vacuum Calorimeter

An accurate series of experiments in which the specific heats of a number of substances at low temperatures were determined was carried out by Nernst and Lindemann (1911) using the method of electrical heating with the apparatus shown in Fig. 17. In dealing with metals the substance acted as its own calorimeter. A cylindrical plug of the metal was made to fit into a hollow cylinder B of the same metal. A platinum heating coil H was wound on paraffin-waxed paper round the plug A and insulated from B by filling the narrow interspace with paraffin wax. The platinum spiral served both as heating coil and as a platinum resistance thermometer by which the temperatures of the substance were obtained. The calorimeter so constructed was then suspended by the connecting leads in a glass vessel V which could be evacuated through the tube *C*. In carrying out an experiment the vessel *V* was placed in a constant temperature bath of ice, liquid air, &c.,

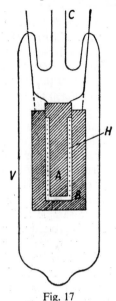

Fig. 17

according to the temperature required, and allowed to stay there for some time until the calorimeter had attained the temperature of the bath. The vessel *V* was then evacuated, and heat was supplied to the specimen by means of the heating coil to raise the temperature by a small amount (about 1°). The specific heat of the substance was then calculated from the observed values of the current and the potential difference applied to the coil as in the previous section. Heat losses are almost entirely eliminated by carrying out the experiment in a vacuum, the only correction required being one for the very small radiation losses.

A modified form of calorimeter (Fig. 18) was used for non-metals. It consisted of a silver vessel round which was wrapped the heating coil. This was covered with a layer of paraffin-waxed paper and then surrounded by a sheet of silver foil to improve the thermal contact and to diminish heat losses. A weighed amount of the substance under test was placed inside the silver vessel and the lid soldered in position. The narrow tube at the top was then closed by a drop of solder to keep the air inside (hydrogen was used at very low temperatures), the presence of the air being required to establish rapidly a condition of thermal equilibrium inside the vessel. The silver calorimeter was then placed inside the glass vessel *V* (Fig. 17) and the experiment carried out as described for good conductors. For very low temperatures the platinum coil was replaced by two coils, one of constantan which was used as the heating coil, and another of lead which served as the thermometer coil.

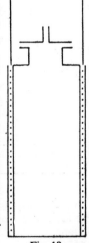

Fig. 18

2.14 Combustion Calorimeter—Calorific Values of Fuels

The determination of the calorific value of a given fuel represents an important problem to the engineer. The calorific value is the quantity of heat evolved when a given mass of the fuel is completely burnt; it is usual to express the result in calories per gram or British thermal units per pound of the fuel used. There are two main types of calorimeter employed for determining this value, the " bomb " type and the " bell " type. In the former type the fuel is burnt by electrical heating in oxygen at high pressure, and this results in instantaneous combustion which takes place with explosive violence. The calorimeter is a cylindrical vessel made of strong steel or other suitable alloy and lined with enamel to prevent oxidation of the metal of the calorimeter. It can be closed with a pressure-tight cover, and the heat of the combustion process is measured by placing the "bomb" in an outer calorimeter containing a known mass of water and recording the rise of temperature produced.

A typical calorimeter of the "bell" type is shown in Fig. 19. In this type of calorimeter the combustion takes place in oxygen under normal

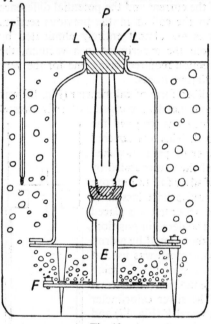

Fig. 19

pressure, and consequently the heat is released more gently over a period of several minutes. To effect the determination of the calorific value of a solid fuel using this calorimeter, a sample of the fuel is carefully ground up in an iron mortar, and an amount of from 1 to 2gm. weighed out in the crucible C. This is then placed in position in the clips as shown, and the glass cover is screwed in position to form an air-tight joint. A rubber bung closes the glass cover at the top, and through this passes the oxygen pipe P which is adjusted so that its lower end is $\frac{1}{2}$" to 1" above the surface of the fuel. Electric leads LL also pass through the bung and connect to a small heating spiral embedded in the fuel. The whole apparatus is immersed in a calorimeter of known water equivalent containing a known mass of water, and a steady stream of oxygen allowed to enter the combustion chamber. The fuel is ignited by connecting a battery to the leads LL, and the hot gases produced by the combustion process pass down the tube E to the water in the calorimeter via the baffle plate F. The rise in temperature of the water is recorded by the thermometer T, and the calorific value of the fuel calculated from

$$\frac{\left(\begin{array}{c}\text{Weight of water} + \text{water equivalent} \\ \text{of apparatus and calorimeter}\end{array}\right) \times \begin{array}{c}\text{Temperature} \\ \text{rise}\end{array}}{\text{Weight of fuel used}}$$

In the case of liquid fuels the crucible and electrical ignition system are replaced by a lamp provided with a wick of asbestos fibre.

TABLE 5

Calorific Value of Fuels

Fuel	Heat value	
	Cal. per gm.	B.Th.U. per lb.
Solid		
Anthracite coal	8,800	15,850
Gas coke	6,000	10,800
Liquid		
Petrol.....................................	11,400	20,520
Paraffin oil	11,200	20,160
Methylated spirit........................	6,400	11,520

2.15 Latent Heat

When a pure crystalline substance is heated a temperature is eventually reached at which it changes sharply into the liquid state. This temperature is known as the *melting point* (or fusing point) of the solid, and for each crystalline substance the temperature at which it passes from the solid to the liquid state or *vice versa* is fixed and definite under given conditions of external pressure. Heat communicated to such a substance at the temperature of its melting point is absorbed by the substance without corresponding rise in temperature until all the substance is melted. The number of calories of heat required to effect the change of state from solid to liquid of 1gm. of the substance without change of temperature is called the *latent heat of fusion* of the substance. Similarly, for the transition of a liquid into the vapour state, heat is absorbed at constant temperature (the *boiling point*) until all the liquid has been converted into vapour. The heat required to convert 1gm. of a substance from the liquid state to vapour without change of temperature is called the *latent heat of vaporisation* of the substance.

In both cases of change of state referred to above the heat energy supplied at the steady temperatures of the melting point and the boiling point of the substance is used up in increasing the potential energy of the molecules (work done against their attractive forces), and by doing work against the external pressure. The latent heat of vaporisation is always greater than the latent heat of fusion since the very much greater volume changes at the higher change of state necessitate the absorption of much more heat to establish this large volume of vapour against the external pressure.

TABLE 6

Latent Heat of Fusion in calories per gm. of
some common substances

Substance	Temperature °C.	Latent Heat
Acetic Acid	4	44
Beeswax	61·8	42·3
Benzene	5·4	30
Bromine......................................	−7·3	16·2
Ice	0	80
Lead	327	5
Mercury	−38·9	3
Napthalene..................................	80	35
Sulphur	115	9

TABLE 7

Latent Heat of Vaporisation in calories per gm. of
some common substances

Substance	Temperature °C.	Latent Heat
Acetic acid..................................	118	94
Alcohol (ethyl)	78·3	205
Benzene	80·1	93
Bromine......................................	61	45·6
Chloroform	61	58
Ether	34·5	88·4
Mercury	358	68
Sulphur	316	362
Water	100	539

2.16 Determination of the Latent Heat of Fusion of Ice

The usual method of determining the latent heat of fusion of ice is by the method of mixtures. Small pieces of dry ice at 0°C. are added to a mass of warm water in a calorimeter, the process being repeated until the temperature of the contents of the calorimeter is as far below room temperature as it previously was above it. The amount of ice added is obtained by finding the increase in weight of the calorimeter and contents at the completion of the experiment. If this mass is M gm. and the temperature of the m gm. of water in the calorimeter of water equivalent m_1 gm. is reduced from t_1°C. to t_2°C., then the latent heat of fusion L of the ice is calculated from the following heat equation:

$$ML + Mt_2 = (m + m_1)(t_2 - t_1) \quad . . . . \quad [9]$$

The main objection to this method of determining the latent heat of fusion of ice is that it is not possible to ensure that the ice is perfectly free from water when transferred to the calorimeter. This difficulty can be avoided by using ice which is at a temperature a few degrees below its melting point. This will ensure that the ice is dry, but equation [9] must then be modified to include a term for the heat required to raise the temperature of the ice to 0°C. before the melting process occurs. This necessitates a value for the specific heat of ice which may be measured by the method outlined in section 2.13.

Another method of finding the latent heat of ice is by the use of Bunsen's ice calorimeter (see section 2.22) using a substance of known specific heat, *e.g.*, water.

2.17 Latent Heat of Fusion of Metals—Experiments of Awbery and Griffiths

Awbery and Griffiths (1926) applied the method of mixtures to determine the latent heat of fusion of metals using the calorimeter illustrated in Fig. 20. The calorimeter was constructed of copper and

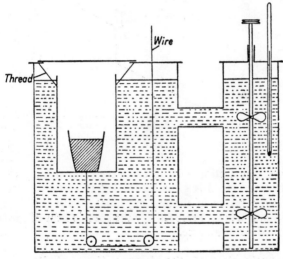

Fig. 20

comprised two chambers communicating as shown. Thorough mixing was ensured by using two propellors driven by a shaft running the length of the small chamber in which was also placed the thermometer. To reduce errors due to loss of heat on transference of the hot body to the calorimeter, Awbery and Griffiths used a large mass (1 to 5kgm.) of the metal which was heated in a crucible to a high temperature by a furnace, the temperature of which was obtained using a platinum, platinum-rhodium thermo-couple. The crucible containing the hot liquid metal was then quickly transferred to a hollow metal vessel suspended by threads from the top of the calorimeter. The lid of the calorimeter was quickly closed and the vessel drawn under the water by the wire attachment shown. This procedure was adopted to avoid errors due to steam losses. A blank experiment using the crucible alone enabled Awbery and Griffiths to calculate the difference between the *total heat* of the metal at the temperature of the furnace and at the final temperature of the experiment (corrected to 20°C. in each case). This difference gives the sum of the heat given out by the mass of liquid cooling to the temperature of the melting point of the solid substance, the heat evolved on solidification, and the further heat given out by the substance in the solid state cooling from the melting point to the final temperature. The experiment was repeated with the metal at different initial temperatures, and a graph drawn showing the total heat in calories per gm. against the temperature, as shown in the case of lead in Fig. 21. The melting point of the metal was obtained by the cooling

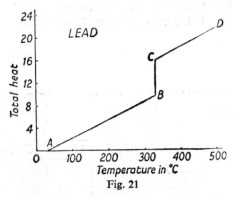

Fig. 21

curve method (see section 9.03(*b*)) using the same thermo-couple. The latent heat of the metal is given by the length of the vertical portion *BC* at the temperature of the melting point, whilst the slopes of the portions *AB* and *CD* give respectively the specific heat of the solid and liquid forms of the substance.

2.18 Determination of the Latent Heat of Vaporisation— Berthelot's Method

A simple method of determining the latent heat of vaporisation of water is to pass a known mass M of steam from water boiling at 100°C. into a calorimeter of water equivalent m_1 containing m gm. of water, and noting the subsequent rise in temperature of the calorimeter and contents. If the initial temperature t_1°C. is raised to t_2°C. by the condensation of the steam, the latent heat L of vaporisation is calculated from the following heat equation:

$$ML + M(100 - t_2) = (m + m_1)(t_2 - t_1)$$

Apart from errors due to radiation losses, the value of L calculated from this equation will only be approximate due to the impossibility of avoiding droplets of water being carried over in the steam supply. The increase in mass M of the calorimeter during the experiment is then not entirely due to steam condensed in the calorimeter, and the value of L obtained will be too low.

A modification of this method using the form of apparatus shown in Fig. 22 was devised by Berthelot with a view to overcoming the disadvantages of the simple method described above. The water is boiled

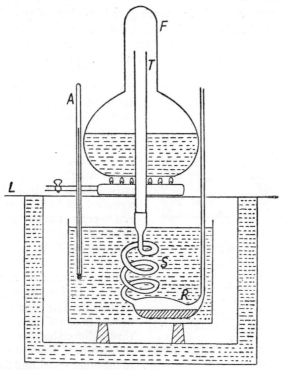

Fig. 22

in a flask F by a circular gas burner as shown, direct transfer of heat to the calorimeter being prevented by covering the calorimeter and protecting water jacket by an asbestos sheet L. The steam passes down a tube T, through the boiling water in the flask, and in this way partial condensation is avoided before the steam enters the calorimeter. The steam condenses in the spiral S immersed in the water in the calorimeter and is collected in a reservoir R. The passage of the steam is continued until a temperature rise of about 5°C. is recorded by the thermometer A when the mass of steam condensed is obtained by removing the spiral and reservoir from the calorimeter and finding the increase in weight. In this way the mass of steam is more accurately obtained than by finding the increase in weight of the relatively large mass of water (subject to evaporation losses during the experiment) in the calorimeter as is done in the simple direct method. Berthelot's apparatus can be applied to the determination by the latent heat of vaporisation of other liquids (of known specific heats), whether or not these liquids are mixable with water.

2.19   Henning's Electrical Method

An accurate method of measuring the latent heat of vaporisation of water was devised by Henning using the method of electrical heating to produce a measured mass of steam. The scheme of Henning's apparatus is given in Fig. 23. The water which was contained in a bronze

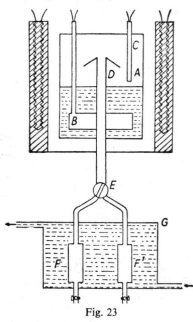

vessel A, fitted with an air-tight lid, was heated by an electrical heating element contained in a ring-shaped metal cylinder B. The temperature of the steam produced was measured by a thermo-couple inserted in the nickel tube C which projected into the vessel A. The steam passed down a tube D which had a conical shield fitted at its upper end to prevent water droplets being splashed into it. By means of a two-way tap E the steam could be directed into either of two condensing vessels F, F' contained in a vessel G through which water circulated. The steam was condensed in one of these vessels until the conditions became steady when the stream of steam was deflected to the other vessel, and the amount condensed in a measured time was

Fig. 23

found by running off the water through the tap provided and weighing it. The steam at the end of this time was switched back to the first condensing vessel.

To reduce heat losses from the vessel A which take place during the course of the experiment Henning surrounded it by an electrically heated ring-shaped oil bath the temperature of which was adjusted so as to be as nearly as possible equal to that of A. Any small remaining heat loss was then eliminated by repeating the experiment with different current and voltage values but with the temperature the same as before. The duration of this second experiment was the same as the first, and consequently, the heat losses were identical in both experiments. Hence if m_1 and m_2 were the masses of steam condensed in a time t seconds in the two experiments, and the voltage and current values were E_1, I_1 and E_2, I_2 respectively

$$m_1 L = \frac{E_1 I_1}{J} t + h$$

$$\text{and } m_2 L = \frac{E_2 I_2}{J} t + h$$

where L is the latent heat of vaporisation, J is the mechanical equivalent of heat in joules per calorie, and h represents the heat losses which occurred in each experiment. Then by subtraction:

$$\left[\frac{E_2 I_2 - E_1 I_1}{J} \right] t = (m_2 - m_1)L$$

$$\text{or } L = \frac{(E_2 I_2 - E_1 I_1)t}{J(m_2 - m_1)} \text{ calories per gm.}$$

The method is applicable to any liquid, and by varying the pressure inside A the boiling point of the liquid can be adjusted, and the latent heat of vaporisation can be determined for a wide range of temperatures.

2.20 Awbery and Griffiths' Continuous Flow Method

The method of continuous flow has been applied by Awbery and Griffiths to the determination of latent heats of vaporisation. The arrangement of their apparatus is shown in Fig. 24. The liquid was boiled in a vessel V heated electrically by inner and outer heating coils H_1 and H_2 which assure a steady rate of production of vapour. The temperature of the vapour so formed was determined by a thermocouple (not shown) which projected through the lid of V into the vapour space above the liquid. Through the base of the boiling vessel a re-entrant tube projected above the liquid surface and carried a silica tube S which extended about 20 inches below the base of V. S was enclosed by a double water jacket C through which water circulated as shown, and the liquid produced by the condensation of the vapour in S was received in a flask F. The rise in temperature of the water stream produced by the condensation of the vapour was recorded by the thermo-couples placed in the inflow and outflow tubes, and another thermo-couple situated at the

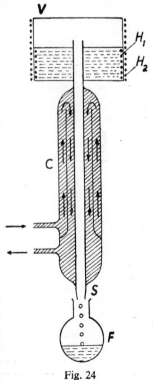

Fig. 24

lower end of S gave the temperature of the liquid as it left the apparatus.

If the temperature of M gm. of water flowing through the condenser per second is raised from θ_1 to θ_2°C. by the condensation of m gm. of vapour of latent heat L calories per gm., and if the boiling point of the liquid is t_1°C. and the temperature of the outflowing liquid is t_2°C., then

$$M(\theta_2 - \theta_1) = mL + ms(t_1 - t_2) \quad \ldots \ldots \quad [10]$$

where s is the mean specific heat of the liquid between the temperatures t_1 and t_2°C. s is determined by a separate experiment when a value of L can be calculated from equation 10.

2.21 Relation between Latent Heat and Boiling Point— Trouton's Rule

An interesting empirical relationship between the latent heat of evaporation and the boiling point of a substance was put forward by Pictet in 1876, and later by Ramsay in 1877 and Trouton in 1884. It is commonly referred to as *Trouton's Rule*. This states that *the ratio of the molecular latent heat of vaporisation of a substance to its boiling point on the absolute scale of temperature is a constant*. Thus if M is the molecular weight of the substance in grams, L the latent heat of vaporisation in calories per gm. and T the boiling point on the absolute scale, then

$$\frac{ML}{T} = \text{constant}$$

For many substances the value of the constant is about 21 calories per °C. Table 8 indicates the limits of applicability of the rule, and it will be seen that there is fair agreement in the case of non-associated vapours, but for associated vapours there are wide departures from the rule.

TABLE 8
Test of Trouton's Rule

Substance	Molecular heat in calories ML	Boiling Point on absolute scale $T°A$	$\dfrac{ML}{T}$
Non-associated vapours:			
Oxygen	1,630	90·1	18·1
Chlorine	4,600	239·5	19·2
Carbon disulphide	6,490	319	20·3
Ethyl ether	6,466	307	21·1
Mercury	14,200	630	22·6
Sodium	23,300	1,155	20·2
Zinc	27,730	1,180	23·5
Associated vapours:			
Ethyl alcohol	9,550	351	27·0
Water	9,710	373	26·0
Formic acid	5,550	374	14·8

2.22 Latent Heat Calorimetry—Bunsen's Ice Calorimeter

Bunsen's ice calorimeter depends on the fact that a change in volume occurs during the liquefaction of ice. Thus the specific volume of ice at 0°C. is 1·0908cc. whilst that of water at the same temperature is 1·0001cc. Hence when one gram of ice is melted there is a contraction in volume of 0·0907cc. Using a value for the latent heat of ice it is thus possible to measure small quantities of heat by the measurement of the volume changes produced in melting corresponding masses of ice. In practice, however, it is usual to calibrate Bunsen's ice calorimeter in such a way that an accurate knowledge of the latent heat of ice is not necessary to the estimation of the heat quantities involved.

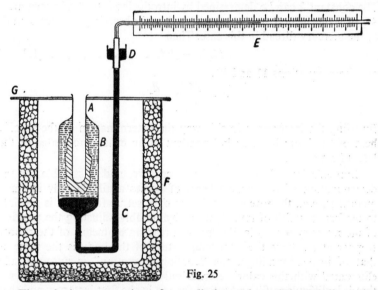

Fig. 25

The calorimeter consists of a small glass tube A (Fig. 25) fused into a wider glass tube B which is provided with a glass stem CD terminating in an iron collar at D. The stem and the lower part of B are filled with pure boiled mercury, whilst the remaining space in B contains pure air-free water. A rubber stopper through which passes the end of a graduated tube E is forced into the end of the tube C, and by adjusting the position of this stopper the extremity of the mercury thread in E can be placed at any suitable position on the scale. To prepare the instrument for use some ether is poured into A and evaporated quickly by bubbling air through it. This causes a shell of ice to form round the outer surface of A as indicated, and when a sufficient layer of ice has thus been produced, the instrument is placed in a double-walled vessel F containing melting ice and the whole covered with a lid G. This arrangement ensures that practically no heat exchange takes place between the instrument and its surroundings. The whole apparatus is left until the temperature is at 0°C. and the mercury thread in E is steady when the instrument is ready for use.

In performing an experiment a little water at 0°C. is placed in A, and a given mass M_1 of a substance of specific heat s which has been heated to some steady temperature θ_1°C. is then rapidly transferred to this water and the recession d_1 of the mercury thread in E is measured. The apparatus should be left until the recession is complete as otherwise the water in B will be above 0°C. If then k represents the quantity of heat corresponding to each division on the scale, we have

$$M_1 s \theta_1 = k d_1 \quad . \quad . \quad . \quad . \quad . \quad [11]$$

The constant k can be determined by introducing into A a known mass M_2 of water heated to a temperature θ_2 and observing the recession d_2 of the mercury thread in this case. Then

$$M_2 \theta_2 = k d_2 \quad . \quad . \quad . \quad . \quad . \quad . \quad [12]$$

and from equations 11 and 12:

$$s = \frac{M_2 \theta_2}{M_1 \theta_1} \cdot \frac{d_1}{d_2}$$

By using the instrument in this way the determination of the specific heat of the body is made independent of an exact knowledge of the latent heat of ice.

Bunsen's ice calorimeter is very sensitive, and is suitable for the determination of the specific heats of bodies available only in small quantities, e.g., the rare earths. When once set up for use it is possible to perform a series of investigations by suitably adjusting the position of the mercury thread in E after each reading by means of the rubber stopper at D. Since the final temperature of the body is the same as that of its surroundings, the "cooling correction" is automatically eliminated with this calorimeter. Disadvantages of the instrument are that it is difficult to fill and set up for use in the first instance, and that the ice formed from a given specimen of water is capable of having slightly different densities. This latter fact limits the use of the instrument for high precision work.

2.23 Joly's Steam Calorimeter

The latent heat of steam may be applied to the determination of specific heats just as the latent heat of ice is applied in ice calorimeters. The steam calorimeter devised by Joly is illustrated in Fig. 26, which shows a section of the steam chamber suitably lagged with non-conducting material, in which is suspended one of the pans of a delicate balance. The solid whose specific heat is required is placed in this pan and its mass determined. The body is allowed to remain in the pan for some time so as to take up the temperature of the chamber, which is determined by an accurate thermometer (not shown) inserted in the steam chamber. Steam is now suddenly admitted to the chamber by a wide steam duct, and the whole chamber becomes immediately filled with saturated vapour. The steam intake is then regulated to a very gentle flow so as to obviate any interference with subsequent weighings by the steam draught. On first entering, the steam condenses upon the body and pan and on the surfaces of the chamber, drops of water being prevented from falling from the roof of the chamber on to the pan by means of the shield shown. The increase in weight of the pan, due to the condensed water, is determined about five minutes after the entry of the steam and the final temperature of the chamber taken. If M is the mass of the body, s its specific heat, w the increase in weight of the pan, c the thermal capacity of the pan and its accessories, L the latent heat of steam, and $\theta_1°$C. and $\theta_2°$C. are the initial and final temperatures of the enclosure respectively, we have

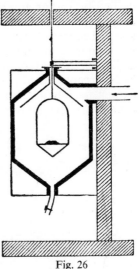

Fig. 26

$$wL = Ms(\theta_2 - \theta_1) + c(\theta_2 - \theta_1) \quad . \quad . \quad . \quad . \quad . \quad [13]$$

The quantity c is determined by a preliminary experiment in which there is no solid in the pan, and thus the specific heat s of the body can be calculated from equation 13.

Precautions are necessary to prevent interference with the weighings, which would otherwise result from the steam condensing and forming a drop of water at the point where the pan suspension wire passes from the chamber. This is effected by closing the tubular opening by a smooth flat disc which has a small central hole through which the suspension wire passes. The disc can move freely in a horizontal plane and so readily adjusts itself to the position of the wire. The small amount of steam still escaping is prevented from condensing on the wire by surrounding it with an electrically heated nickel or platinum spiral immediately above the hole.

Advantages of the method are that the final temperature of the body is the same as that of its surroundings, and hence there is no "cooling correction" to be applied, and that the readings taken consist of weighings which permit of determinations of great accuracy. Small quantities of substances can be dealt with, and the method is applicable to liquids as well as to solids. A special form of the calorimeter by which the specific heat of gases can be determined is described in section 7.05.

2.24 Dewar's Liquid Oxygen Calorimeter

Another form of calorimeter using latent heat values has been devised by Dewar and applied to the determination of specific heats of substances at low temperatures. Instead of finding the mass of vapour that will condense on a cold body to bring its temperature up to that of the vapour as in Joly's apparatus, Dewar applied the reverse process and found the mass of liquid evaporated at its boiling point when a hot body is cooled to the temperature of the liquid. Dewar used liquid oxygen and liquid hydrogen as his calorimetric substances, the arrangement of his liquid oxygen calorimeter being shown in Fig. 27. A vacuum

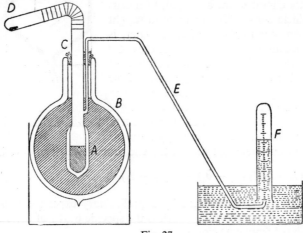

Fig. 27

vessel A of capacity between 25 and 50cc. containing liquid oxygen is placed inside a larger vacuum vessel B of 2 or 3 litres capacity, which also contains liquid oxygen. The vessel A is provided with a wide stem C, to which is attached by a length of flexible rubber tubing a glass tube D containing a known mass of the specimen whose specific heat is required. By tilting the tube the specimen can be introduced into the liquid air contained in A, some of which is vaporised by the heat given out on the substance being cooled to the temperature of the liquid air. The gas thus generated passes out via a narrow outlet tube E, and is collected over water or oil in a graduated receiver F.

As used by Dewar the method was a comparative one in which the volume of oxygen released as above was compared with that liberated by a given mass of lead in an experiment performed immediately afterwards. The thermal capacities of the two bodies are proportional to the volume of oxygen collected in each case, and since the specific heat of lead was accurately known at low temperatures, that of the specimen was readily calculated. This method of dealing with the observations eliminated many sources of error, and obviated the need for an accurate evaluation of the latent heat of vaporisation of liquid air. It is interesting to note that instead of liquid oxygen Dewar used "old" liquid air from which most of the nitrogen had evaporated, and his results covered a long list of specific and atomic heats of metals over different low temperature ranges.

2.25 Experimental Results—Dulong and Petit's Law

A survey of the results of the specific heats of elements in the solid state, made as long ago as 1819 by Dulong and Petit, led to an important generalisation known as Dulong and Petit's law. They found that for a large number of elements the product of the atomic weight and specific heat was approximately constant. This product is known as the *atomic heat* of the substance, and Dulong and Petit's empirical law may be stated as follows: "*For elements in the solid state the atomic heat is a constant and equal to 6·4 calories per gram atom per °C.*" (The value of the constant here stated is that obtained by Regnault in taking the mean value for 32 elements.)

The extent to which the law is true is shown in table 9, which gives the atomic heats of a number of solid elements based on specific heat values taken at ordinary temperatures at constant pressure. It will be seen that the law is not exact and that for certain exceptions, notably the elements carbon and boron, the deviations from the law are very marked, but these are less at higher temperatures.

TABLE 9

Element	Atomic Weight	Specific Heat	Atomic Heat
Aluminium	27·0	0·212	5·72
Arsenic	74·9	0·083	6·22
Boron	10·8	0·307	3·32
Cadmium	112·4	0·055	6·13
Carbon	12·0	0·160	1·92
Copper	63·6	0·091	5·79
Gold	197·2	0·031	6·11
Iron	55·8	0·110	6·12
Lead	207·2	0·030	6·21
Nickel	58·7	0·109	6·40
Silicon	28·1	0·182	5·11
Silver	107·9	0·056	6·04
Tin	118·7	0·054	6·31
Zinc	65·4	0·92	6·02

The general agreement for a large number of substances is, however, sufficiently close to justify the view that the regularity of the atomic heats of the elements observed by Dulong and Petit is not merely fortuitous, but is perhaps a general statement of some more detailed and exact law relating to specific heats of elements in the solid state.

The classical theory of the equipartition of energy due to Boltzman (see section 7.16) gives some theoretical justification to Dulong and Petit's generalisation. If the ideal solid is pictured as a space lattice with the atoms vibrating about fixed equilibrium positions in the lattice, the atoms may be considered to have three degrees of vibrational freedom corresponding to the component velocities referred to the three standard directions, the total kinetic energy of the substance due to these velocities being dependent on the temperature. For simple harmonic vibrations it may be shown that the mean kinetic energy is equal to the mean potential energy, and hence for the lattice as a whole in which the atoms are capable of all displacements at any instant, the total kinetic energy is equal to the total potential energy. Each form of energy is capable of three degrees of freedom, making six in all, and by Boltzmann's theorem of the equipartition of energy each degree of freedom has an associated energy equal to $\frac{1}{2} RT$ ergs per gram atom where R is the gas constant in ergs per °C. Hence it follows that the total internal energy U of the substance is given by,

$$U = 3RT \text{ ergs} \quad \cdots \quad \cdots \quad [14]$$

On differentiating equation 14 with respect to temperature we obtain the change in the internal energy for a change of 1 °C. in the temperature, or in other words, the atomic heat of the substance in work units. Thus the atomic heat

$$\frac{dU}{dT} = 3R \text{ ergs}$$

$$= \frac{3R}{J} \text{ calories}$$

$$= 5.96 \text{ cal. per gm. atom per °C.}$$

Since we are concerned only with changes in the internal energy of the substance the value so deduced is the atomic heat at constant volume. The results considered by Dulong and Petit were based on the experimental values of the specific heats of the substances which were all obtained under conditions of constant pressure. These values may be corrected to constant volume by the application of thermodynamical principles when values are obtained which are lower by 2 to 4% approximately than the corresponding values at constant pressure in the case of solid substances. It would thus appear that the empirical law of Dulong and Petit is in large measure substantiated on theoretical grounds.

2.26 Variation of Specific Heat with Temperature— The Quantum Theory

The values of the atomic heats given in table 9 are based on determinations of the specific heats of the elements at ordinary temperatures, that is at temperatures about 50°C. The systematic work of Nernst and Lindemann and others at low temperatures shows, however, that the specific heat of a substance is by no means constant in value but decreases as the temperature falls, attaining very low values as the temperature approaches the absolute zero. Thus Dulong and Petit's law ceases to be even approximately true as the temperature is reduced materially below O°C., and may only be considered to represent the facts at high temperatures when the atomic heats of the solid elements approach the value stated in the law. The experimental investigations on the variations of specific heats of solids with temperature may be summed up as follows:

(a) The specific heat increases with temperature such that at high temperatures the atomic heat at constant volume approaches the value required by Dulong and Petit's law.

(b) At low temperatures the specific heat decreases, converging to zero as the temperature approaches the absolute zero.

(c) For a range of temperature immediately above the absolute zero the variation of the specific heat is proportional to the cube of the absolute temperature (*Debye's* T^3 *law*).

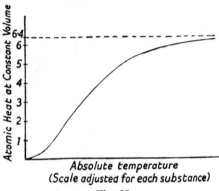

Absolute temperature
(Scale adjusted for each substance)

Fig. 28

The general shape of the curve (see Fig. 28) representing the temperature changes in the atomic heat is the same for all substances, and by a suitable choice of the value of a parameter θ (different for each substance), the curves for all substances may be made to coincide. Thus the atomic heat at constant volume for all simple solid substances as a function of the absolute temperature may be expressed as

$$\text{Atomic heat} = f\left[\frac{T}{\theta}\right]$$

It is thus clear that the classical interpretation of the specific heats of solids, based on the law of equipartition of energy as given in the previous section, does not adequately represent the facts. The classical theory, which is based on the ordinary laws of mechanics, fails to predict any temperature variation in the specific heat of a solid element, and leads to a constant result of 5·96 for the atomic heat at constant volume. It would thus appear that the laws of classical mechanics break down in this instance, and that a modification of the principle of partition is required at low temperatures.

The anomaly was satisfactorily explained by Einstein in 1907, who applied the principles of the quantum theory to the specific heats of solid bodies. This theory, which was introduced by Planck in 1901 in an endeavour to explain the facts of thermal radiation (see section 12.14), postulates that the energy of a system having a fundamental periodicity of v can only change by certain discrete amounts or "quanta" which are integral multiples of hv, h being a universal constant (*Planck's constant*) of value $6·62 \times 10^{-27}$ erg. sec. The justification for the quantum theory rests on its successful explanation of widely different phenomena in the field of Physics, such topics as the photo-electric effect, line spectra, &c., in addition to those mentioned in this section, having been elucidated by this radical departure from the principles of classical dynamics.

In applying the hypothesis to solids, Einstein supposed that the vibrating atoms of the solid could only have energy value of o, hv, $2 hv$, &c., and accordingly continuous energy changes, as assumed by the classical theory were not possible between the atoms of the substance. On the basis of these assumptions, Einstein showed that the energy associated with each degree of freedom of an atom, instead of having a value of $\frac{1}{2} kT$ (where k is Boltzmann's constant, namely, the value of the gas constant calculated for a single atom), must be expressed in the form

$$\frac{1}{2} \frac{hv}{e^{\frac{hv}{kT}} - 1} \qquad \ldots \ldots \ldots \quad [15]$$

This reduces to its classical value of $\frac{1}{2} kT$ if the temperature is high enough or if the characteristic frequency of the atomic oscillation is sufficiently small. An expression for the atomic heat at constant volume derived from equation 15 is of the form,

$$\text{Atomic heat} = 3R \frac{x^2 e^x}{(e^x - 1)^2} \qquad \ldots \ldots \ldots \quad [16]$$

where $x = \frac{hv}{kT}$, which approaches the classical value of $3R$ as $x \to o$, that is, at high temperatures. Debye subsequently improved Einstein's theory and obtained a modified expression of equation 16 which shows close quantitative agreement with the experimental data.

QUESTIONS.   CHAPTER 2

1. Describe carefully some method of determining the specific heat of a liquid.
The temperature of a mass m gm. of a liquid A is 20°C., of a mass $2m$ gm. of a liquid B is 15°C., and of a mass $3m$ gm. of a liquid C is 6°C. On mixing A and B the resulting temperature is 17°C., on mixing B and C the resulting temperature is 10°C. What is the resulting temperature on mixing A and C, assuming that there is no chemical action between the liquids?   (L.)

2. Explain the terms specific heat, capacity for heat, water equivalent.
A copper ball weighing 200gm. and of specific heat 0·1 is heated to the temperature of boiling water and is transferred rapidly to a calorimeter containing water, the temperature of which is thereby raised from 15°C. to 20°C. The ball is removed from the calorimeter, heated in an oil bath and again transferred to the calorimeter, the temperature of which now rises from 18°C. to 28°C. What is the temperature of the oil bath?   (C.)

3. Define specific heat and point out how the numerical value in any particular case depends on the units of mass and temperature involved.
A mass of 750gm. of hot mercury is mixed with 100gm. of cold water in a beaker and the temperature rise is 2·5°C. The experiment is repeated, but this time the

100gm. of water is heated and poured into the 750gm. of cold mercury contained in the same beaker. The temperature then rises 10·0°C. If, in each experiment, the difference in the initial temperatures of the hot and cold liquids is the same, determine the specific heat of mercury. (L.)

4. Sketch the apparatus you would use to determine the specific heat of copper by the method of mixtures and explain briefly the precautions you would take to increase the accuracy of the result. (Deal with each precaution in a separate short paragraph.)

When 500gm. of mercury at 50·0°C. are poured into a vacuum flask containing 90gm. of water at 15°C. the temperature of the mixture is 19°0·C. When 90gm. of water at 50°C. are poured into 500gm. of mercury at 15·0°C. contained in the same flask the temperature of the mixture is 38·0°C. Find the specific heat of mercury and the effective thermal capacity of the vacuum flask. (N.)

5. When a quantity of liquid bismuth at its melting point is transferred to a calorimeter containing oil the temperature of the oil rises from 12·5°C. to 27·6°C. When the experiment is repeated with all the circumstances the same except that the hot bismuth is solid, the temperature of the oil rises to 18·1°C. The specific heat of bismuth is 0·032 cal. gm.$^{-1}$ degC.$^{-1}$. What is its latent heat of fusion? (Melting point of bismuth is 271°C.) (N.)

6. Describe how to determine the specific heat of a liquid by the method of electrical heating, illustrating your description with diagrams and pointing out any precautions which should be taken to increase the accuracy of the work.

Heat is supplied at the rate of 8·0 watts to 200gm. of a liquid in a calorimeter until the temperature of the liquid becomes constant. The rate of cooling at this temperature, when the supply of heat is stopped, is 0·95°C. per min. If the thermal capacity of the calorimeter is 12 cal. per degC., what is the specific heat of the liquid? (N.)

7. Define *specific heat*, and explain how you would measure the specific heat of ice at a temperature a few degrees below its melting point.

At low temperature the specific heat C per gram-molecule of sodium chloride varies with absolute temperatue T as follows:

$$C = k \left(\frac{T}{\theta}\right)^3$$

When $K = 460$cal. mole^{-1} deg. K^{-1} and $\theta = 280°K$. One gram-molecule of sodium chloride is initially at a temperature of 10°K.; if 0·25cal. is supplied, what is the final temperature? (Camb. Schol.)

8. The specific heat of a certain liquid varies slightly with temperature. Describe how it could be measured at any chosen temperature between 0° and 100°C.

When 100g. of a liquid at 100°C. are mixed with 100g. of water at 20°C. the temperature of the mixture is 80°C. If the experiment is repeated with the initial temperature of the liquid changed to 40°C., the temperature of the mixture is 60°C. What can you deduce from these results? Assume that the specific heats of the liquid, of water and of the mixture are independent of temperature, and that there is no loss of heat by cooling. (Camb. Schol.)

9. How can the specific heat of a liquid be found by a continuous flow method? What are the advantages and disadvantages of this method?

An electric kettle (rated accurately at 2½kW.) is used to heat 3 litres of water from 15°C. to boiling point. It takes 9½min. How much heat has been lost? Outline briefly the explanation of this loss of heat. (S.)

10. Define *latent heat of vaporization* and write a critical account of a method for determining its value for water.

200gm. of water in a copper calorimeter of mass 100gm. had a temperature of 10°C. Steam was passed in for one minute and the temperature rose to 45°C. During the next two minutes it fell to 43°C. Calculate the mass of steam condensed.

(Latent heat of steam = 540 cal. per gm. Specific heat of copper = 0·1 cal. per gm. per degC.) (S.)

11. Give a sketch of Bunsen's ice calorimeter explaining how you would use it to find the specific heat of a solid.

If a solid of mass 3gm., specific heat 0·095 cal. gm.$^{-1}$ and temperature 105°C. is dropped into the ice calorimeter, find the travel produced in the capillary tube of diameter 1·6mm. if the latent heat of melting of ice is 80 cal. gm.$^{-1}$ and the density of ice is 0·917 gm. cm.$^{-3}$ (L.)

12. Give an account of a method, based on latent heat calorimetry, of determining the specific heat of a solid of which a small quantity only is available.

When a solid, of mass 5gm. and specific heat 0·22 cal. gm.$^{-1}$ degC.$^{-1}$, is heated to 100°C. and then dropped into the inner tube of a Bunsen's ice calorimeter, the meniscus in the capillary tube, bore 1·0mm., is observed to move through 16cm. Calculate the density of ice at 0°C. The latent heat of fusion of ice is 80 cal. gm.$^{-1}$ (L.)

13. Describe Bunsen's ice calorimeter and the method of using it.

Ten grams of copper at 100°C. are dropped into the inner tube of a prepared Bunsen's ice calorimeter. How far will the end of the mercury thread move in the capillary tube which is of 1mm. diameter? Estimate the smallest quantity of heat which could be detected with this apparatus.

(Density of ice at 0°C. = 0·917 gm. cm.$^{-3}$ Latent heat of fusion of ice = 80 cal. gm.$^{-1}$ Specific heat of copper = 0·095 cal. gm.$^{-1}$ degC.$^{-1}$) (L.)

14. Define *thermal capacity* and hence define *specific heat*.

Draw labelled diagrams showing (a) a steady (or continuous) flow method to determine the mean specific heat of a liquid over a small range in temperature, (b) an electrical heating method to determine the latent heat of vaporization of ethyl alcohol at its normal boiling point (about 80°C.).

For *one* of the above experiments describe and explain how a correction for the exchange of heat between the calorimeter and its surroundings may be made. (L.)

15. Define *specific heat*. Describe how you would determine the specific heat of a liquid by the method of cooling, and give the theory of the method.

The density of a certain liquid at 0°C. is ϱ, its coefficient of expansion is α, and its specific heat is σ. Energy is supplied electrically at the rate of W watts to a quantity of the liquid which initially occupies a volume V at 0°C. Obtain an expression for the volume occupied by the liquid when energy has been supplied for t seconds. (O.)

16. Describe an electrical method for the determination of the latent heat of vaporization of a liquid.

Give two advantages of the method over the method of mixtures.

When water boils at its normal boiling point each gm. of water produces 1,650cm.^{3} of steam. If the latent heat of vaporization is 540 cal. gm.$^{-1}$, what fraction of the energy involved in the water-to-vapour conversion is used in doing work against the external pressure? What happens to the rest of the energy needed for the conversion?

(Assume that 1 atmosphere = 10^{6} dyne cm.$^{-2}$ and 1 calorie = 4·2 joules.) (N.)

17. A closed calorimeter containing a quantity of hot oil rests on a non-conducting stand. What factors determine (a) the rate at which heat is lost, (b) the rate at which the oil cools? Write down an equation relating the rates (a) and (b).

An iron oven of mass 10kgm. is heated by a thermostatically controlled 200watt heating element which is switched on at 60°C. and off at 65°C. If the interval between switching off and on is 4 minutes, find that between switching on and off.
(Specific heat of iron, 0·105 cal. gm.$^{-1}$ degC.$^{-1}$. It may be assumed that the rate of cooling of the oven is uniform over the range 65°C. to 60°C.) (N.)

18. Define *latent heat of vaporization*.
With the aid of a labelled diagram of the apparatus describe an experiment to determine the latent heat of vaporization of water at the prevailing boiling point by a method of electrical heating. Give two advantages of this method over the method of mixtures.
In an experiment to determine the latent heat of vaporization of liquid nitrogen, 5·0gm. of copper at 20°C. were dropped into liquid nitrogen at its normal boiling point (−196°C.). The nitrogen evaporated was collected and occupied 1·38 litre at 20°C. and 770mm. of mercury pressure. Assuming that the mean specific heat of copper is 0·075 cal. gm.$^{-1}$ degC.$^{-1}$ and that the density of nitrogen at S.T.P. is 1·25 gm. litre^{-1}, calculate the latent heat. (N.)

19. Describe an experiment to determine the specific heat of glass using a glass stopper. Explain the procedure which should be adopted to lessen experimental errors.
An immersion heater of 10watts output produces a rise of temperature of 1·4degC. per minute in 160gm. of oil contained in a vacuum flask. When the mass of oil is increased to 200gm. and the output of the heater raised to 20 watts the temperature of the oil rises 2·4degC. per minute. Calculate the specific heat of the oil and the thermal capacity of the flask and heater. (N.)

20. The specific heat of gallium metal is 0·33 joule gm.$^{-1}$ degC.$^{-1}$ Explain carefully how this result may be determined experimentally. Indicate the sources of error in your method and estimate the accuracy which could be achieved.
(Melting point of gallium = 30°C.)
A ball of gallium is released from a stationary balloon, falls freely under gravity and on striking the ground it just melts. Calculate the height of the balloon assuming that the temperature of the gallium just before impact is 1°C. and that all the energy gained during its free fall is used to heat the gallium on impact. Why are the conditions specified in this problem unrealistic?
(Latent heat of fusion of gallium = 79 joule gm.$^{-1}$) (O. and C.)

21. What do you understand by the *specific heat* of a substance? Describe how you would measure the specific heat of a sample of rock, describing the precautions that you would take to obtain an accurate result.
A room is heated during the day by a 1kW electric fire. The fire is to be replaced by an electric storage heater consisting of a cube of concrete which is heated overnight and is allowed to cool during the day, giving up its heat to the room. Estimate the length of an edge of the cube if the heat it gives out in cooling from 70°C. to 30°C. is the same as that given out by the electric fire in 8 hours.
(Density of concrete = 2·7 gm. cm.$^{-3}$; specific heat of concrete = 0·85 joule gm.$^{-1}$ degC.) (O. and C.)

22. Describe a method for the determination of the latent heat of steam. What errors are likely to arise in the experiment, and how may they be reduced to a minimum?
What becomes of the energy which is absorbed by a liquid when it vaporizes without change of temperature?
When water boils at a pressure of one atmosphere each gramme of water produces 1,650c.c. of steam. How much work is done against atmospheric pressure, and how much heat will be required to supply this work? (O. and C.)

23. Give the theory and practice of determining by experiment the *latent heat of fusion* of ice. What precautions must be taken to minimise errors in the experiment? A pond is covered with a layer of ice 2·4cm. thick. The air temperature just above the ice is −1°C. Calculate approximately how long it will take for the thickness of the ice to increase by 0·2cm., given that the density of ice is 0·9 g. cm.$^{-3}$, its thermal conductivity is 5×10^{-3} cal. cm.$^{-1}$ sec.$^{-1}$ degC.$^{-1}$ and its latent heat of fusion is 80 cal. g.$^{-1}$ (A.)

24. What are the advantages and disadvantages of continuous flow calorimetry? Describe how this method has been applied to determine **either** (*a*) the latent heat of steam, **or** (*b*) the specific heat of a gas at constant pressure.

A liquid flows at a steady rate along a tube containing an electric heating element. When the power supplied to this heater is 10 watts and the rate of flow of the liquid is 50 gm. per min., the outflow temperature of the liquid is 5degC. higher than the inflow temperature. When the rate is doubled, the power supplied to the heater must be increased to 19 watts for the outflow temperature to be unchanged. If the inflow temperature is the same in each experiment, calculate a value for the specific heat of the liquid. (Assume that 4·2 joules = 1 calorie).   (N.)

25. Describe experiments you would carry out to find approximately what proportion of the heat lost by a cooling body is lost by radiation. Confine yourself to the two cases: (*a*) the cooling of a brass ball after being heated in a Bunsen burner, (*b*) the cooling of a calorimeter after being filled with boiling water.

Two identical calorimeters (water equivalent = 10gm.) containing 50c.c. of water and paraffin, respectively, are heated to 60°C. and allowed to cool. The water takes 2 minutes to cool from 50°C. to 45°C. How long will it take the paraffin to cool through the same range?

(Specific heat of paraffin = 0·52 cal. per gm. per degC. Specific gravity of paraffin = 0·8.)   (S.)

26. Some substance was placed in a container and heated to 75°C. The container was then held in a retort stand and allowed to cool in a space where the temperature was kept at 15°C. The following table gives the temperature of the substance at various times:

Time in min.	0	1	2	2½	3	11	12	13	14	15	17
Temp. in °C.	64	50	40	37	40	40	40	35	31	28	24

Plot the graph of temperature against time and explain its shape.

Calculate the specific heat of the substance at 50°C. and its latent heat at 40°C. if its specific heat at 30°C. is 0·6 cal. gm.$^{-1}$ deg.C.$^{-1}$ Assume Newton's law of cooling and neglect the water equivalent of the container.   [S.]

27. Explain the physical principles of a domestic refrigerator employing an evaporating liquid and give a labelled diagram showing its essential components. How may the temperature of the main storage compartment be regulated and what factors determine the lowest attainable temperature?

A certain refrigerator converts water at 0°C. into ice at a maximum rate of 5 gm. per minute when the exterior temperature is 15°C. Assuming that the rate at which heat leaks into the refrigerator from its surroundings is proportional to the temperature difference between the exterior and interior and is 2·5 watt degC.$^{-1}$, what is the maximum exterior temperature at which this refrigerator could just maintain a temperature of 0°C. in the interior?

(Latent heat of fusion of ice = 330 joule gm.$^{-1}$)   (O. and C.)

28. What is Newton's law of cooling?

A calorimeter containing a hot liquid is placed inside an enclosure whose walls are at 10°C., and cools from 80°C. to 60°C. in 10 minutes. How long will it take to cool from 60°C. to 40°C. if Newton's law of cooling holds? (Camb. Schol.)

29. An electric kettle containing 900gm. of water initially at room temperature (10°C.) begins to boil 10 minutes after it is switched on to a constant power supply. On switching off the power, the cooling is found to be initially at the rate of 6°C. per minute. If the heat capacity of the kettle is equivalent to 100gm. of water, at what rate was the power consumed, assuming Newton's law of cooling? (Camb. Schol.)

30. Describe and explain how you would determine the specific heat of a liquid.

A calorimeter of water equivalent 10gm. contains 70gm. of a substance in the liquid state at its melting point, and the substance is found to solidify completely in 21 minutes. A similar calorimeter containing 80gm. of water cools at the rate of 1·5°C. per minute at the same temperature, the room temperature being the same in both cases. What is the latent heat of fusion of the substance? (N.)

31. Distinguish between *rate of loss of heat* and *rate of fall of temperature*.

A copper calorimeter of mass 100gm., containing 150gm. of water at 70°C., loses heat at the rate of 88 cal. per minute. If the water is replaced by an equal volume of turpentine at 70°C. what will be (a) the rate of loss of heat, (b) the rate of fall of temperature?

(Density of turpentine = 0·87gm. per c.c. Specific heat of turpentine 0·42. Specific heat of copper 0·1.) (C.)

32. State Newton's law of cooling and describe how you would verify it experimentally.

500gm. of sulphur, contained in a test tube of negligible thermal capacity is initially at a temperature of 150°C. Assuming Newton's law of cooling to be obeyed find the total time for the sulphur to cool down to 20°C. if solidification commences after 10 minutes.

(Melting point of sulphur = 115°C. Specific heat of liquid sulphur = 0·235 cal. gm.$^{-1}$ °C. Specific heat of solid sulphur = 0·163 cal. gm.$^{-1}$ C. Latent heat of fusion for sulphur = 9 cal. gm.$^{-1}$. Temperature of surrounding = 10°C.)

(Oxford Schol.)

33. Explain what is meant by Newton's law of cooling and give examples to show (a) when the law applies, and (b) when it does not apply.

Assuming that the resistance of a lamp filament is proportional to its absolute temperature, show that the increase of the resistance caused by passing a current i through the filament is proportional to i^2 for very small currents and to $i^{2/3}$ for very large currents. (Camb. Schol.)

H.T.–3+

CHAPTER III

EXPANSION OF SOLIDS

3.01 Introductory

Almost all bodies increase in size in consequence of a rise in temperature. This effect is a matter of common experience and has wide practical application. The expansion of metals, though small, is of particular concern to the constructional engineer. For example, a large steel bridge is capable of appreciable contractions and expansions due to fluctuations of temperature, and for this reason is provided with rocking posts to avoid strains in the structure which would otherwise result. For similar reasons there are expansion bends in long lengths of metal piping, gaps are left in railway lines, and clocks and watches require compensation, &c. Clearly therefore it is a matter of first importance that the small increments in length, area, and volume, resulting from temperature changes, should be capable of precise evaluation for different substances. These changes in size are calculated from the temperature changes, and the original dimensions by reference to the various expansion coefficients. The definition of these coefficients, and the manner in which they are measured, is outlined in the subsequent sections of this chapter.

3.02 Linear Expansion

In considering the linear expansion of a substance, it is clear that increase in length must be proportional to the original length of the specimen. Thus if a rod of length 1 metre expands by 2mm. when its temperature is raised by a given amount, a rod of 2 metres will expand by 4mm. under the same conditions. In addition, the amount of expansion will depend on the change of temperature, and experiment has shown that it is very nearly proportional to the temperature increase. Thus if l_o is the length of a rod at 0°C., and l_t its length at t°C., then the amount of expansion is

$$l_t - l_o = \alpha l_o t$$

from which

$$l_t = l_o (1 + \alpha t) \quad . \quad . \quad . \quad . \quad . \quad . \quad [17]$$

where α is a constant which is numerically equal to the amount by which unit length increases for unit temperature rise.

This constant, which may be defined as the fractional increase in length per degree rise in temperature, is known as the *coefficient of linear expansion.* It is independent of the length of the specimen, but its value will alter with different temperature scales.

TABLE 10

Linear expansion coefficients for various substances per °C.

Substance	$\alpha \times 10^6$	Substance	$\alpha \times 10$
Pure metals:		Invar......................	1·0
Aluminium	25·5	Phosphor-Bronze.........	16·8
Copper	16·7	Solder (2Pb. 1Sn.)	25·0
Gold	13·9	Steel	11·0
Iron	12·0	*Non-metals:*	
Nickel.....................	12·8	Brick	9·5
Platinum	8·9	Ebonite.....................	c. 70·0
Silver.....................	18·8	Glass (flint)	7·8
Alloys:		(soda)...............	8·5
Brass	18·9	Porcelain	3·0
Constantan	17·0	Silica (fused) 0–30°C. ...	·42

It is implicit in equation 17 that equal increments of temperature cause equal amounts of expansion. Experimental investigations over a wide range of temperature (measured on the gas scale) show that the exact dependence of length on temperature is given by a formula of the type:

$$l_t = l_o (1 + at + a't^2 + a''t^3)$$

where the constants a, a', a'' have numerical values for pure metals of the order of 1×10^{-5}, 10^{-11}, 10^{-14} respectively. In view of the smallness of the coefficients a' and a'' in relation to a, the more approximate relationship given by equation 17 is sufficient for normal work. It must be realised however that a does in fact vary slightly with temperature, and any experimentally derived value is the mean value of the coefficient between the two temperatures concerned. The actual value of the linear coefficient of expansion at any particular temperature may be specified as follows. If δl is the small increment of length l corresponding to a small change δt in temperature at the temperature t, then

$$\delta l = al\delta t$$

$$\text{or } a = \frac{1}{l} \cdot \frac{\delta l}{\delta t}$$

giving a limiting value at the temperature t of

$$a = \frac{1}{l} \cdot \frac{dl}{dt}$$

It is not always convenient to measure the expansion of a rod from an initial temperature of 0°C. In actual practice, if l_{t_1}, l_{t_2} represent the lengths of a rod at temperatures t_1 and t_2 respectively, it is usual to use the following equation from which to calculate a:

$$l_{t_2} = l_{t_1} \{1 + a (t_2 - t_1)\}$$

This equation is sufficiently accurate for most practical purposes, and is justified on account of the values of a being so small. Thus

$$l_{t_2} = l_0 (1 + at_2)$$
$$l_{t_1} = l_0 (1 + at_1)$$

hence $\dfrac{l_{t_2}}{l_{t_1}} = \dfrac{1 + at_2}{1 + at_1}$

or $\quad l_{t_2} = l_{t_1} (1 + at_2)(1 + at_1)^{-1}$
$$= l_{t_1} (1 + at_2)(1 - at_1 + a^2 t_1^2 - \; . \; .)$$
$$= l_{t_1} (1 + at_2 - at_1 - \; . \; .)$$
$$= l_{t_1} \{1 + a (t_2 - t_1)\}$$

ignoring terms involving a to the second and higher powers.

The error involved in using the more approximate equation can readily be shown in any actual case to be very much smaller than the experimental error of the various measurements involved.

3.03   Superficial and Volume Expansions

In general a rise in the temperature of a body results in equal relative expansion in the three standard directions; hence the area of a plate, and the volume of a block of material, both increase with temperature. The relative increase in area per degree rise in temperature is known as the *coefficient of areal* or *superficial expansion* (β). Thus if A_t, A_o are the areas at temperatures of $t°$ and $0°C$. respectively, then

$$\beta = \frac{A_t - A_o}{A_o t}$$

or $A_t = A_o (1 + \beta t)$ . . . . [18]

Similarly the *coefficient of cubical expansion* (γ) is defined as the relative increase in volume per degree rise in temperature. Hence if V_t, V_o are the volumes at $t°$ and $0°C$. respectively, then

$$\gamma = \frac{V_t - V_o}{V_o t}$$

or $V_t = V_o (1 + \gamma t)$ . . . . [19]

As with the linear coefficient of expansion, both β and γ show slight variation with temperature, but for normal work equations 18 and 19 represent the state of affairs to a sufficient degree of accuracy.

A relation between the three coefficients can readily be established as follows. If a square plate of side l_o has its temperature raised by $t°C.$, then each side will become $l_o (1 + at)$ units in length. Then

$$A_t = l_o^2 (1 + at)^2$$
$$= A_o (1 + 2at + a^2 t^2)$$

Since a is small, we may ignore the term involving a^2, hence

$$A_t = A_0 (1 + 2at)$$

Comparing this with equation 18, it will be seen that

$$\beta = 2a$$

Similarly, a cube of side l_0 whose sides will increase in length to $l_0 (1 + at)$ when the temperature is raised by $t°C$, will have volume at $t°C$. of

$$V_t = l_0^3 (1 + at)^3$$
$$= V_0 (1 + 3at + 3a^2t^2 + a^3t^3)$$
$$= V_0 (1 + 3at)$$

ignoring the last two terms.

On comparing this equation with equation 19, we have

$$\gamma = 3a$$

Thus to a close approximation, the ratio between the three coefficients is

$$a : \beta : \gamma = 1 : 2 : 3$$

Since it is a matter of some difficulty to measure β and γ directly, it is usual only to determine a and to obtain the other two coefficients from the relationships given.

3.04   Expansion of Crystals

It is well to remember that the above only applies to *isotropic* bodies, that is, to that class of bodies whose properties are the same in all directions. For *anisotropic* bodies, such as crystals, the expansion coefficients are different in different directions. The calculation of the volume of such bodies at any particular temperature is in general a difficult matter. It is, however, possible to cut the crystal with faces parallel to three mutually perpendicular axes, called the *principal axes of dilation*, such that the faces still remain at right angles when the temperature is raised. The coefficients of linear expansion in these directions are called the principal coefficients of expansion, and if for a given specimen of dimension l_0', l_0'', l_0''' these coefficients are a', a'', a''' then

$$V_t = l_0' (1 + a't) \, l_0'' (1 + a''t) \, l_0''' (1 + a'''t)$$
$$= l_0' l_0'' l_0''' \{1 + (a' + a'' + a''')t\}$$
$$= V_0 \{1 + (a' + a'' + a''')t\}$$

to the first degree of small quantities. Thus for such crystals

$$\gamma = a' + a'' + a'''$$

Uniaxal crystals possess an axis of crystalline symmetry. The physical properties of these crystals are the same in a plane perpendicular to this axis and hence they have only two principal coefficients of expansion. With certain crystals negative coefficients of expansion are obtained along one or other of the principal axes, although measurements usually show that the volume coefficient γ is a positive quantity.

TABLE 11

Expansion of Crystals

	a' ‖ axis	a'' $+$ axis	$\gamma = a' + 2a''$
Quartz	$7 \cdot 5 \times 10^{-6}$	$13 \cdot 7 \times 10^{-6}$	$34 \cdot 9 \times 10^{-6}$
Iceland spar	$25 \cdot 1 \times 10^{-6}$	$-5 \cdot 6 \times 10^{-6}$	$13 \cdot 9 \times 10^{-6}$

3.05   Measurement of Linear Coefficient of Expansion

It is a matter of some importance that the linear coefficients of expansion of substances should be accurately measured. With the specimen in the form of a bar or tube, this involves the measurement of the length of the bar, the rise in temperature during the experiment, and the increase in length of the bar consequent on this rise of temperature. The first two measurements present no great difficulty, but the measurement of the actual expansion, which for a metre-long bar subject to a rise of temperature of up to 100°C. is only of the order of 1mm., requires special treatment. To measure such a small increment of length directly calls for the use of vernier microscopes and micrometer screw gauges, but before the standard methods using these instruments are described, a method based on the magnification of the small change in length will be given.

3.06   Optical Lever Method

This is a sensitive method, and if adequate precautions are taken to ensure that the apparatus is firm and rigid, is capable of giving quite accurate results in ordinary laboratory practice. The specimen in the form of a metal tube (Fig. 29) closed at both ends and fitted with inlet

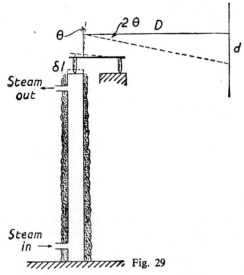

Fig. 29

and outlet tubes at the sides for steam circulation, is supported in a vertical position with the lower end standing on a firm base. The tube should be provided with a heat insulating jacket of cotton wool or felt. (If the specimen is provided in the form of a rod, it can be heated by enclosing it in a suitable steam jacket.) An optical lever of standard design is supported in a horizontal position with its rear leg on the upper end of the metal tube, and its two front legs placed on a firmly clamped metal block or piece of plate glass placed on a level with the top of the tube. A lamp and scale, arranged for vertical displacement, are fixed a measured distance D from the mirror of the lever and the position of the spot of light on the scale is read. Steam is now passed through the tube, and the expansion of the metal causes the mirror to be tilted and produces a deflection of the spot on the scale. The position of the spot is read at the full deflected position on completion of the expansion. The perpendicular distance between the back and front legs of the optical lever is now carefully measured, as also is the length of the metal tube when cold (when the spot should have returned to its original undeflected position). The tube may be assumed to be at air temperature (which is measured by a thermometer suspended near it) when cold, and the steam temperature is obtained from tables having ascertained the atmospheric pressure. From these observations the coefficient of expansion is calculated as follows.

Let the length of the tube be L cm. and the expansion δL cm. when the temperature is increased from $t°C$. to $T°C$. Then if b is the perpendicular distance between the back leg and the line through the two front legs, the angle through which the mirror is tilted is $\dfrac{\delta L}{b}$ radians. Now if the deflection of the spot of light on the scale at a distance D cm. from the mirror is d cm., the angle through which the beam is deflected is $\dfrac{d}{D}$ radians. This angle is twice the angle through which the mirror is tilted.

$$i.e., \quad \frac{d}{D} = 2\,\frac{\delta L}{b} \text{ or } \delta L = \frac{bd}{2D}$$

Thus, since the linear coefficient of expansion (a)

$$= \frac{\text{increase in length}}{\text{original length} \times \text{temperature change}}$$

we have

$$a = \frac{bd}{2DL\,(T-t)} \text{ per } °C.$$

3.07 Screw Gauge Method

This is another simple laboratory method by which a good value for the linear coefficient of expansion can be determined. The specimen in the form of a tube AB (Fig. 30) about a metre in length, is supported

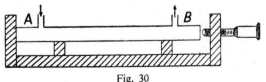

Fig. 30

in a horizontal position on a firm base. The end A of the tube is in contact with a fixed stop, whilst a screw gauge makes contact with the other end B as indicated in the diagram. The reading of the screw gauge is taken with the tube at the temperature of the room, and the gauge is then screwed well back to permit of the expansion of the tube when heated by passing steam through it. The gauge is then screwed up to make contact with the fully expanded tube, and the increase in length of the tube for the measured temperature rise is obtained from the difference of the two readings. The value of the linear coefficient of expansion is then calculated as before from

$$\frac{\text{increase in length}}{\text{original length} \times \text{temperature change}}$$

Another method embodying the screw gauge principle is to use a spherometer with the tube in the vertical position described for the optical lever method. The three legs of the spherometer rest on a firm ebonite table cut with a central hole to permit the expansion of the tube. The centre leg is screwed down through the hole to make contact with the end of the tube and the reading taken. It is then screwed back to allow for the expansion, on completion of which a further reading is taken with the centre leg in contact with the tube. In this way the amount of expansion of the tube is obtained from which α is calculated.

3.08 Comparator Method

The modern standard precision methods for the accurate measurement of the expansion of substances are given in this and the subsequent two sections. The comparator method is the one used by the International Bureau of Weights and Measures for the standardisation of measuring rods, and is shown in Fig. 31. The amount of expansion of

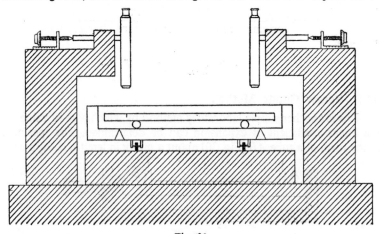

Fig. 31

a metal rod can be very accurately measured by this method by comparison with a standard rod maintained at a fixed temperature. The rod of material under investigation is contained in a double-walled vessel and surrounded with water the temperature of which is thermostatically controlled. The water is kept very efficiently stirred throughout the experiment, and its temperature recorded by accurate thermometers. The rod is supported freely on rollers in the vessel to permit expansion in both directions. A standard rod is contained in a similar vessel whose temperature is kept constant at 0°C. Both rods have fine scratches made near each end so that the distance between the scratches measured on the rods is about one metre. These scratches are viewed through two microscopes supported vertically by massive stone pillars. The microscopes are fitted with cross wires and are capable of horizontal movement in a direction parallel to the rods by means of micrometer screws. The vessels containing the rods are supported on a table resting on wheels so that the rods can be brought under the microscopes in turn. Readings are taken with the cross wires focused on the scratches on the rods, firstly with the standard and the specimen at the temperature of melting ice, and then with the specimen at a succession of controlled temperatures. Finally, the standard is again brought under the microscopes to check whether any movement of the support has occurred during the experiment. In this way, by adding together the two outward movements of the microscopes, the expansion of the specimen relative to the standard bar can be obtained for any required temperature range.

3.09 Henning's Tube Method

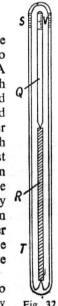

A convenient and accurate method of obtaining the expansion of a specimen in the form of a rod relative to fused silica is provided by Henning's method (Fig. 32). A fused silica tube T contains the specimen R, which rests with its lower end supported by a pointed projection also of fused silica. On the top of the specimen is placed a pointed rod of fused silica Q which has a scale V engraved on its upper end. Another scale S is engraved on the outer tube which is cut away at the top to enable the movement of V against S to be read (by a microscope). In measuring the expansion of the specimen for a given temperature range, the tube is immersed up to a level corresponding to a point mid-way up Q in a water bath, first at the low temperature and then at the higher temperature, the temperatures of the water bath being thermostatically controlled. The expansion of the rod against silica is then read directly from the relative displacement of the two scales V and S.

To obtain the absolute expansion of R it is necessary to allow for the expansion of the silica, and this is usually

Fig. 32

3*

determined optically. Since the expansion of silica is very small **the** correction can be applied very accurately.

3.10 Fizeau's Method

This is a very sensitive method enabling the expansion of small specimens to be measured over a very wide range of temperature. It is particularly suitable for crystals, and information regarding the expansion coefficients along the various principal axes has been very helpful in the study of the crystalline state. Since it is unusual for a crystal to have dimensions exceeding 1cm. the expansions to be measured are of

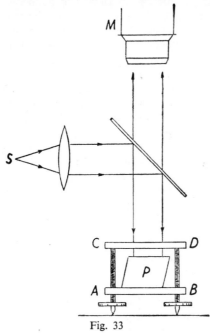

Fig. 33

the order of 10^{-5}cm. or less. These small changes in length are measured in Fizeau's method in terms of the wavelength of light by an interference method.

The specimen under test is cut into a plate P (Fig. 33) with parallel faces, and is from 1 to 10mm. thick. The specimen is then placed on a plane metal disc AB, which is supported by three metal screws passing through it. An optically smooth glass plate CD rests on the top of these screws, which can be adjusted so as to bring P very close to CD, thereby obtaining a very thin air film of nearly uniform thickness between them. Monochromatic light from a cadmium or mercury vapour source S, after being formed into a parallel beam by passing through a lens, is caused to fall normally on this thin air film. Interference occurs between the lower surface of the plate CD and the upper surface of the specimen P, and a set of approximately parallel fringes is seen through the microscope M. If the lower surface of CD is slightly convex a Newton's rings system of interference fringes is formed. A change of temperature will now cause an alteration in the thickness of the air film equal to the difference in expansion between the specimen and the projecting part of the metal screws. The change in thickness is measured by observing the displacement of the fringes relative to lines engraved on the lower surface of CD—a displacement of one fringe clearly corresponding to an alteration of thickness of the air film equal to half a wavelength of the light used. Since a displacement of $\frac{1}{5}$th of a fringe width can readily be detected, changes in the thickness of the air film of the order of 6×10^{-6}cm. can be measured using light of wavelength of the order 6×10^{-5}cm.

With the Newton's rings system of fringes, rings appear to expand outwards as the thickness of the air film decreases or to contract, and disappear towards the centre of the system with increase in thickness of the film. The number of fringes passing a fixed mark is observed when the temperature of the specimen is raised. This enables the difference in expansion of the specimen and screws to be obtained. The expansion of the screws for the same range of temperature is determined by removing the specimen and taking observations of the fringes formed by interference between the lower surface of *CD* and the upper surface of *AB*. From the two readings the expansion of the specimen can be found.

3.11 Results and Applications—Grüneisen's Law

An interesting generalisation, which provides an important test of any theory of the solid state, was found empirically by Grüneisen as a result of the measurements of the coefficients of expansion of isotropic metals at different temperatures. Grüneisen found that the ratio of the coefficient of expansion of a metal to its specific heat at constant pressure is constant at all temperatures. Some results taken from Robert's "Heat and Thermodynamics" illustrating Grüneisen's law for certain metals are given in table 12.

TABLE 12
Grüneisen's Law

Metal	Temperature °C.	$a \times 10^6$	$\dfrac{a \times 10^6}{c_p}$
Aluminium	−173	13·6	107
	−100	18·2	109
	0	23·0	110
	100	24·9	112
	300	29·0	119
	438	29·8	112
Copper	−87	14·1	174
	0	16·1	177
	100	16·9	180
	400	19·3	179
	600	20·9	182
Platinum	−150	7·4	269
	−100	7·9	268
	0	8·9	280
	100	9·2	277
	875	11·2	267

3.12 Compensation of Clocks and Watches

Some of the consequences of expansion have already been mentioned. It is clear that the expansion of metals, though small, has profound effects which frequently call for special consideration. A further instance of this may be given in the change in length of metal scales and tapes. Barometric heights read against brass scales need correction for temperature changes (see section 4.11), whilst the difficulty of errors in surveying tapes has been overcome by the use of *invar steel* (a nickel-steel alloy containing 36 per cent nickel) which has a negligibly small expansion coefficient.

Timing mechanisms, such as pendulum clocks and watches, are affected by changes in temperature, and compensating devices are necessary if gains or losses in the indicated times are to be avoided.

The need for such a compensating device in the case of pendulum clocks is illustrated by the following example relating to an iron pendulum of coefficient of expansion 12×10^{-6} per deg. C. keeping correct time at 0°C. Assuming the formula for a simple pendulum (where l_0 is the length of the simple pendulum equivalent to the rigid clock pendulum), the time lost per day at a temperature of 20°C. can be calculated as under, where T_0 and T_{20} are the oscillation periods at 0°C. and 20°C. respectively,

$$T_0 = 2\pi \sqrt{\frac{l_0}{g}}$$

$$T_{20} = 2\pi \sqrt{\frac{l_0(1 + 0.000012 \times 20)}{g}}$$

$$= 2\pi \sqrt{\frac{l_0}{g}}(1 + 0.000012 \times 20)^{\frac{1}{2}}$$

$$= T_0\{1 + \tfrac{1}{2} \times 0.000012 \times 20 - \tfrac{1}{8}(0.000012 \times 20)^2 + \ . \ .\}$$

$$= T_0(1 + \tfrac{1}{2} \times 0.000012 \times 20)$$

ignoring second and higher order terms. Hence the fractional increase in the period of the pendulum

$$= \frac{T_{20} - T_0}{T_0} = \tfrac{1}{2} \times 0.000012 \times 20$$

$$= 0.00012 \ ^*$$

This means that a 1 second interval at 0°C. actually takes 0.00012 seconds longer at 20°C., or the clock loses 0.00012 seconds for each second time interval. Thus in one day the clock will lose

$$24 \times 60 \times 60 \times 0.00012 = 10.368 \text{ seconds.}$$

Compensating devices of pendulum clocks depend on the differential expansions of two metals. The principle can be understood by reference

---

* or using the calculus,

$$T = 2\pi\sqrt{\frac{l}{g}} \ \therefore \ \frac{dT}{dl} = \frac{2\pi}{\sqrt{g}} \times \tfrac{1}{2}l^{-\frac{1}{2}} = 2\pi\sqrt{\frac{l}{g}} \cdot \frac{1}{2l} = \frac{T}{2l} \ \text{or} \ \frac{dT}{T} = \tfrac{1}{2}\frac{dl}{l} = \tfrac{1}{2}\alpha t$$

to Fig. 34 where l_1 and l_2 are the lengths of two metal rods A and B, having coefficients of expansion of a_1 and a_2 respectively. For oscillations of constant time the centre of gravity of the bob G must always be at the same distance below the point of support O. In other words, the expansion of the rod A downwards must be balanced by an identical expansion of the rod B upwards. Thus for a temperature change of $t°$C.

$$l_1 a_1 t = l_2 a_2 t$$

$$\text{or } \frac{l_1}{l_2} = \frac{a_2}{a_1}$$

Hence for compensation the lengths of the two rods must be in the inverse ratio of their coefficients of expansion.

If A is of iron ($a_1 = 0.000012$) and B of brass ($a_2 = 0.0000189$), then

$$\frac{l_1}{l_2} = \frac{0.0000189}{0.000012} \doteq \frac{3}{2}$$

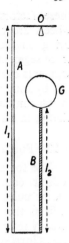

Fig. 34

Harrison's gridiron pendulum (Fig. 35) is constructed on this principle, the relative lengths of the iron (heavy lines) to brass (unshaded) being given by the above ratio. More modern practice is to construct clock pendulums of very low expansion coefficient alloys such as *invar* or *elinvar*.

In watches the mechanism is controlled by a balance wheel and hair spring. The time of oscillation is given by $T = 2\pi \sqrt{\dfrac{I}{c}}$ where I is the moment of inertia of the balance

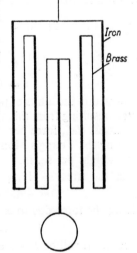

Iron

Brass

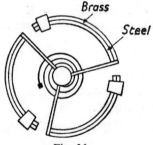

Brass

Steel

Fig. 36   Fig. 35

wheel about its axis of rotation, and c is a constant depending on the elastic properties of the hair spring. A rise in temperature causes I to increase due to the expansion of the radial arms, and weakens the elasticity of the spring; consequently the watch loses time due to both these factors. To compensate for these effects, the rims of balance wheels are made from bimetallic strips—the more expansible metal being on the outside. Fig. 35 shows such a balance wheel made in three segments from brass and steel strips. With rise in temperature the segments curl inwards, thus reducing the effective radius of the wheel, and by a suitable adjustment of the movable weights, complete compensation can be obtained for both the expansion of the wheel and the change in elasticity of the hair spring.

3.13   Forces Due to Expansion

The forces resulting from the expansion and contraction of metal rods are enormous. Rigidly braced structures are subject to tremendous strains unless adequate provision is made to allow for the expansion, such as the gaps in railway lines, and the rocking posts for bridges already referred to in the first section of this chapter. Some idea of the forces involved may be gained from the following numerical example.

Consider a length L feet of steel with an area of cross section of 5 sq. in. subject to a temperature variation of 30°C. The coefficient of linear expansion of steel is $11{\cdot}0 \times 10^{-6}$ per deg. C., hence the increase in length for this temperature change is

$$L \times 0{\cdot}000011 \times 30 \text{ feet.}$$

The rod is thus strained by an amount:

$$\text{Strain} = \frac{\text{increase in length}}{\text{original length}}$$

$$= \frac{L \times 0{\cdot}000011 \times 30}{L}$$

$$= 0{\cdot}00032$$

Now within the limits of perfect elasticity (which apply in this case), the stress resulting from this strain is, by Hooke's law:

Stress (*i.e.*, force per unit area of cross section) $= E \times$ strain
E being the Young's Modulus of the material which for steel is 13,500 tons weight per sq. in.

Thus the stress $= 13,500 \times 0{\cdot}00033$

$$= 4{\cdot}455 \text{ tons per sq. in.}$$

Hence the force due to expansion of the specimen is

$$22{\cdot}275 \text{ tons weight.}$$

A result, it will be observed, which is independent of the length of the rod.

If the rod is prevented from contracting a contractile force of the same magnitude is called into play. This contractile force of metals finds many useful applications, among which may be mentioned the production of steam-proof joints by riveting boiler plates with red-hot rivets, and the "shrinking" of iron tyres on cart wheels.

QUESTIONS. CHAPTER 3

1. Explain what is meant by the linear coefficient of expansion. How may it be determined in the case of a solid?

Two equal bars, 50cm. long, one of brass and the other of iron, are joined together at one end and a needle 1mm. in diameter and carrying a pointer is clipped between their free ends. When the bars are heated the needle rotates through 10°. What is the temperature interval through which they were heated? The coefficients of linear expansion of brass and iron are 1.8×10^{-5} per degC. and 1.2×10^{-5} per degC. respectively. (O. and C.)

2. It is required to fit a brass ring of internal diameter 19·95cm. on a wheel of diameter 20·00cm. Through what range of temperature must the ring be heated so that it just slips on to the wheel?

(Coefficient of linear expansion of brass = 0·000018 per degC.) (N.)

3. Describe an experiment to find the coefficient of linear expansion of a metal.

The coefficient of linear expansion of brass is 0·000018 per degC. What is the coefficient of cubical expansion and why?

The coefficient of linear expansion of iron is 0·000012 per degC. What lengths of brass and iron rods must be taken so that when heated together their difference is 1ft. at all ordinary temperatures? (N.)

4. A piece of aluminium wire 1mm. in diameter, at room temperature, is threaded through a circular hole in a nickel plate. What must be the diameter of the hole in the plate, at room temperature, in order that the area of the annular aperture surrounding the wire may be independent of temperature?

(Assume that the coefficient of linear expansion of aluminium is 0·0000225 per degC. and that of nickel 0·0000144 per degC.) (N.)

5. An aluminium disc 8cm. in diameter at 15°C. just fits in a circular hole in a steel plate at 100°C. What is the area of a gap between them at 15°C.? The coefficient of linear expansion of aluminium is 0·000017 and of steel 0·000012 per degC. (L.)

6. Describe an accurate method of determining the linear coefficient of expansion of a solid.

A thermal delay switch consists of two strips of different metals rigidly held together and fixed at one end to a support. When this bimetallic strip is heated it bends, because of the different expansions of the two metals, and the free end can make contact, e.g., to complete some circuit. Heat, at the rate of 4 calories per second, is supplied to such a bimetallic strip (made of copper and iron) whose thermal capacity is 1 cal. per degC. Each component of the strip is 5cm. long and 1mm. thick and the distance separating the free end from a contact is 1mm. when the strip is cold. Find the time taken for contact to be established from the moment heating starts.

Coefficients of linear expansion of copper and iron = 17×10^{-6} per degC. and 12×10^{-6} per degC. respectively. Assume that mechanical stresses play no part in the process and that there is no loss of heat from the strip by radiation or any other process. (Camb. Schol.)

7. Derive a relationship between the coefficients of linear and cubical expansion for an isotropic solid.

Describe fully a method employing an optical lever for determining the coefficient of linear expansion of a metal rod.

Explain carefully how the balance wheel of a watch is constructed to reduce the effects of change of temperature. (A.)

8. Define (a) mean coefficient of linear expansion between 0°C. and t°C., (b) true coefficient of linear expansion at t°C.

The length l_t of a metal rod at t°C. is related to its length l_0 at 0°C. by the equation $l_t = l_0(1 + 11 \times 10^{-6} t + 7 \times 10^{-9} t^2)$. Calculate the mean coefficient of expansion of the metal between 0°C. and 100°C. and the true coefficient at 200°C.

Explain how the expansion of solids is involved in (a) the construction of clocks and watches, and (b) the manufacture of electric lamps.   (W.)

9. Describe a method of determining the coefficient of linear expansion of a metal in the form of a rod, explaining how possible errors are eliminated.

Give a brief account of the chief methods of compensating watches and pendulum clocks in order to avoid errors due to change of temperature.

A clock with an iron pendulum keeps correct time at 15°C. What will be the error, in seconds per day, if the room temperature is 20°C.?

(The coefficient of linear expansion of iron is 0·000012 per degC.)   (N.)

10. Distinguish between (a) the mean coefficient of linear expansion of a rod between 0°C. and θ°C., and (b) its coefficient of linear expansion at θ°C.

Give an account of a comparator method for measuring the mean coefficient of linear expansion of a steel rod.

Three thin rods are joined together at their ends to form an equilateral frame ABC; another rod connects A to P, the mid-point of BC. The linear coefficient of expansion for the materials of the rods AB and AC is α; that for BC is β. Show that for a small rise θ in the temperature of the system there will be no tendency for its sides to buckle provided the coefficient of linear expansion for the material of the rod is $\frac{1}{3}(4\alpha - \beta)$.

(N.B. Terms in θ^2 may be neglected.)   (L.)

11. Explain the construction and action of two devices which make use of the difference of thermal expansibility of two solids.

A clock controlled by a brass pendulum keeps correct time at 15°C. If the clock is used uncorrected to time a pendulum on a day when the temperature is 20°C., find the apparent percentage error in the value of g. Assume that the coefficient of linear expansion of brass is 0·0000186 per degC.   (N.)

12. Define the coefficient of volume expansion of a substance and show that its value is three times the coefficient of linear expansion.

A temperature-compensated simple pendulum consists of a glass tube of negligible mass, containing mercury. Calculate what fraction of the length of the tube the mercury should occupy in order that the distance of its centre of gravity from the upper end of the tube may be independent of temperature.

(Coefficient of linear expansion of glass = 7×10^{-6} per degC.; coefficient of volume expansion of mercury = $1·8 \times 10^{-4}$ per degC.)   (Oxford entrance)

13. Describe a standard method by means of which the coefficient of linear expansion of metal bars has been accurately determined.

A steel wire, of diameter 1·00mm., supports a load which is sufficient to keep the wire taut, the temperature being 20°C. If the temperature falls to 0°C. what additional load would be required to restore the length of the steel wire to its initial value?

(Young's modulus for steel = $2·00 \times 10^{12}$ dynes per cm.^{2} Coefficient of expansion of steel = 0·000010 per degC.)   (W.)

14. A steel wire 8 metres long and 4mm. in diameter is fixed to two rigid supports. Calculate the increase in tension when the temperature falls 10°C.

(Linear coefficient of expansion of steel 12×10^{-6} per degC. Young's modulus for steel 2×10^{12} dynes per sq. cm.)   (O. and C.)

15. Describe an accurate method of measuring the coefficient of linear expansion of a substance in the form of a rod.

A uniform rod of 2 sq. mm. cross section is heated from 0° to 20°C. Find the force which must be exerted over its ends to prevent it expanding. Find also the energy stored in unit volume. The coefficient of linear expansion of the rod may be taken as $1 \cdot 2 \times 10^{-5}$ per degC. and Young's modulus as 10^{12} dynes per sq. cm. (L.)

16. Describe briefly a method for measuring the thermal expansion coefficient of a liquid. What causes thermal expansion in solids and liquids?

A thermograph consists of a bimetallic strip wound in a helix. One end is clamped firmly, and the free end actuates the pen mechanism. The bimetallic strip consists of two tapes, each 0·5mm. thick, welded back to back. One tape is of German silver whose linear expansion coefficient α is 18×10^{-6} per°C. and the other is of nickel steel for which α is 9×10^{-6} per °C. If the free end of the helix is to rotate by 1° per °C. change in temperature, how long must the strip be? (Camb. Schol.)

17. Describe briefly an accurate method of measuring the thermal expansion of a solid.

The pendulum OA of a clock consists of a thin steel rod suspended at O, with a cylindrical vessel of mercury resting on a small thin platform fastened to A. At 15°C. the length OA is 100cm. and the height of the mercury in the vessel is 12cm. The coefficient of *linear* expansion of steel is $10 \cdot 0 \times 10^{-6}$, of mercury 60×10^{-6} and of the material of the vessel 15×10^{-6} per °C. If the clock goes correctly at 15°C., calculate the percentage error in its timing at 25°C.

(You may assume that the device acts as a simple pendulum of length equal to the distance from O to the centre of gravity of the mercury, and that the base of the vessel has negligible thickness). (Camb. Schol.)

18. Describe *two* methods of measuring the linear coefficient of thermal expansion of a substance.

In a measurement of this kind, the temperature measured at the centre of a bar of length $2L$ is T_0, but at a point distant x from the centre the temperature is $T_0(1 - ax^2)$ where a is small. Find the error in the value of the coefficient of expansion calculated on the assumption that the bar is of uniform temperature T_0.

(Oxford Schol.)

CHAPTER IV

EXPANSION OF LIQUIDS

4.01 Introductory

In dealing with the expansion of liquids we are concerned only with volume changes. Liquids have no definite shape of their own, but conform to that of the containing vessel. Since the container as well as the liquid expands on being heated, it is clear that the measured increase in volume of the liquid is not the actual volume increase for the particular temperature range concerned, but merely its apparent or relative change compared with the expansion of the vessel. We may therefore define two coefficients of expansion (a) the *real or absolute coefficient* which is the actual increase in volume of unit volume per degree rise in temperature, and (b) the *apparent* or *relative coefficient* which is the increase in volume of unit volume per degree rise in temperature when the liquid is heated in an expansible vessel.

Most liquids expand somewhat irregularly and consequently the coefficient of expansion is not a constant. Mercury, it will be remembered, however, expands fairly regularly and is therefore suitable for thermometry. It is usual to distinguish between the *zero coefficient* and the *mean coefficient*. The former is the increase in volume compared with the volume at 0°C. for each degree rise in temperature. Thus if V_o is the volume of the liquid at 0°C. and V_t its volume at t°C., then the zero coefficient of expansion is

$$\gamma = \frac{V_t - V_o}{V_o t}$$

The mean coefficient of expansion is the increase in volume expressed as a fraction of the original volume per degree rise in temperature. Thus if V_1 and V_2 are the volumes at t_1°C. and t_2°C. respectively, then the mean coefficient of expansion is

$$\frac{V_2 - V_1}{V_1 (t_2 - t_1)}$$

In giving the mean coefficient of apparent expansion of a liquid it is necessary to specify the nature of the material of the containing vessel against which the measurements are made, and also the temperature range over which the readings are taken.

4.02 Relation between the Real and Apparent Coefficients

Suppose the liquid to be contained in some graduated vessel such as a volume dilatometer which reads correctly at 0°C. Let the observed volume at $t°$C. be V_1, then the apparent coefficient of expansion γ^1 will be given by

$$\gamma' = \frac{V_1 - V_o}{V_o t}$$

Now if g is the coefficient of cubical expansion of the containing vessel, then the real volume of the V_1 graduation at $t°$C. will be $V_1(1 + gt)$ since each unit volume of the vessel has become $(1 + gt)$ units at $t°$C. Thus the real coefficient of expansion of the liquid is given by

$$\gamma = \frac{V_1(1 + gt) - V_o}{V_o t}$$

$$= \frac{V_1 - V_o}{V_o t} + \frac{V_1}{V_o} \cdot g$$

$$= \gamma' + (1 + \gamma' t) g$$

$$= \gamma' + g + \gamma' g t$$

Ignoring the third term which contains the product of two very small quantities, we may write

$$\gamma = \gamma' + g$$

Thus for all but the most accurate work we may say that the real coefficient of expansion of a liquid is the sum of its apparent coefficient and the coefficient of cubical expansion of the material of the vessel.

TABLE 13

Coefficients of expansion of various liquids (mean coefficients for a range of temperature round 18°C.).

Liquid	Coefficient of expansion $\times 10^5$ (per °C.)
Alcohol (methyl)	122
„ (ethyl)	110
Ether (ethyl)	163
Glycerine....................................	53
Mercury	18·2
Olive oil	70
Paraffin	90
Turpentine	94
Toluene	109
Xylol	101
Water— 5°–10°	5·3
10°–20°	15·0
20°–40°	30·2

4.03 Change of Density with Temperature

Since the volume of a given mass of liquid changes with temperature, it is clear that the density of the liquid will also vary with temperature.

Consider a mass M of liquid whose real coefficient of expansion is γ. Let the volumes of the liquid at 0°C. and t°C. be V_o and V_t respectively, and ρ_o and ρ_t the corresponding densities. Then

$$\rho_o = \frac{M}{V_o} \text{ and } \rho_t = \frac{M}{V_t}$$

therefore, since M is constant

$$V_o\rho_o = V_t\rho_t$$

but $\quad V_t = V_o(1 + \gamma t)$

hence $V_o\rho_o = V_o(1 + \gamma t)\rho_t$

or $\quad\quad \rho_o = \rho_t(1 + \gamma t) \ . \ . \ . \ . \ . \ . \ .$ [20]

It is thus evident that measurements of the density of a liquid at different temperatures will yield a value for γ.

Relative Methods for Determination of the Apparent Coefficient of Expansion

4.04 Dilatometer Method

This is a simple direct method most suitable for volatile liquids, and for taking a continuous set of readings over a range of temperature. The dilatometer consists of a glass bulb to which is attached a calibrated stem. The volume of the bulb is determined by weighing it first empty and then full of mercury. From the weight of mercury filling the bulb and the density of the mercury the volume of the bulb can be calculated. The stem is calibrated in a similar way by weighing a measured length of a mercury thread. The liquid under test is now placed in the dilatometer so as to fill the bulb and the lower part of the stem. The volume V_o of the liquid at 0°C. is then found by placing the bulb and lower stem in ice shavings and reading off the position of the liquid thread against the graduations on the stem. On raising the temperature to t°C. the volume V_t of the liquid can be similarly found, when the mean apparent coefficient of expansion of the liquid can be calculated from $\dfrac{V_t - V_o}{V_o t}$

4.05 Weight Thermometer Method

This method, together with the one immediately following, is based on the variation of density of a liquid with temperature. The measurements involved in both methods are determinations of mass, and as such measurements can be taken with much greater accuracy than measurements of volume, these methods are more reliable than the one given above. The weight thermometer (Fig. 37) consists of an elongated glass or fused silica bulb with an elongated neck of narrow bore bent twice at right angles. The vessel is first weighed empty, and then it is completely filled with the liquid by alternately heating and cooling it with the open end of the neck under the surface of the liquid contained in a small beaker. The liquid should be well boiled before introduction into the weight thermometer so as to drive out any dissolved air, thus preventing the appearance of air bubbles in the thermometer. When full the weight thermometer is again weighed when the mass M_o of liquid filling it at the low temperature can be found. The bulb of the thermometer is now completely immersed in a water bath, and the temperature steadily raised to $t°C$. The thermometer is kept at this

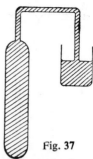

Fig. 37

temperature until the liquid has ceased to expand and there is no further issue from the orifice. It is then removed, dried, and re-weighed, when the mass M_1 of liquid filling it at the higher temperature can be found. The coefficient of expansion γ of the liquid is then calculated as follows: Let the volumes of the vessel at the lower and higher temperatures be V_o and V_t respectively, and let the corresponding densities of the liquid be ρ_o and ρ_1.

Then $\rho_o = \dfrac{M_o}{V_o}$

and $\rho_1 = \dfrac{M_1}{V_1} = \dfrac{M_1}{V_o(1 + gt)}$

where g is the coefficient of cubical expansion of the vessel.

Now $\rho_o = \rho_1(1 + \gamma t)$ (from equation 20)

$$\therefore \frac{M_o}{V_o} = \frac{M_1(1 + \gamma t)}{V_o(1 + gt)}$$

from which

$$\gamma = \frac{M_o - M_1}{M_1 t} + \frac{M_o}{M_1}g \quad \ldots \ldots \ldots \quad [21]$$

This gives the real coefficient of expansion; if however we neglect the expansion of the vessel, the apparent coefficient γ' is given by

$$\gamma' = \frac{M_o - M_1}{M_1 t} \quad \ldots \ldots \ldots \ldots \quad [22]$$

or, in words:

$$\gamma' = \frac{\text{mass expelled}}{\text{mass remaining} \times \text{temperature change}}$$

A specific gravity bottle may be used as a simple laboratory form of weight thermometer, and has the added advantage that it is much more easily filled than the standard form of weight thermometer described above. Another form convenient to use is the pyknometer shown in Fig. 38. The instrument is filled to a fixed mark B on one of the capillary

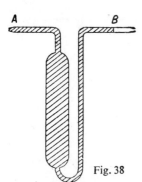

Fig. 38

arms by applying suction and drawing in the liquid by the narrow orifice at A. Application of blotting paper to A during the course of the experiment will absorb the overflow liquid and ensure that the liquid is kept at the mark B. The necessary weighings are readily taken for insertion in equation 22, from which a value of γ' may be obtained.

Another interesting use of the weight thermometer is to determine the coefficient of cubical expansion of a solid in the form of a powder. The expansion of the liquid against the solid can be found in terms of the relative mass of liquid expelled when the solid is immersed in the liquid contained in the weight thermometer. Then from previous measurements of the expansion of the liquid when filling the weight thermometer alone, it is possible to calculate the expansion of the solid.

4.06 Matthiessen's Method

This method applies the principle of Archimedes to the determination of the density of a liquid at two different temperatures, the coefficient of expansion being calculated from the relation between density and temperature already established. A suitable sinker for the experiment is shown in Fig. 39. This consists of a glass or silica bulb hermetically sealed and suitably loaded with mercury or lead shot so that it just sinks in the liquid at the low temperature. By thus using a large sinker of low mean density the apparent loss in mass to actual mass of the sinker is increased with consequent increase in the sensitivity of the method.

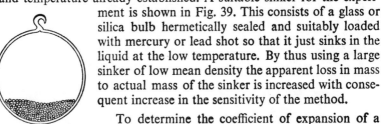

Fig. 39

To determine the coefficient of expansion of a liquid, the sinker is first weighed in air, and then again when completely immersed in the liquid at the temperature of the room. The liquid is now heated, and a determination of the mass of the sinker is again made when it is immersed in the liquid at some higher temperature. This is best done by suspending the sinker from the balance hook by a fine wire passing through a hole in the base (the pan is removed to permit of this). The purpose of this is to avoid interference by convection currents with the counterpoise of the balance. The counterpoise position can be conveniently obtained by placing a weight which is just too light on the balance pan, and as the liquid cools noting the temperature at which counterpoise is exact. The liquid should be kept continuously stirred throughout the process. Proceeding in this way the upthrust can be found at a number of temperatures as the liquid cools, enabling the mean coefficient of expansion to be found over several ranges of temperature.

The theory of the method is as follows:

Let M = the weight of the sinker in air.

$\quad M_o$ = its weight when immersed in the liquid at the low temperature.

$\quad M_1$ = its weight when immersed in the liquid at the higher temperature.

$\quad t$ = the difference in the temperatures of the cold and hot liquid.

If V_o and V_t represent the volumes of the sinker at the low and higher temperatures respectively, and ρ_o and ρ_1 are the corresponding densities of the liquid, then by Archimedes' principle

Upthrust at low temperature $= M - M_o = V_o\rho_o$

Upthrust at higher temperature $= M - M_1 = V_1\rho_1$

$$\therefore \rho_o = \frac{M - M_o}{V_o}$$

and $$\rho_1 = \frac{M - M_1}{V_1} = \frac{M - M_1}{V_o(1 + gt)}$$

where g is the coefficient of cubical expansion of the sinker.

Now $\rho_o = \rho_1(1 + \gamma t)$

hence $$\frac{M - M_o}{V_o} = \frac{M - M_1}{V_o(1 + gt)} \cdot (1 + \gamma t)$$

i.e., $$\gamma = \left(\frac{M_1 - M_o}{M - M_1}\right)\frac{1}{t} + \left(\frac{M - M_o}{M - M_1}\right)g \quad \ldots \ldots [23]$$

Thus, ignoring the expansion of the sinker, we have for the apparent coefficient of expansion γ'.

$$\gamma' = \left(\frac{M_1 - M_o}{M - M_1}\right)\frac{1}{t}$$

or, in words:

$$\gamma' = \frac{\text{difference between upthrusts at the low and higher temperatures}}{\text{upthrust at high temperature} \times \text{temperature change}}$$

It will be seen from equations 21 and 23 that a knowledge of g, the cubical coefficient of expansion of the weight thermometer and the sinker respectively, enables a value of the real coefficient of expansion of the liquid to be obtained by these two latter methods. The difficulty is, however, that g has to be determined on a specimen of the material, and cannot conveniently be done on the vessel itself. Accordingly, as the heat treatment received during the making of the vessel often results in change of physical properties of the material, other methods which do not involve the determination of the expansion of the vessel are used for finding the real coefficients of expansion of liquids.

Absolute Methods for Determining the Real Coefficient of Expansion of a Liquid

4.07 Dulong and Petit's Experiment

The first direct determination of the real coefficient of expansion of a liquid was carried out by Dulong and Petit in 1817. Their method, which has been extended and modified by later workers, is based on the hydrostatic equilibrium of two columns of liquid. The pressure exerted by a column of liquid is dependent only on the vertical height of the column and the density of the liquid, and is not affected by the shape or size of the containing vessel. Hence measurements of the heights of two columns of liquid balanced when they are at different temperatures enables the real coefficient of expansion of the liquid to be obtained directly. Dulong and Petit's method is illustrated by the following laboratory form of their apparatus shown in Fig. 40.

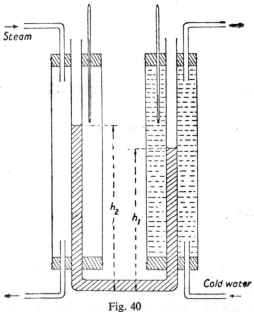

Fig. 40

The limbs of a glass U-tube are surrounded by wide glass tubing bored to take thermometers and inlet and outlet tubes as shown. Steam from a steam heater is passed through one jacket whilst cold water passes through the other. This is continued until the levels of the liquid in the two limbs of the U-tube are constant. The heights h_1 and h_2 of the cold and hot columns are then read. If the temperatures of the two columns are t_1 and t_2 respectively, and ρ_1 and ρ_2 the corresponding densities of the liquid, then since the columns are hydrostatically balanced, the two columns exert equal pressures:

i.e., $h_1\rho_1 = h_2\rho_2$

or $\dfrac{\rho_2}{\rho_1} = \dfrac{h_1}{h_2}$

now $\dfrac{\rho_2}{\rho_1} = \dfrac{1 + \gamma t_1}{1 + \gamma t_2}$ by equation 20.

∴ $\dfrac{h_1}{h_2} = \dfrac{1 + \gamma t_1}{1 + \gamma t_2}$

or $\gamma = \dfrac{h_2 - h_1}{h_1 t_2 - h_2 t_1}$

In Dulong and Petit's original experiment the cold column was surrounded by melting ice, and the hot column was contained in an oil bath which was heated by a furnace built round it. The temperature of the oil bath was taken by an air thermometer and a mercury weight thermometer. The readings indicated by the two thermometers showed divergencies at high temperatures, and the final calculation was based on the readings of the air thermometer. A cathetometer was used for measuring the heights h_1 and h_2 of the liquid columns both of which protruded slightly above their containing jackets. Their measurements were made with mercury in the U-tube. An accurate determination of the real coefficient of mercury is valuable as this liquid can then be used to determine the expansion of the containing vessels used in the methods described in sections 4.05 and 4.06.

4.08 Regnault's Experiments

The simple arrangement adopted by Dulong and Petit is unsatisfactory for the following reasons:

(*a*) Convection currents along the bottom of the U-tube, and the necessity of the mercury columns protruding above the containing jackets so as to be visible, result in unequal heating of the mercury columns.

(*b*) The quantity $h_2 - h_1$ is not capable of accurate determination since the two limbs are widely separated.

(*c*) The condition of true hydrostatic equilibrium of the two columns is invalidated to some extent due to the mercury surfaces being at different temperatures, with consequent different capillary depressions resulting from the variation of surface tension with temperature.

Regnault (1847) modified Dulong and Petit's original method in two experiments designed to obviate the above objections. His first experiment is shown diagrammatically in Fig. 41. The mercury is contained in two wide vertical arms connected by two narrow horizontal tubes. A small hole E in the upper tube enables the mercury to attain atmospheric pressure, and the mercury surfaces are separated in the lower

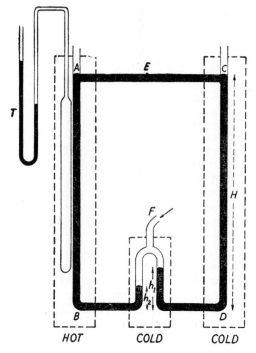

Fig. 41

horizontal tube by applying pressure from a pump to the inverted U-tube at F as indicated. The arm AB is surrounded by an oil bath the temperature of which is taken by a gas thermometer T. The other three arms are maintained at the same temperature by circulating cold water round them. If ρ_o and ρ_t are the densities of the cold and hot mercury and the equilibrium heights are as given in the diagram, then if P is the pressure in excess of the atmospheric pressure applied at F, we have

$$H\rho_t = h_2\rho_o + P \text{ for equilibrium on the left hand limb,}$$

and $H\rho_o = h_1\rho_o + P$ for equilibrium on the right hand limb, thus eliminating P

$$H\rho_o - H\rho_t = (h_1 - h_2)\,\rho_o$$

i.e., $\qquad \dfrac{\rho_o}{\rho_t} = \dfrac{H}{H - (h_1 - h_2)} = 1 + \gamma t$

hence $\qquad \gamma = \dfrac{h_1 - h_2}{(H - h_1 + h_2)\,t}$

Regnault's second experiment is shown in schematic form in Fig. 42. The vertical limbs CD and EF are connected by narrow flexible iron

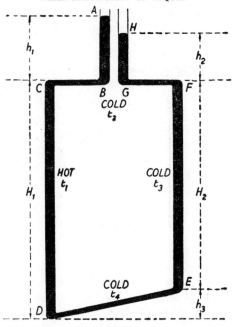

Fig. 42

tubing to permit independent expansion of the limbs (greatly exaggerated in the figure). The tops of the limbs are connected by narrow tubing, adjusted so as to be accurately horizontal, to two wide vertical tubes AB and HG placed side by side in a cold water bath at a controlled temperature. The temperature of the limb EF is similarly controlled, whilst the limb CD is surrounded with an oil bath and heated. The tube DE is exposed to the air, and its mean temperature is taken as being the same as the air temperature. Let the heights and temperatures in the steady state be as indicated in the diagram. Then, since the pressures at A and H are equal, by equating pressures at the level through D we have

$$h_1\rho_2 + H_1\rho_1 = h_2\rho_2 + H_2\rho_3 + h_3\rho_4$$
$$\text{or } (h_1 - h_2)\rho_2 + H_1\rho_1 = H_2\rho_3 + h_3\rho_4$$

ρ_1, &c., being the densities of the mercury at the corresponding temperatures t_1, &c.

Now $\rho_0 = \rho_1(1 + \gamma t_1) = \rho_2(1 + \gamma t_2) = $ &c.

$$\therefore \frac{h_1 - h_2}{1 + \gamma t_2} + \frac{H_1}{1 + \gamma t_1} = \frac{H_2}{1 + \gamma t_3} + \frac{h_3}{1 + \gamma t_4}$$

All the quantities in this equation are accurately determinable with the exception of t_4. This temperature is involved in a small order term

84 EXPANSION OF LIQUIDS

containing h_3, which contributes very little to the final result. Consequently an approximate value for t_4 does not prevent an accurate evaluation of γ from the equation.

4.09 Experiments of Callendar and Moss

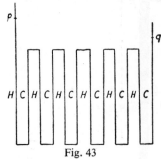

Fig. 43

In Regnault's second method the length of the mercury columns was $1\frac{1}{2}$ metres. Callendar and Moss (1911) extended the method so as to increase the effective lengths of the columns. This was done by using six pairs of hot and cold columns connected in series as shown (Fig. 43). The length of each column was 2 metres, and hence the difference in height $p - q$ was eight times that obtained by Regnault. All the hot columns were arranged side by side in a constant temperature bath, the cold columns being similarly arranged. The temperatures of the baths were taken by platinum resistance thermometers, the wires of which extended the full lengths of the columns, thus giving the mean value of the temperature of the columns. In this way Callendar and Moss carried out determinations of the expansion of mercury between the temperatures of 0° and 300°C. Selected values from their results are given in table 14.

TABLE 14

Real coefficient of expansion of mercury

Temperature range °C.	$\gamma \times 10^8$
0—30	18,095
50	18,124
75	18,164
100	18,205
200	18,406
300	18,657

4.10 Anomalous Expansion of Water

Most liquids expand fairly uniformly over moderate temperature ranges. Water, however, shows marked exceptions to this general rule, its coefficient of expansion increasing rapidly with temperature (see table 13). In addition it exhibits anomalous behaviour between 0 and 4°C. by contracting as the temperature is raised in this interval. The variation in volume of a given mass of water can be studied using a volume dilatometer with a graduated stem of fine bore. The expansion bulb contains an amount of mercury equal to one sixth of its volume to compensate for the expansion of the glass, and thus the readings of the dilatometer register the actual volume change of the mass of water contained in it. By placing the instrument in a water bath with a $\frac{1}{10}$ °C. thermometer and slowly adding ice, the volume can be read off for a series of temperatures down to 0°C. A graph (Fig. 44) of the volume

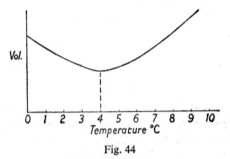

Fig. 44

against temperature shows a fairly steady contraction from room temperature to 4°C., after which the water expands as the temperature is reduced to 0°C. Water has thus a temperature of maximum density at 4°C. This fact, together with the fact that there is a sudden expansion as water freezes at 0°C., is of profound importance in nature. With wintry conditions the coldest layers of a pond or lake are those at the surface, and when these layers freeze the ice so formed remains on the surface. Without this freakish behaviour of water the pond would freeze solid from the bottom upwards greatly to the detriment of aquatic life!

The existence of a maximum density for water is strikingly demonstrated by Hope's experiment. A tall narrow vessel containing water is surrounded by an annular tray at its middle (Fig. 45). A freezing mixture cools the water, and the readings of the two thermometers T_1 and T_2 are taken at suitable intervals. Temperature-time graphs for these two thermometers are then drawn (Fig. 46), which show that T_2 falls very rapidly until its temperature is approximately 4°C. when it remains steady, while T_1 falls at first slowly and then very quickly to 0°C. The temperature of

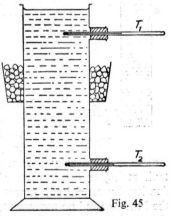

Fig. 45

maximum density is given by the point of intersection of the two curves.

The explanation of the curves is as follows. When the water initially at room temperature (say 10°C.) is first cooled by the freezing mixture, contraction takes place and the cooled liquid falls, affecting the reading of the lower thermometer. This process goes on until T_2 records approxi-

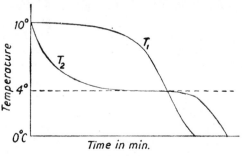

Fig. 46

mately 4°C., when the water expands on further cooling. Thus after 4°C. the cooler water is less dense, and continues to rise until the top layers record a temperature of 0°C. and freezing commences.

A more accurate method of finding the temperature of maximum density is that of Joule and Playfair. Their apparatus (Fig. 47) consisted of two tall cylinders a, a connected at the bottom by a tube b fitted with

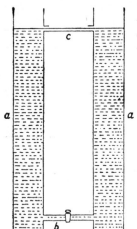

Fig. 47

a stop-cock, and at the top by an open trough or channel c. The principle of the method depends on the fact that for a small range of temperature the expansion curve for water is symmetrical about the temperature of maximum density and consequently, if the densities of the water in the two tanks is the same for two different temperatures, the mean of these will be the temperature of maximum density. To determine when equality of density was obtained, a small glass bead was floated in the channel c, and the temperatures of the tanks adjusted so that one was slightly below and the other slightly above 4°C. On opening the stop-cock a difference in density between the water in the two columns was evidenced by a movement of the bead in the channel towards the tank of higher density since convection currents will be set up from the tank at higher density through the stop-cock to the tank at lower density. By careful adjustment temperatures were obtained for which the bead showed no movement, the mean of these temperatures giving the temperature of maximum density. This temperature was found by Joule and Playfair to be 3·95°C. for a series of readings over a range of temperature near this point.

The following table gives for various temperatures the density and specific volume of water in gm. per millilitre and in millilitres respectively. The *litre* is defined as the volume occupied by one kilogram of pure air-free water at the temperature of maximum density, and under a pressure of one standard atmosphere. Clearly therefore an accurate knowledge of the temperature of maximum density is required to establish this practical unit of volume. The temperature now generally accepted is given by the relation:

t max. $= 3 \cdot 98 - 0 \cdot 0225 (p - 1)°$C. units

p being the pressure measured in atmospheres. Defined as above the litre is found to be 1,000·028cc., and for practical convenience flasks, cylinders, &c., are graduated in millilitres (ml) and litres (l). This discrepancy arises from the fact that the standard unit of mass, namely the International Prototype Kilogram (which is the mass of a cylinder of platinum-iridium alloy, and a copy of the original Borda platinum kilogramme) has not exactly the same mass as that of a cubic decimetre of pure water at the temperature of its maximum density as was originally intended. Later measurements revealed the slight discrepancy, but the platinum standard kilogram was adhered to.

TABLE 15

Densities in gm. per millilitre and specific volumes in millilitres of water at various temperatures.

Temperature °C.	Density	Specific volume
0	·99987	1·00013
2	·99997	1·00003
4	1·00000	1·00000
6	·99997	1·00003
8	·99988	1·00012
10	·99973	1·00027
12	·99953	1·00047
14	·99928	1·00072
16	·99897	1·00103
18	·99862	1·00138
20	·99823	1·00177
30	·99567	1·00433
40	·9922	1·0079
50	·9881	1·0120
60	·9832	1·0171
70	·9778	1·0227
80	·9718	1·0290
90	·9653	1·0360
100	·9584	1 0434

4.11 Barometric Correction for Temperature

It is customary to express the pressure of the atmosphere in terms of the height of a column of mercury at 0°C., which exerts an equivalent pressure. The barometric height taken at some temperature $t°C$. will require correcting on account of the expansion of the scale, and also for the change of density of mercury with temperature. These two temperature corrections are effected as follows:

(a) *Scale correction.*—Let H be the observed height at $t°C$. measured against a scale graduated at $0°C$. and with a linear coefficient of expansion of l. Clearly at $t°C$. the scale divisions will not be true, each cm. marked on the scale being in reality $(1 + lt)$ cm. Thus the H observed divisions at $t°C$. will actually be $H(1 + lt)$ cm. Let this height corrected for scale expansion be H_t, then

$$H_t = H(1 + lt) \qquad . . . . \qquad [24]$$

(b) *Correction for density of mercury.*—The pressure due to a column of liquid at a given place depends on the vertical height and the density of the liquid in such a way that the product of height and density is constant. Thus with a mercury barometer the fluctuations in the density of the mercury, as the temperature changes, will cause corresponding changes in the height of the barometer. If the barometric heights are H_o and H_t at temperatures of $0°C$. and $t°C$., then

$$H_o\rho_o = H_t\rho_t$$

where ρ_o and ρ_t are the corresponding densities of the mercury.

i.e., $H_o = H_t \dfrac{\rho_t}{\rho_o}$

Now $\rho_o = \rho_t (1 + \gamma t)$

γ being the real coefficient of expansion of mercury,

and $H_t = H(1 + lt)$ from equation 24 above.

hence $H_o = H(1 + lt)(1 + \gamma t)^{-1}$

or, ignoring second and higher order terms

$$H_o = H\{1 - (\gamma - l)\, t\}$$

To obtain the standard barometric reading from the above temperature corrected reading H_o it is further necessary to apply corrections for latitude and height above sea level. These corrections are required to reduce the value of the acceleration of gravity g to its standard value, which is that measured at sea level in a latitude of $45°$. Tables showing these corrections are given in Kaye and Laby's book of Constants.

4.12 Thermostats

A thermostat or thermo-regulator is a device for maintaining a water bath or oven at a given fixed temperature. A common laboratory form is shown in Fig. 48. An elongated bulb T contains a liquid with a high coefficient of expansion such as toluene or alcohol. This liquid is in contact with a mercury column which extends almost up to the nozzle C of a tube connected to the main gas supply at A. As the temperature in the vicinity of T rises, the toluene expands pushing up the mercury, which closes the nozzle thus cutting off the gas supply and preventing further heating. To ensure that the gas supply shall not be completely extinguished, a by-pass tube B with adjustable tap is provided, which supplies just sufficient gas to keep the burner alight. The fall of temperature which follows this action causes the toluene to contract, and the gas supply through C is re-established. In this way the temperature of the bath in which the toluene tube is immersed can be controlled to within a degree or so. The temperature at which the bath is to be maintained can be regulated at will by adjustment of the initial gap between the mercury surface and the nozzle C.

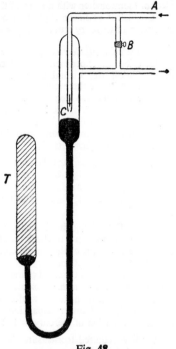

Fig. 48

The range for which liquid thermostats can be used is limited by the boiling point of the expanding liquid, which in the case of toluene is 111°C. For higher temperature control it is customary to use devices which depend for their action on the expansion of metals. In the bimetallic thermoregulator the principle of action depends on the change of a compound strip of two such metals as brass and invar steel when subject to temperature changes. Details of design of these appliances vary widely, but the object is to cause the movement of the compound strip to open suitably adjusted electric contacts controlling the heating circuit when the temperature rises above a certain point, and conversely to close the contacts and switch in the heating supply again as the temperature falls.

QUESTIONS. CHAPTER 4

1. The volume of the bulb of a glass dilatometer up to the zero graduation of the stem, which is of uniform bore, is 10,000 times the volume corresponding to one division of the graduated stem. The liquid in the dilatometer stands at the graduation mark 12, when the instrument is at 0°C., and at 162, when the temperature of the instrument is raised to 100°C. Find the mean coefficient of absolute expansion of the liquid in the dilatometer for the range 0°C. to 100°C.

(Coefficient of cubical expansion of glass is 0·000025 per 1°C.)

Describe and give the theory of a method of determining *directly* the coefficient of absolute expansion of a liquid. (W.)

H.T.–4+

2. The volume of the mercury in a mercury-in-glass thermometer is 0·30c.c. at 0°C. and the distance between the fixed points is 20·0cm. Find the area of cross-section of the bore. Find also the volume at 0°C. of the alcohol in an equally sensitive alcohol thermometer with a stem of the same bore.

(Coefficient of linear expansion of glass = 0·000008 per degC.; coefficients of expansion of mercury and alcohol = 0·000182 and 0·00110 per degC. respectively).

(N.)

3. Define *coefficient of apparent expansion* of a liquid.

Describe an experiment to determine the coefficient of apparent expansion of paraffin oil over a suitable range.

Two centigrade thermometers, one mercury and the other alcohol, have the same distance between the degree markings. The volume of the mercury in the bulb and up to the 0°C. mark is 1¾ times the corresponding volume of the alcohol. Compare the radii of the bores of the two thermometer stems, given that the coefficients of apparent expansion of mercury and alcohol in glass are 0·00015 and 0·00105 per degC. respectively.

(N.)

4. A cylindrical glass tube contains mercury at 0°C. What will be the percentage change (*a*) in the height of the mercury column, (*b*) in the pressure exerted by the mercury at the bottom of the tube, on raising the temperature by 50°C.? (Coefficients of cubical expansion of glass and mercury are 0·000024 and 0·00018 per °C. respectively.)

5. (*a*) Describe, giving full experimental details, how you would use a constant volume air thermometer to check the 50°C. mark on a mercury-in-glass thermometer. Make **two** criticisms of the simple constant volume air thermometer.

(*b*) The volume of mercury in a mercury-in-glass thermometer is 0·35cm.^{3} and the distance between the fixed points is 25·0cm. Find the area of cross section of the bore of the stem in mm.^{2} Assume that the bore is uniform and explain why the expansion of the stem can be neglected in the calculation.

(Coefficient of linear expansion of glass = 9×10^{-6} degC.$^{-1}$ Coefficient of absolute expansion of mercury = 182×10^{-6} degC.$^{-1}$)

(N.)

6. It is required to measure the temperature of an oven which is maintained at about 300°C. Compare the advantages and disadvantages for this purpose of (*a*) the mercury-in-glass thermometer, (*b*) the platinum resistance thermometer, (*c*) the thermocouple.

A mercury-in-glass thermometer contains 14g. of mercury. Calculate the rise in the level of mercury in the capillary tube on raising the temperature of the thermometer bulb from 0°C. to 100°C., given that the capillary bore diameter is 0·4mm., the volume of the glass bulb is 1cm.^{3}, the coefficient of cubical expansion of mercury is 18×10^{-5} degC.$^{-1}$, the coefficient of linear expansion of the glass is 10^{-5} degC.$^{-1}$ and the density of mercury at 0°C. is 13·6 g. cm.$^{-3}$ To simplify the calculation assume that the whole of the mercury is heated and that the capillary bore diameter remains constant but not the volume of the thermometer bulb.

(A.)

7. A solid of density 0·88 gm. per c.c. floats in a beaker of olive oil of density 0·92 gm. per c.c. The coefficient of linear expansion of the solid is $1·2 \times 10^{-}$ per °C. and the coefficient of cubical expansion of olive oil is 7×10^{-4} per °C. Through what range of temperature must the beaker and its contents be raised before the solid just sinks in the oil?

8. Show the connexion between the real and apparent coefficients of expansion of a liquid, and describe in detail **one** method of measuring the latter.

A sealed glass bulb contains 29c.c. of mercury and 1c.c. of air at standard temperature and pressure. Find the pressure and volume of the enclosed air at 50°C., neglecting the vapour pressure of mercury at this temperature.

(Coefficient of linear expansion of glass = 0·000008 per °C. Coefficient of real expansion of mercury = 0·00018 per °C.) (W.)

9. A glass bottle is filled with a given liquid at 0°C., and when its temperature is raised to 40°C. 0·12gm. of the liquid is expelled. On now raising the temperature to 100°C. a further 0·17gm. of the liquid is expelled. Calculate (a) the mass of the liquid originally in the bottle, (b) the real coefficient of expansion of the liquid. Take the coefficient of cubical expansion of glass as 0·00001 per °C.

10. Describe how you would use a specific gravity bottle to find the coefficient of expansion of paraffin oil relative to glass between 0° and 50°C.

A specific gravity bottle contains 44·25gm. of a liquid at 0°C. and 42·02gm. at 50°C. Assuming that the coefficient of linear expansion of the glass is 0·00001 per degC., (a) compare the densities of the liquid at 0°C. and 50°C., (b) deduce the coefficient of real expansion of the liquid. Prove any formula employed. (N.)

11. Describe an experiment to determine how the density of a liquid varies with its temperature.

A sinker made of silica (whose expansion may be neglected) weighs 110·505gm. in air; 60·515gm. in water at 0°C. and 61·000gm. in water at 50°C. Determine the mean coefficient of expansion of water between 0°C. and 50°C. (N.)

12. Describe how you would measure in a laboratory the coefficient of absolute expansion of turpentine. Work out the formula from which you would calculate the coefficient.

Explain how the coefficient of absolute expansion is related to the coefficient of expansion relative to glass. (O.)

13. Describe and explain how you would use a weight thermometer to find the absolute coefficient of expansion of paraffin oil between 0°C. and 50°C., supposing the coefficient of volume expansion of the container to be known. Derive the formula from which your result would be calculated.

A weight thermometer contains 36·75gm. of a high boiling point liquid when filled at 0°C. and 32·50gm. when filled at 200°C. Assuming the coefficient of volume expansion of glass to be 25×10^{-6} degC.$^{-1}$, compare the densities of the liquid at 0°C. and 200°C. (L.)

14. What is meant by the *apparent coefficient of expansion* of a liquid and how would you determine it for a liquid such as turpentine?

A glass sinker of mass 150gm. is weighed in a liquid at 10°C. What is its apparent weight? It is then placed in the same liquid at 100°C. and weighed immediately. If it is weighed again after ten minutes will the apparent weight increase or decrease and by how much?

(Density of glass at 10°C. = 2·0 gm. per c.c. Density of liquid at 10°C. = 0·90 gm. per c.c. Mean coefficient of cubical expansion of glass 10–100°C. = 24×10^{-6} per degC. Mean coefficient of expansion of liquid 10–100°C. = 9×10^{-4} per degC.) (S.)

15. Describe how a hydrostatic balance may be used to investigate the expansion of water between 0°C. and 70°C., and explain how you would determine the density of water at any given temperature.

A sealed glass bulb of mass 15·6gm. floats on the surface of a liquid at 0°C., and at this temperature the external volume of the bulb is 20cm.³ and the density of the liquid is 0·8 gm./cm.³ If the liquid is gently heated, at what temperature will the bulb just become completely submerged?

(Coefficient of real expansion of the liquid = $9·0 \times 10^{-4}$ per degC.; coefficient of linear expansion of glass = $8·3 \times 10^{-6}$ per degC.) (W.)

16. How would you investigate the change of volume of a mixture of ice and water, initially at 0°C. as the temperature is gradually raised to 50°C? Draw a rough volume-temperature graph to illustrate the results you would expect to obtain.

A block of metal, which has a volume of 25·00c.c. at 20°C., weighs 66·25gm. in air, 45·75gm. when immersed in a certain liquid at 20°C., and 46·63gm. when immersed in the same liquid at 80°C. Calculate the mean coefficient of expansion of the liquid over the range 20–80°C. given that the coefficient of *linear* expansion of the metal is $2·55 \times 10^{-5}$ per °C.   (O. and C.)

17. Describe a method determining directly the absolute coefficient of expansion of a liquid.

A compensated pendulum consists of an iron rod of negligible mass, to which, at a distance of 1 metre from the knife edge, is fastened a hollow iron cylinder of length 16cm., internal diameter 5cm., and mass 800gm. This cylinder contains 3,200gm. of mercury. Find the change in distance of the centre of mass of the pendulum from the knife edge for a 1°C. rise in temperature.

(Coefficient of linear expansion of iron = 10^{-5} per degC.; coefficient of volume expansion of mercury = 2×10^{-4} per degC., and the density of mercury = 13·6 gm. per c.c.)   (O. and C.)

18. Describe an experiment by which the true coefficient of expansion of a liquid may be accurately determined.

A mercury barometer has a brass scale, which is correct at 0°C. The barometer reads 74·85cm. of mercury, the temperature of the surrounding air being 20°C. Reduce this reading to 0°C.

(Coefficient of cubical expansion of mercury = 0·000181; coefficient of cubical expansion of brass = 0·000054).   (N.)

19. A thread of liquid, whose coefficient of real expansion is *a* per degC., occupies a length $l_0$cm. in a capillary tube when the temperature is 0°C. What is the true length of the thread at t°C. (*a*) if the expansion of the tube is negligible, (*b*) if the coefficient of linear expansion of the material of the tube is λ per degC.?

What, in case (*b*), is the apparent length of the thread at t°C. if it is read from a scale etched on the tube and correct at 0°C.?   (N.)

20. Define the *coefficient of thermal expansion* of a liquid. Find an expression for the variation of the density of a liquid with temperature in terms of its expansion coefficient.

Describe, without experimental details, how the coefficient of thermal expansion of a liquid may be determined by the use of balanced columns.

A certain Fortin barometer has its pointers, body and scales made from brass. When it is at 0°C. it records a barometric pressure of 760 mm. Hg. What will it read when its temperature is increased to 20°C. if the pressure of the atmosphere remains unchanged?

(Cubical expansion coefficient of mercury = $1·8 \times 10^{-4}$ degC.$^{-1}$; linear expansion coefficient of brass = 2×10^{-5} degC.$^{-1}$)   (O. and C.)

21. Describe, with diagrams, some form of thermostat. Explain how it works and point out *two* practical uses for which it is suitable.   (O.)

CHAPTER V

EXPANSION OF GASES AND THE GAS LAWS

5.01 Introductory

The problem of measuring the expansion of gases is complicated by the fact that the volume of a fixed mass of gas is dependent on the pressure as well as on its temperature. A gas will distend and completely fill any space into which it is introduced, and consequently it is meaningless to assign a particular volume to the gas at a fixed temperature unless the pressure at which the volume is measured is fully specified. Thus in dealing with gases, three variables require consideration, namely volume, pressure, and temperature, and three relationships are possible between these quantities as each in turn is maintained constant whilst the connection between the remaining two variables is examined. These three relationships have been the subject of extensive experimental investigation for different gases, and the results of these investigations are expressed in three generalisations known as the *gas laws*. These laws are discussed in turn in the paragraphs that follow, and for the simple laboratory experiments by which the gas laws are usually demonstrated, the reader is referred to elementary textbooks on the subject.

5.02 Boyle's Law

This law relates the pressure and volume when the temperature is kept constant, and may be stated as follows: *The volume of a given mass of gas is inversely proportional to the pressure if the temperature is kept constant*,

thus $p \propto \dfrac{1}{v}$ if the temperature is constant,

or $p \times v = $ constant . . . . [25]

The compressibility of gases was first studied by Robert Boyle in 1662, and the results of his work were later independently confirmed by E. Mariotte in 1676. Their work was extended by later workers, and the results of observations made on a whole range of gases are summed up by the statement given above. It is well to remember, however, that Boyle's law is a wide generalisation indicating the general behaviour of all gases subject to pressure variation at constant temperature. As will be seen later the law is subject to marked deviations in certain instances, particularly under conditions of high compression and for gases at temperatures near to their critical temperatures. The law is only strictly obeyed for gases at low pressures and high temperatures, although for ordinary gases under normal conditions of temperature and pressure sufficiently close agreement exists to warrant the use of equation 25 in routine numerical work.

5.03   Charles' Law

This law deals with the expansion of gases at constant pressure. The variation of the volume of a gas with temperature at constant pressure was investigated by J. A. C. Charles in 1787 and later by L. J. Gay-Lussac in 1802. In its simplest form the law may be stated as follows:

For a given mass of any gas heated at constant pressure, a given increase in the temperature always produces a constant relative increase in the volume of the gas.

Thus is v_0 is the volume of a given mass of gas at 0°C. and v_t its volume at t°C., then

$$v_t = v_0 (1 + a_p t) \quad \ldots \ldots \ldots \quad [26]$$

where a_p is the *coefficient of increase of volume of the gas at constant pressure*. The earlier work of Charles and Gay-Lussac was followed by extensive investigation by Regnault, who determined the value of a_p for a wide range of gases. It was found that for all gases the coefficient of expansion at constant pressure was constant and equal to 0·00367 or $\dfrac{1}{273}$ per deg. C. Therefore for all gases

$$v_t = v_0 \left[1 + \frac{1}{273} \cdot t \right]$$

and Charles' law may be restated as follows: *At constant pressure the volume of a fixed mass of gas increases by $\dfrac{1}{273}$ of its volume at 0°C. for each degree C. rise in temperature.*

Under normal conditions there is fairly close agreement with Charles' law, especially with the so-called "permanent" gases. The extent of this agreement can be seen from the constancy of the values of a_p (table 16) obtained by Heuse and Otto (1929) in their experiments to check the validity of the gas laws. At low pressures and high temperatures a gas approximates more and more closely to its ideal behaviour, and under these conditions a_p for all gases tends to a limiting value. Recent work by W. H. Keesom (1934) indicates that this limiting value is 0·003661.

5.04   Law of Pressures

If the volume of a gas is maintained constant, its pressure increases regularly with the temperature. As will be seen shortly, for a gas obeying Boyle's law, the coefficient of increase in pressure at constant volume is numerically the same as the volume coefficient at constant pressure. Accordingly the law of pressures may be stated as follows:

For a given mass of gas heated at constant volume, the pressure increases by $\dfrac{1}{273}$ of its pressure at 0°C. for each degree C. rise in temperature.

Thus if p_t and p_o are the pressures of the given mass of gas at t°C. and 0°C. respectively, then

$$p_t = p_o (1 + a_v t)$$

where a_v is *the coefficient of increase of pressure of the gas at constant volume.*

This law follows from Charles' law on the assumption that the gas also obeys Boyle's law. Let v_o, p_o be the volume and pressure of a given mass of gas at 0°C. and v_t, p_t be the corresponding values at t°C. Then, if the temperature of the gas is increased from 0°C. to t°C. whilst the pressure remains constant at p_o,

$$v_t = v_o (1 + a_p t)$$

Let the pressure of the gas be increased to p_t until the volume is v_o with the temperature constant throughout at t°C. Then by Boyle's law

$$p_o v_t = p_t v_o$$

Eliminating v_t from the above equations we have

$$\frac{p_t v_o}{p_o} = v_o (1 + a_p t)$$

or $\qquad p_t = p_o (1 + a_p t)$ . . . . [27]

which establishes the law in question.

From equations 26 and 27 it will be seen that if the gas obey's Boyle's law then the coefficient of increase of pressure at constant volume a_v is identical with the coefficient of increase in volume at constant pressure a_p. The extent to which this is true for the gases helium, hydrogen, and nitrogen is revealed in table 16, which gives the mean coefficients between the ice and steam points for the three gases at different initial pressures.

TABLE 16

Pressure in cm. of mercury		39·02	53·31	72·72	99·45
Helium	a_v	0·0036597	0·0036594	0·0036587	0·0036579
	a_p	0·0036611	0·0036602	0·0036611	0·0036604
Hydrogen.............	a_v	0·0036604	0·0036604	0·0036593	0·0036589
	a_p	0·0036617	0·0036613	0·0036620	0·0036621
Nitrogen	a_v	0·0036664	0·0036668	0·0036699	0·0036734
	a_p	0·0036673	0·0036671	0·00366709	0·0036674

5.05 Absolute Zero and Absolute Scale of Temperatures

Considering a gas that strictly conforms to Charles' law we have

$$v_t = v_o(1 + a_p t)$$

where t is the temperature of the gas in degrees centigrade at which its volume is v_t. This equation may be rewritten as follows:

$$v_t = v_o a_p \left[\frac{1}{a_p} + t \right]$$

$$= v_o a_p T \quad \cdots \cdots \cdots \cdots \cdots \quad [28]$$

where $T = \dfrac{1}{a_p} + t \quad \cdots \cdots \cdots \cdots \cdots \quad [28a]$

Equation 28 defines a new temperature T which is such that the volume of the gas is directly proportional to it. The volume is clearly zero when T is zero, that is (from equation 28a) when $\left[\dfrac{1}{a_p} + t \right]$ is zero, or when $t = -\dfrac{1}{a_p} = -273 \cdot 16°C$. (with $a_p = 0 \cdot 0036608$). Since it is not possible to contemplate a negative volume for a gas, the temperature at which the ideal gas would occupy zero volume suggests that it is the lowest temperature attainable. This temperature is known as the *absolute zero*, and the scale of temperature obtained by dis-placing the zero to this point is referred to as the *absolute scale of temperature*. It is clear that both the absolute and centigrade scales are measured in the same units, the absolute temperatures being obtained by adding $273 \cdot 16$ to the temperatures as measured on the centigrade scale.

The pressure exerted by a gas when heated at constant volume can in a similar way be made to define the absolute zero of temperature.

Thus $p_t = p_o(1 + a_v t)$

$$= p_o a_v \left[\frac{1}{a_v} + t \right] = p_o a_v T$$

where $T = \dfrac{1}{a_v} + t$

For an ideal gas which conforms exactly with Boyle's law it has been shown that $a_v = a_p$, and hence it is seen that the scale of temperature defined above is identical with the absolute scale established on the basis of Charles' law. Thus the pressure of an ideal gas varies directly as the absolute temperature, and such a gas exerts zero pressure at the absolute zero of temperature.

The ideal gas cannot be realised in practice, although gases such as nitrogen, hydrogen, and helium conform very closely to the ideal state, particularly at low pressures. Thus absolute temperatures cannot be precisely evaluated experimentally, but for practical purposes the absolute scale can be established using a standard constant volume gas thermometer (see section 1.08) containing hydrogen or helium. The value of the absolute zero of temperature can be obtained by recording the pressures of the gas at the ice and steam points and extrapolating to zero pressure as indicated in Fig. 49. The value so obtained will,

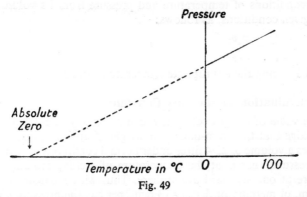

Fig. 49

however, differ slightly from its value on the ideal gas scale. The Joule-Kelvin effect which measures the departure of an actual gas from the ideal state can be used, however, to provide the necessary correction between the observed and true values of the absolute zero. A method of correcting gas thermometer readings is given in section 5.11.

5.06   The Ideal Gas Equation

As already established, the ideal gas is one which conforms exactly to the laws of Boyle and Charles, which may be written respectively as follows:

$$pv = \text{const.}$$

and $\dfrac{v}{T} = \text{const.}$ where T is the absolute temperature.

Thus for a given mass of an ideal gas we may combine the two above relationships to obtain,

$$\frac{pv}{T} = \text{const.} \quad . \; . \; . \; . \; . \; . \; . \; . \; . \; . \; . \; . \; . \quad [29]$$

If the given mass considered is the gram-molecular weight or *mole*, the constant appearing in equation 29 is, on the assumption of Avagadro's hypothesis, the same for all gases (assumed perfect). This universal constant is known as the *gas constant* and is given the symbol R. Equation 29 can then be written as

$$pv = RT \quad . \; . \; . \; . \; . \; . \; . \; . \; . \; . \; . \; . \; . \quad [30]$$

4*

This equation embodies the three laws covering the behaviour of an ideal gas, and is called the *ideal gas equation*. The behaviour of real gases in the main approximates closely to that of the ideal gas, and consequently equation 30 can be used in calculations dealing with real gases, particularly if the changes in the variables are small. Thus the equation may be used to obtain the volume of a given mass of gas under stated conditions of temperature and pressure from its volume under other given conditions, as follows:

$$\frac{p_1 v_1}{T_1} = \frac{p_2 v_2}{T_2}$$

T_1 and T_2 being the appropriate temperatures in *absolute* units.

5.07   Evaluation of the Gas Constant

The value of the gas constant R can be readily calculated from the knowledge that the gram-molecular weight of any gas (assumed ideal) occupies a volume of 22·4 litres under normal conditions of temperature and pressure, that is at the temperature of melting ice and under a pressure of one standard atmosphere. Thus with a barometric height of 76cm. of mercury of density 13·6gm. per c.c. the pressure p is

$$76 \times 13·6 \times 981 \text{ dynes per sq. cm.}$$

and since $T = 273·2°A$

and $v = 22400$ c.c.

then $R = \dfrac{pv}{T} = \dfrac{76 \times 13·6 \times 981 \times 22400}{273·2}$

$= 8·31 \times 10^7$ ergs per gm. molecule per °C.

As evaluated above the units of R are in ergs since the product pv has the dimensions of energy or work. Since $4·185 \times 10^7$ ergs are equivalent to 1 calorie, the value of R in heat units is

$$R = \frac{8·31 \times 10^7}{4·185 \times 10^7} = 1·986 \text{ calories per gm. molecule per °C.}$$

or approximately 2 calories per mole per °C.

It is as well to remember that the above value for R relates only to the gram-molecular weight of the gas, in which case it is a constant for all gases. Its value will be different if masses other than the mole are considered. A specially important case is that dealing with 1 gram of a gas. The calculation proceeds as before using a value for v of 22400c.c. divided by the molecular weight of the gas in grams. Clearly, therefore, the value of the gas constant for unit mass can be derived from the value of R obtained above by dividing it by the molecular weight (M) in grams. Thus:

$$R' \text{ (gas constant per unit mass)} = \frac{R}{M}$$

and clearly the value of R' will be different for different gases. In the case of oxygen ($M = 32$), its value will be

$$R' \text{ (for 1 gm. of oxygen)} = \frac{8\cdot31}{32} \times 10^7$$

$$= 2\cdot597 \times 10^6 \text{ ergs per gm. per °C.}$$

For any other mass (m) of a gas of molecular weight (M) the value of the gas constant will be $\dfrac{m}{M} R$.

5.08 Deviations from the Gas Laws

In the foregoing sections of this chapter the behaviour of gases has been described in terms of certain generalisations which lead to the relationship $pv = RT$. This equation relates to the properties of an ideal gas, and the extent to which it represents the actual behaviour of real gases has been the subject of extensive experimental investigation since the early part of the nineteenth century. Despretz in 1827 carried out experiments on the compressibility of different gases, comparing their behaviour with that of air. He placed equal amounts of the gases in a set of barometer tubes placed side by side in a closed cistern containing mercury which was fitted with screw plungers. On raising the pressure by screwing in the plunger, the gases were seen to be unequally compressed. Gases which are easily liquefied, such as carbon dioxide, ammonia, &c., were found to be much more compressible than air and the other "permanent" gases, and it was clear that in their case at least marked deviations existed from Boyle's law.

At high pressures the volume of the gas becomes too small to be measured accurately, and in 1847 Regnault made careful experiments using the apparatus shown in Fig. 50, in which this difficulty is overcome. A quantity of the pure dry gas was introduced into a strong glass tube T, which contained mercury at its lower end and was connected to an open mercury manometer M inserted into the same cistern of mercury. More mercury could be forced up M and into T by means of a plunger P, and throughout the experiment the temperature of the gas in T was kept constant by means of a water jacket. At the start of the experiment the mercury was adjusted to be on the same level in T

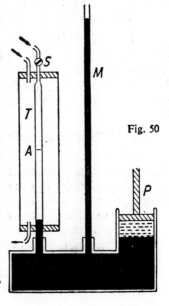

Fig. 50

and M when the gas was at atmospheric pressure p_1. Mercury was then pumped into T until it rose to a mark A, at which the volume of the gas was halved. The pressure p_2 required to do this was read off, and if Boyle's law was obeyed then $\frac{2p_1}{p_2}$ should equal unity. The stop-cock S was now opened and more gas pumped into T until the mercury was depressed to its original level, when the gas was at some pressure p_3. S was now closed and the gas compressed as before until its volume was halved, when the value of $\frac{2p_3}{p_4}$ should again equal unity if Boyle's law was true for the gas. Proceeding in this way Regnault was able to show that all gases except hydrogen were more compressible than if they obeyed Boyle's law, the product pv for these gases, instead of remaining constant, decreasing as the pressure applied increased.

Amagat (1880) extended Regnault's work, and investigated the behaviour of gases at very high pressures. He employed a steel pressure tube several hundreds of feet in length situate in the vertical shaft of a mine. The tube connected with a strong glass tube of small bore via a mercury chamber fitted with a screw plunger. The glass tube was first carefully calibrated by examining the behaviour of nitrogen over a wide range of pressures, and the behaviour of other gases was then compared with that of nitrogen. The glass tube was water-jacketed to ensure a constant controlled temperature.

Amagat represented his results graphically by plotting the product pv for the gas against p to obtain a set of isothermals. His general results may be summarised as follows:

(i) For all gases at low pressures the product pv decreases as p increases.

(ii) As the pressure increases the product pv attains a minimum and then shows a steady increase with p.

Fig. 53 (section 5.10) shows a typical set of p, pv isothermals for a gas which will be considered more fully later.

5.09 Holborn's Experiments

In a series of experiments commenced in 1915 by Holborn and his co-workers, the difficulties of the accurate measurement of the volumes of gases under high compression and of obtaining reliable values of the high pressures used, were overcome by measuring the mass of gas required to fill a given volume at various pressures which were determined by a pressure balance specially devised for the purpose. The instrument used is shown diagrammatically in Fig. 51. The pressure of the gas is transmitted via a tube A containing oil to a piston C which fits accurately into a cylindrical hole bored in a firmly clamped metal block B. A frame F, with attached weights E, is supported by a screw D

resting on the upper end of the piston, and the weights are adjusted until the total weight of E, F, D and C is just balanced by the thrust of the oil on the lower end of C. Measurements are always taken with the piston in a fixed position where the cross sectional area a has been accurately measured by calibration against a mercury manometer at a fairly low pressure. To reduce friction a fairly slow stream of oil continually emerges between the piston and cylinder, and further to reduce friction the piston is caused to rotate slowly whilst measurements are being taken. The pressure communicated by the gas to the piston when balance is obtained is $\frac{mg}{a}$ where m is the mass of the parts E, F, D and C.

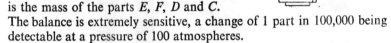

Fig. 51

The balance is extremely sensitive, a change of 1 part in 100,000 being detectable at a pressure of 100 atmospheres.

The scheme of Holborn's apparatus is shown in Fig. 52. The apparatus was first of all completely evacuated of air with all the taps open. The tap T_2 and the taps S_1, S_2 . . . were then closed and the gas under investigation introduced into a long narrow stout glass vessel A via the connecting tube D until the required pressure, as measured by the pressure balance, was attained. The tap T_1 was then closed. The glass vessel A was surrounded by a strong steel vessel B containing mercury, which was brought to the same pressure as the gas in A by introducing compressed air via the tube C at the same time as the gas was pumped into A. This was done to prevent the glass vessel bursting, and also to avoid changes in its volume on account of the high pressures of the gas inside it. The vessel B in its turn was surrounded by a bath (not shown) for maintaining the temperature constant at any desired value. The volume of the vessel A and the connecting tubes up to the taps T_1 and T_2 having been previously determined, the volume of the gas at the measured pressure was therefore accurately known. It was now necessary to find the mass of this given volume of gas. To do this the gas was introduced into the low pressure side of the apparatus by opening T_2 and the gas allowed to fill a sufficient number of the flasks F_1, F_2, &c., until the pressure, recorded by the mercury manometer M, was nearly atmospheric. The temperature of the gas was obtained from the constant temperature bath surrounding the flasks, and the volume of the gas was found by adding together the volumes of A, the connecting tubes, and those of the flasks filled—all these volumes having been previously

calibrated. Having thus determined the volume, pressure, and temperature of the gas, the volume of the gas was reduced to *N.T.P.* and its

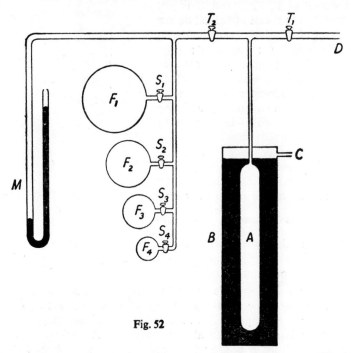

Fig. 52

mass calculated. From this the volume occupied by a given mass of the
gas at the high pressure was obtained. The whole apparatus was now
exhausted and the experiment repeated at a different pressure. Continuing in this way the volume occupied by a given mass of gas for a
range of pressure values at different temperatures was obtained.

5.10 General Nature of the Results

The work of the many investigators on the compressibility of gases
can be summed up by the series of curves shown in Fig. 53. The curves
are a family of isothermals obtained by plotting the product *pv* against
p for a range of temperatures increasing from the bottom of the
diagram. The general nature of the curves is the same for all gases, and
hence no scales have been marked on the figure. It is seen that at high
temperatures the curves commence with an initial positive slope and
show a gradual rise as the pressure increases. At lower temperatures the
initial slope is negative, and the curves slope downwards to attain a
minimum value, after which they follow a gradual upward trend. The
locus of the minima of the curves is marked by the dotted curve *M*.

For temperatures lower than the *critical temperature* (see section 10.02) there is a sudden vertical drop due to the change from the gaseous to the

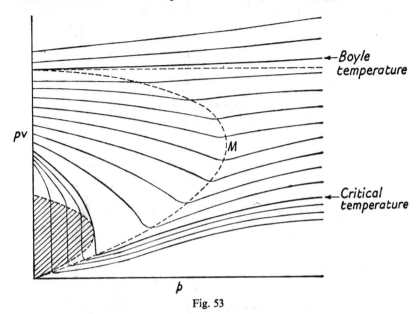

Fig. 53

liquid state. When liquefaction is complete the curve rises steadily as the pressure increases. The shaded area in the diagram represents the zone of liquefaction.

In order to represent analytically the behaviour revealed by gases in these experiments, Kamerlingh Onnes introduced the following empirical relation:

$$pv = A + Bp + Cp^2 + Dp^3 + \quad . \quad . \quad . \quad [31]$$

where $A, B, C, D, \ldots$ are constants which are characteristic of a given gas. The values of the constants vary with temperature, and for a given curve it is possible to evaluate them so as to specify completely the state of the gas for the temperature represented by the curve. The coefficients are called *virial coefficients* and their magnitudes decrease rapidly from A along the series. Thus for nitrogen at 100°C. the relative order of magnitude of the coefficients $A, B, C \ldots$ are $1 : 10^{-3} : 10^{-6}$. The first virial coefficient A is equal to RT (for 1 gram molecule of the gas). The second coefficient B is of great importance in dealing with individual gases, and it is found to vary in exactly the same way for all gases. It has negative values at low temperatures which increase through zero to positive values at high temperatures. The temperature at which it is zero is of particular importance and is called the *Boyle temperature* (see section 10.05). At this temperature Boyle's law is closely obeyed for a wide range of pressures from zero upwards, since the terms Cp^2, Dp^3, &c., are unimportant except at high pressures. Thus for nitrogen, which has a Boyle temperature of 50°C., the deviations from Boyle's law are less than one part in 1000 for pressures extending up to 19 atmospheres.

The value of B can be obtained by differentiating equation 31 and putting $p = o$. That is,

$$B = \left[\frac{d(pv)}{dp} \right]_{p\,=\,o}$$

Hence by extrapolating the curve to zero pressure and taking the slope of the tangent to the curve at this point, B can be evaluated as indicated in Fig. 54.

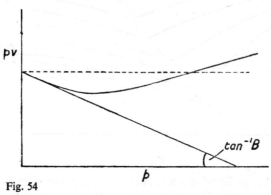

Fig. 54

A consideration of the general nature of the results of the experiments on the compressibility of gases with a view to their interpretation suggests strongly the existence of inter-molecular forces. The fact that below the Boyle temperature gases are more compressible than the ideal gas is highly suggestive of the existence of inter-molecular attractions which are the equivalent of an added external pressure. This in effect reduces the external pressure necessary to produce a given compression, and thus the product pv is lower than predicted by the ideal gas equation. The rise in the graph that occurs at high pressures can be explained on the assumption that the molecules have a finite size which reduces the effective volume of the gas and results in an over estimate of the product pv. At temperatures above the Boyle temperature the higher values of pv would suggest that at these temperatures there exists a force of repulsion between the molecules. The question of these molecular forces will be further discussed in chapter X on the continuity of state.

5.11 Correction of Gas Thermometer Readings

It has already been mentioned that the scale of temperature given by a gas thermometer containing hydrogen or helium at low pressure is the closest practical realisation of the centigrade thermodynamic scale of temperatures, and that the scale obtained in this way is the one to which all other scales are ultimately referred. Since however all gases show departures from the gas laws in varying degree it follows that different gases will give scales of temperature which show deviations from one another. Nevertheless it is possible to correct the readings of a given gas thermometer to the ideal or thermodynamic scale from a knowledge of the virial coefficients for the gas. The theory of these corrections is as follows:

Consider a constant volume gas thermometer in which the constant volume of the gas is V_0. Using the virial expansion (equation 31) for temperatures of 0°C. (ice point), 100°C. (steam point), and t°C. we have, using appropriate suffixes to indicate the values of the pressure and virial coefficients at the corresponding temperatures,

$$p_0 v_0 = A_0 + B_0 p_0 + C_0 p_0^2 + \ldots$$

$$p_{100} v_0 = A_{100} + B_{100} p_{100} + C_{100} p_{100}^2 + \ldots$$

$$p_t v_0 = A_t + B_t p_t + C_t p_t^2 + \ldots$$

Now the gas temperature t_g is defined by the equation

$$t_g = 100 \frac{p_t - p_0}{p_{100} - p_0}$$

thus using the virial equations given above as far as the term containing the second virial coefficient (terms beyond this can be ignored at the pressures used in gas thermometry), we have

$$t_g = 100 \frac{A_t - A_0 + B_t p_t - B_0 p_0}{A_{100} - A_0 + B_{100} p_{100} - B_0 p_0}$$

$$= 100 \left[\frac{A_t - A_0}{A_{100} - A_0} \right] \left\{ \frac{1 + \dfrac{B_t p_t - B_0 p_0}{A_t - A_0}}{1 + \dfrac{B_{100} p_{100} - B_0 p_0}{A_{100} - A_0}} \right\}$$

Applying the binominal theorem and ignoring second and higher order terms, this becomes

$$t_g = 100 \left[\frac{A_t - A_0}{A_{100} - A_0} \right] \left\{ 1 + \frac{B_t p_t - B_0 p_0}{A_t - A_0} - \frac{B_{100} p_{100} - B_0 p_0}{A_{100} - A_0} \right\}$$

Now the ideal gas scale temperature (t) would be obtained from the readings of the gas thermometer at infinitely low pressure, that is

$$t = \lim. \ 100 \frac{p_t - p_0}{p_{100} - p_0} \text{ as } p \to o$$

$$= 100 \frac{A_t - A_0}{A_{100} - A_0} \text{ (since the volume is constant).}$$

Hence we have

$$t_g = t \left\{ 1 + \frac{B_t p_t - B_0 p_0}{A_t - A_0} - \frac{B_{100} p_{100} - B_0 p_0}{A_{100} - A_0} \right\}$$

from which the necessary correction $t - t_g$ is seen to be

$$t \left\{ \frac{B_{100} p_{100} - B_0 p_0}{A_{100} - A_0} - \frac{B_t p_t - B_0 p_0}{A_t - A_0} \right\}$$

The various corrections are obtained from a knowledge of the virial coefficients (see previous section) and corresponding gas pressures by using a method of successive approximations similar to that used in correcting the platinum resistance temperatures (see section 1.09). The correction to be applied to any observed gas scale reading can then be obtained by interpolation from a graph of the corrections calculated in this manner against gas temperatures.

5.12   Production and Measurement of Low Pressures

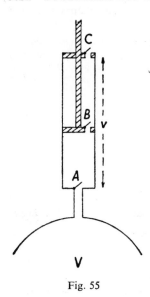

Fig. 55

To conclude this chapter the various devices which depend for their action on Boyle's law to produce and measure low pressures will be discussed. Fig. 55 shows the principle of *Smeaton's exhaust pump*. As the piston is pulled up, air from the vessel to be evacuated opens the valve A and enters the barrel of the pump. On the subsequent downstroke valve A closes, and the air as it is compressed opens valve B and so enters the space above the piston. On the next upstroke this air is forced out via valve C when the pressure in the barrel is increased above that of the atmosphere. The purpose of valve C is to relieve the piston of the thrust of the atmosphere after the first upstroke. After the first upstroke a volume V of air at an initial pressure P becomes $V + v$ at some pressure P_1, where by Boyle's law,

$$PV = P_1 (V + v)$$

thus $\quad P_1 = P \cdot \dfrac{V}{V + v}$

If the pressure falls to P_2 after the end of the second upstroke we have similarly

$$P_1 V = P_2 (V + v)$$

or $\quad P_2 = P_1 \cdot \dfrac{V}{V + v} = P \left[\dfrac{V}{V + v} \right]^2$

Proceeding in this way it will be seen that the pressure P_n at the end of the nth upstroke is

$$P_n = P \left[\dfrac{V}{V + v} \right]^n$$

Apart from the question of valve leakage, it is not possible to obtain a high degree of vacuum with this pump as, after a time, due to the "dead space" above B it will not be possible to compress the air to a pressure sufficiently high for it to open C. The pump has been improved by Fleuss who sealed all the valves with oil, and so eliminated the "dead space" above the piston. Modified in this way the piston exhaust pump is capable of reducing the pressure to about $\frac{1}{50}$ mm. of mercury.

A more efficient pump working on the same principle as the piston pump is that designed by *Toepler,* one form of which is shown in Fig. 56. The action of the pump is as follows. As the mercury reservoir *R* is raised, the pump chamber *A* gradually fills with mercury and the air in it is forced out via the long capillary tube *B*. This process is continued until *A* is completely filled, when the float *F* fitting into the ground glass joint *G*, leading to the apparatus to be exhausted, will have been pushed home. On lowering *R*, *F* opens, and air from the apparatus rushes into *A* which would otherwise become a vacuum. On again raising *R* the volume of this air trapped in *A* when the mercury rises to *P* is pushed out via *B*, the rising mercury acting like a piston. Proceeding in this way a fairly high vacuum (pressures of the order of 10^{-5} mm. of mercury) can be obtained. The production of vacua by this pump is however a slow and cumbersome process, and in modern practice high vacua are obtained by using mercury vapour pumps backed by rotary oil pumps.

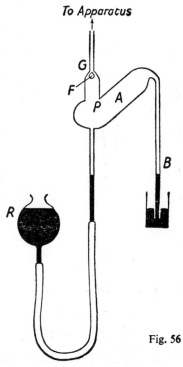

Fig. 56

The measurement of the low pressures produced by these means calls for specially devised manometers. The ordinary liquid in glass manometer is useless for this purpose on account of the difficulties arising from the vapour pressure of the liquid used in the manometer, and a mercury manometer is far too insensitive at these low pressures. Accordingly it is necessary to use a different principle in the construction of low pressure manometers, and the *McLeod gauge* (Fig. 57), described below, depends on Boyle's law for its use. A narrow capillary tube *AB* of known bore is connected to a large bulb *C* whose volume is accurately known. The bulb is connected by a length of clean rubber tubing to a mercury reservoir *R*, and a long vertical tube T_1 connects the gauge at *D* to the apparatus, the pressure in which is to be measured. The tube T_1 is side-tracked by a capillary tube T_2 exactly similar to *AB*, the purpose of this being to eliminate capillary action from affecting the readings of the pressure difference subsequently to be taken. On connecting the instrument to the apparatus the level of mercury is

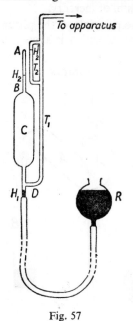

at H_1, and there will be a very small pressure p in the tube T_1 and the vessel C (p will be assumed constant throughout the measurement, the volume of the apparatus being considered very much greater than that of the gauge). On raising R a volume V (equal to the volume of the vessel C and the capillary tube AB) of gas is cut off at the pressure p, and this gas is compressed into the capillary tube until the mercury level is at some point H_2. The mercury has also risen in T_1 and T_2 of course, its level always being above that in AB, and it is usual to raise R until the mercury in T_2 is level with A at H_2. The gas in the capillary is now under a pressure of $(p + h)$ cm. of mercury where h is the distance AH_2. If α is the cross sectional area of the capillary, the volume of the gas is αh, whence by Boyle's law,

$$pV = (p + h)\alpha h$$

Fig. 57

$$\therefore \ p = \frac{\alpha h^2}{V - \alpha h}$$

or more approximately $p = \dfrac{\alpha h^2}{V}$, ignoring αh compared with V (in

practice the gas is compressed about 1000 times). The order of magnitude of the pressures that can be measured by this gauge can be illustrated by the following numerical example:

If $V = 200$c.c. and the diameter of the capillary is $0\cdot1$cm., then $p = 3\cdot9 \times 10^{-4}$ h^2mm. of mercury. For $h = 1$mm., $p = 3\cdot9 \times 10^{-6}$mm. of mercury, and hence the lowest practical limit is of the order 10^{-5}mm.

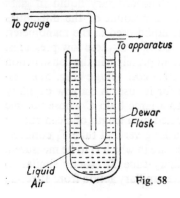

Fig. 58

The McLeod gauge is unreliable when used with gases such as air or carbon dioxide, and cannot be used at all with easily condensible vapours. Its readings are rendered erratic by the entry of moisture into the instrument, and to obviate this a liquid air trap (Fig. 58) may be used with the gauge. However, unlike low pressure gauges of the heat conductivity type (such as the Pirani instrument) and the decrement type, the McLeod gauge gives direct readings of the pressure, and in consequence is widely used as a calibrating instrument.

Another standard type of low pressure gauge will be discussed in the chapter on the Kinetic Theory of Gases (see section 8.20).

QUESTIONS. CHAPTER 5

1. Describe how you would test the accuracy of Boyle's law for pressures between ½ and 2 atmospheres. Have tests been made at very much higher pressures? If so, give a brief account of the method and results.

The volume of a Torricellian vacuum is 5c.c. and its length is 10cm. If 6c.c. of air at the pressure of the atmosphere, which is 75cm., are introduced into the barometer tube, by how much will the mercury column fall? (N.)

2. Give an account of Boyle's law for gases.

If the product of the pressure and the volume of a certain mass of gas at 0°C. is 10^8ergs, what is its value at 100°C.?

Explain why the product of pressure and volume is expressed in ergs. (L.)

3. Describe two forms of barometer and discuss their relative merits for two practical purposes.

A mercury barometer is known to be defective and to contain a small quantity of air in the space above the mercury. When an accurate barometer reads 770mm. the defective one reads 760mm. and when the accurate one reads 750mm. the defective one reads 742mm. What is the true atmospheric pressure when the defective barometer reads 750mm.? (Camb. Schol.)

4. Explain the corrections which have to be made to the readings of a Fortin barometer to obtain the absolute pressure of the atmosphere. A little air has leaked into a barometer tube 100cm. long. The mercury stands at the 70cm. mark when the tube is vertical and at the 78cm. mark when the tube is inclined at 30° to the vertical. What is the atmospheric pressure in millimetres of mercury? (Oxford Schol.)

5. Describe with diagrams, the construction and mode of action of a pump that will reduce the pressure of the air in a vessel to 0·001mm. of mercury. How are pressures of this order of magnitude measured?

A piston pump of effective volume 200cc. is used to exhaust a vessel of volume 1 litre. How many complete strokes will be required to reduce the pressure of the air in the vessel to one-hundredth of its initial value? (Neglect the volume of the connecting tubes, etc., and assume that the temperature remains constant.) (O. & C.)

6. State *Boyle's law* and describe how its validity for dry air at ordinary pressures and room temperature may be investigated experimentally. In what circumstances is the law found to be inaccurate?

The top end of a uniform barometer tube and the surface of the mercury within it stand at 100cm. and 76cm. respectively above the level of the mercury in the reservoir. Hydrogen is introduced into the vacuum above the mercury causing it to be depressed to 35cm. above the level in the reservoir. Helium is then added till the level is the same as that of the reservoir. Calculate the pressure then exerted by the hydrogen in the mixture.

(Changes of the mercury level in the reservoir may be neglected.) (L.)

7. Describe and explain how you would introduce dry air into a Boyle's law apparatus and then use the apparatus to investigate the validity of Boyle's law for air over a pressure range from 0·5 to 1·5 atmospheres.

Sketch the graphs you would expect to obtain if after carrying out the above experiment, you plotted (a) the pressure as ordinate against the reciprocal of volume as abscissa, and (b) the logarithm of the pressure against the logarithm of the volume.

How would you expect the graph (a) to be modified if the " air " used contained just enough water to saturate the volume occupied when the pressure of the " air " is atmospheric?   (L.)

8. State *Boyle's law* and *Charles' law*, and show how they lead to the gas equation $PV = RT$. Describe an experiment you would perform to measure the thermal expansion coefficient of dry air.

What volume of liquid oxygen (density 1·14 gm. cm.$^{-3}$) may be made by lique-fying completely the contents of a cylinder of gaseous oxygen containing 100 litres of oxygen at 120 atmospheres pressure and 20°C.? Assume that oxygen behaves as an ideal gas in this latter region of pressure and temperature.

(1 atmosphere = 1·01 × 10^6 dynes cm.$^{-2}$; gas constant = 8·31 joules degC.$^{-1}$ mole^{-1}; molecular weight of oxygen = 32·0.)   (O. and C.)

9. Describe an experiment to determine how the volume V of a fixed mass of gas maintained at constant pressure varies, as the temperature t, measured with a mercury-in-glass thermometer, is raised from 0°C. to 100°C. Draw a diagram of the apparatus used.

Sketch the (V, t) graph obtained in the experiment.

Indicate and explain how the form of the graph would change if (a) half the mass of the same gas were used, (b) the same mass of another gas were added to the first, (c) a small quantity of water vapour were introduced into the gas at 100°C., such that when the temperature was lowered saturation occurred at 50°C. Assume that the vapour behaves as an ideal gas above 50°C.   (N.)

10. Describe the methods by which you would experimentally measure (a) a very small pressure, (b) a very great pressure.

A straight tube of uniform bore, of which the capacity is 100 cubic inches and the length 2 feet, is closed at one end and sunk into the sea, the closed end being uppermost. Find how far the tube must be sunk below the surface of the sea in order that 80 cubic inches of water may be forced into it. The specific gravity of sea-water is 1·03, and that of mercury 13·6; the mercury barometer stands at 30".

(W.)

11. Describe an experiment to verify the relation between the volume of a fixed mass of gas and its temperature, the pressure being kept constant.

Assuming that a barrage balloon has a volume of 20,000 cub. ft. when filled with hydrogen at 17°C., the pressure being atmospheric, determine the volume of hydrogen, measured at a pressure of 200 atmospheres and a temperature of 5°C., needed to fill the balloon at 17°C.   (N.)

12. Describe a single instrument suitable for the measurement of both very high and very low temperatures.

Two 1-litre glass flasks contain dry air at 27°C. and are sealed hermetically one at each end of a capillary tube with a bore of 1sq. mm. cross section containing a short mercury index. What change in temperature of one of the flasks would move the index 1cm.?   (W.)

13. Define (a) coefficient of increase of pressure at constant volume, (b) co-efficient of increase of volume at constant pressure and show that these coefficients are equal in the case of a perfect gas.

A glass bulb is fitted with a narrow tube open to the atmosphere. Calculate the fraction of the original mass of air in the bulb which is expelled when the temperature of the bulb is raised from 10°C. to 100°C.

14. What do you understand by the equation of an ideal gas?

Two vessels, of volume 10c.c. and 20c.c. connected by a tube of negligible volume, are filled with air. When the smaller vessel is at a temperature of 0°C. and the larger at 200°C., the pressure of the gas is 1 atmosphere. Find an expression for the total mass of gas in the system, and hence or otherwise calculate the pressure when the vessels are at a common temperature of 0°C.

(Absolute zero of temperature = −273°C. Density of air at 0°C. and 1 atmosphere pressure = 1·3gm. per litre.)   (Oxford entrance)

15. Describe a constant volume air thermometer and explain how you would measure the temperature of a liquid given an uncalibrated thermometer of this type.

The spherical bulb of a constant volume air thermometer is made of a metal with a coefficient of linear expansion of 26×10^{-6} per °C. and has a radius of 5cm., the wall thickness being negligible. The pressure in the bulb is 760mm. of mercury at 0°C. and 1030mm. at 100°C. Find the coefficient of expansion of air.

(Camb. Schol.)

16. State the experimental evidence that many gases obey closely the relation $PV = AT$ when P is the pressure, V the volume, and T the temperature of the gas, and A is a constant. In what circumstances is it possible to let A equal R, a constant which is the same for all gases?

Find the value of A which is appropriate to 1gm. of argon from the following data: when a high-pressure cylinder, initially evacuated, which has a volume of 15 litres, is filled at 15°C. with argon at a pressure of 120 atmospheres it is found to increase in mass by 3·1kg.

(1 atmosphere = $1·01 \times 10^6$ dynes cm.$^{-2}$)   (O. and C.)

17. Describe with a suitable diagram the McLeod gauge for measuring low gas pressures.

Derive from first principles the equation for this gauge in which the gas pressure p is given in terms of the bulb volume V, the volume v per unit length of the closed capillary and the length h of the gas column compressed in the closed capillary. Assume the reading is taken by raising the mercury level in the comparison capillary to be opposite the top end of the closed one.

If the diameter of the closed capillary tubing is 1mm, the bulb (plus closed capillary) volume is 300cm.^{3} and the total length of the closed capillary is 10cm., what are the maximum and minimum pressures recordable by this gauge presuming that the minimum recordable value of h is 1mm.?   [A.]

18. Describe and explain in detail how you would measure the pressure of the residual gas in a vessel which had been evacuated to a pressure of about 10^{-3}mm. of mercury.

Discuss the advantages and disadvantages of the type of gauge that you describe.
[A.]

19. Describe, with a labelled diagram, a mechanical rotary vacuum pump capable of exhausting a vessel to a pressure of about 0·01mm. of mercury and explain how it operates. Draw the typical characteristic curve of pumping speed against intake pressure for such a pump. Comment briefly on the uses of rotary pumps. (A.)

CHAPTER VI

MECHANICAL EQUIVALENT OF HEAT

6.01 Early Theories—the Caloric Theory

As already indicated in chapter II it is not necessary to formulate any particular theory as to the nature of heat in making measurements of heat exchanges or of temperature. It was merely stated that the agent of temperature change and the resulting changes in the physical condition of a substance was due to "something" which we called heat being communicated to the body. The facts of experimental science can, however, be better correlated and comprehended if there is available some theory by means of which they can be explained. Indeed theory and experiment go hand in hand, a theory formulated to describe existing facts often pointing the way to the discovery of new facts. Any theory propounded to explain the accumulated data of observation and experiment must, however, fit all the known facts; it must provide a complete and consistent description, and be capable of further experimental tests as the field of practical science is extended. Should any discrepancies arise in this process, the theory must be modified in such a way as to cover all the experimental data, or if this is not possible, it must be replaced by some other theory which can satisfactorily do so. In the early part of the present chapter we shall consider in historical sequence the speculations and theories regarding the nature of heat, and examine their validity as the field of experiment steadily expands.

From the earliest times there have been two rival theories regarding the nature of heat. Certain of the Greek philosophers supposed heat to be due to the rapid vibrations of the molecules of a body. This was but a pure speculation entirely divorced from experiment, and the first rational attempt to found the theory on observation seems to have been made by Francis Bacon (1561–1626). He based his theory of heat on experiments in which heat was produced by friction and percussion, and after mature deliberation concluded that "heat is motion." These ideas, however, were only held by the small minority, and the theory which gained most favour and continued to do so until well into the nineteenth century was the "caloric" theory.

According to this theory heat was a subtle, self-repellent fluid called "caloric" which was generally held to be imponderable. The fluid was invisible and completely filled the interstices of matter. A body became cooler as some of the fluid left it, whilst a gain of caloric by other bodies caused them to become warmer, the total quantity of caloric, which was indestructible, being invariable in any exchange process.

The expansion of bodies resulting from an increase in heat content was readily explained by adherents of the theory as a natural conse-quence of the self-repellent characteristic of the fluid, which would inevitably cause further distention of the substance. Conduction of heat was also explained by the self-repellent nature of the fluid, which property would cause a flow of caloric from the hotter to the colder of two bodies placed in thermal contact. The difference in the specific heats of substances was explained by the various degrees of attraction of different substances for caloric, and latent heats were accounted for by supposing that some of the caloric entered into combination with the particles of the substance when it became inactive and thus did not affect a thermometer. Thus, water = ice + latent heat, and steam = water + latent heat. It can thus be seen that the fluid theory of heat explained many of the facts of observation satisfactorily to the philosophers of the time and who accordingly deemed it to be adequate.

Difficulties arose however when the theory was required to explain the production of heat by percussion and friction. In order to do this the adherents of the theory supposed that in the former case caloric was squeezed out of the substance when hammered resulting in the release of "sensible heat." Heat produced by friction was accounted for by postulating that the capacity of heat of the abraded particles was lower than that of the solid substance from which they had come, the release of the heat thus brought about by the friction process causing the observed rise of temperature. No attempt seems to have been made to check the arbitrary nature of this latter assumption by an actual experiment, and it was the work of Rumford on this very point which later served to overthrow the caloric theory.

6.02 Rumford's Experiments

Count Rumford (Benjamin Thomson) may be said to be the insti-gator of the experiments which were to lead to the true theory of heat. In 1798 Rumford was engaged in the boring of brass cannon at Munich, and was much impressed by the excessive amount of heat generated by the boring process—the supply of heat being apparently inexhaus-tible. The possibility that the heat had been released on account of the lowering of the thermal capacity of the borings as suggested by the calorists was considered by Rumford. On heating equal masses of the borings and metal from the same block to equal temperatures however, and subsequently immersing them separately in equal volumes of water at the same temperature, the same temperature rise occurred in each case. This then would seem to dispose of the explanation put forward by the calorists, as in fact the thermal capacities of the solid and finely divided substances were clearly equal.

Rumford was convinced that the heat could not have been produced at the expense of the borings, but to establish the point further he repeated the boring process using a blunt borer, when he again found that very large quantities of heat were produced although the weight of the borings was in this case very small. The borings therefore could be eliminated as the source of heat. The mass of metal comprising the remainder of the apparatus, being in its original condition and therefore having the same specific heat, could not provide the explanation of the heat gained by all parts of the apparatus during the boring process. There remained one possibility, namely that the heat had its origin in the air surrounding the apparatus. To test this, Rumford excluded the air by enclosing the cylinder in a deal wood box which was filled with water. The boring process was commenced with the temperature of the water at 60°F., and after 2½ hours the water boiled. He argued that the heat in this case could clearly not have come from the water whose only change was to *gain* heat, and concluded that the supply of heat must have been obtained from the effort expended in the boring process. His experiments thus showed that heat was not a material substance and laid the foundation for the dynamical theory of heat. Rumford's conclusions were stated in a paper published in 1798, in which he says:

"In reasoning on this subject we must not forget that most remarkable circumstance, that the source of the heat generated by friction in these experiments appeared evidently to be inexhaustible.

"It is hardly necessary to add that anything which any insulated body or system of bodies can continue to furnish without limitation cannot possibly be a material substance, and it appears to me to be extremely difficult, if not impossible, to form any distinct idea of anything capable of being excited and communicated in the manner the heat was excited in these experiments except it be *motion*."

6.03 Davy's Experiments

An experiment that was to make the caloric theory completely untenable was performed by Davy (later Sir Humphry Davy) in 1799. Davy arranged for two pieces of ice to be rubbed together by a clockwork mechanism, the whole apparatus being placed under the receiver of an air pump. It was observed that the ice melted at the surfaces of contact and continued to do so as the rubbing process was continued. Now as we have seen the calorists admitted that heat is required to melt ice, and in this case the necessary heat could not have had its origin in the extrusion of caloric from the ice block to produce water of a lower caloric content, as it was well known at that time that water had in fact a specific heat almost twice as great as ice! If indeed the experiment was performed by Davy in this manner* it is difficult to see why the

---

* Certain doubts as to whether in fact Davy did carry out the experiment and succeed in melting ice in the manner invariably described have been raised recently by Prof. Andrade. For Andrade's discussion on these points the reader is referred to *Nature*, March 9, 1935.

caloric theory should continue to have any adherents from that time. Many however remained unconvinced, and the theory was clung to tenaciously until well into the nineteenth century, to the time of Joule's classical work on fluid friction. In fact it would appear that Davy himself did not quite comprehend the significance of his experiment, as it was not until 1812 that he made a definite statement of his opinion that heat was a form of molecular energy.

6.04 Joule's Experiments

James Prescott Joule (1818–89), convinced that heat was a form of energy, began his series of brilliant experiments in 1840 which was to put an end to any lingering doubts that existed in the matter, and the decade that followed saw the modern dynamical theory of heat firmly developed. Joule set out to establish the equivalence between heat and other forms of energy. He argued that if heat was a form of energy, then when a given quantity W of work or other form of energy disappeared to produce heat, a definite quantity H of heat should be produced, the relationship between the two quantities being such that

$$\frac{W}{H} = \text{constant } (J).$$ The constant J is known as the *mechanical equivalent of heat*. The numerical value of J will depend on the units in which W and H are measured, but otherwise it will have a constant value independent of the manner in which the transformation of the energy is carried out. Joule not only succeeded in showing this, but obtained highly concordant values of J in a series of remarkable experiments extending over a period of nearly 40 years.

Joule's early experiments consisted in the generation of heat by churning a given mass of water in a suitable vessel by means of a system of paddles driven by a pair of descending weights. His apparatus is shown diagrammatically in Fig. 59. A known mass of water contained in a calorimeter A of determined thermal capacity, was churned by paddles moving in the sections of the calorimeter as shown. Four fixed vanes projecting inwards from the walls of the calorimeter (Fig. 59(b)) prevented the mere rotation of the water by the paddles. Loss of heat from the calorimeter by conduction was prevented by placing it on a wooden grating (not shown) which touched the floor only at the four corners, and by connecting the axis of the paddle to the driving mechanism through a boxwood cylinder C. The paddles were rotated by permitting two large weights M_1 and M_2 to fall vertically through a given distance, the motion of the weights being communicated to the paddles via two large pulleys BB and the cylinder D as shown. The amount of heat generated as the weights fell once to the ground was not sufficient to cause a rise of temperature in the water sufficiently great to be measured accurately, but by removing the clutch pin P the paddle

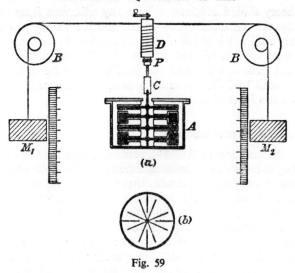

Fig. 59

system could be disconnected from the cylinder D, and the weights wound up again by the handle provided for the process to be repeated.

If m_1 and m_2 represent the masses of the weights which fall n times through a distance h, the total potential energy lost by the weights is

$$n (m_1 + m_2) gh$$

and if the weights reach their lowest point with a velocity v, they will possess a total kinetic energy of $\frac{1}{2} (m_1 + m_2) v^2$, and hence the energy W imparted to the cylinder is

$$W = n (m_1 + m_2)(gh - \tfrac{1}{2}v^2) \quad . \quad . \quad . \quad . \quad [32]$$

If M is the total water equivalent of the calorimeter and its contents, and the temperature is observed to rise from t_1 to t_2 during the experiment, the heat produced is $M (t_2 - t_1)$ and the mechanical equivalent of heat is given by the equation:

$$J = \frac{n (m_1 + m_2) \left[gh - \dfrac{v^2}{2} \right]}{M (t_2 - t_1)}$$

A correction was made by Joule for the work done against friction external to the water. He did this by disconnecting the cylinder D from the paddle and winding the cord round D in such a way that as one weight fell the other was raised. A small mass μ was placed on one of the weights until the velocity of descent was the same as that observed during the experiment. The amount of energy used in overcoming this friction during the experiment was thus $n \mu gh$, and this was subtracted from the energy as expressed in equation 32 to give the corrected value of W. A further correction was made for radiation losses from the calorimeter during the experiment, and Joule gave as his result for J the value 772ft.-lb. per British Thermal Unit. Joule used specially constructed mercury thermometers sensitive to $\frac{1}{200}$°F., but they were not corrected to the gas scale. When this correction is applied, Joule's result is very nearly 778ft.-lb. per B.Th.U.

6.05   Joule's Later Experiments on Fluid Friction

In 1878 at the request of the British Association, Joule repeated his earlier work on fluid friction with an improved form of apparatus. In these later experiments the work expended in churning the water was measured in a more satisfactory manner by using a calorimeter suspended so as to be free to rotate with the paddle and measuring the couple required to be applied to the calorimeter just to prevent motion. With this arrangement the work done in churning the water is given directly, without the need for correcting for friction external to the calorimeter as previously.

The scheme of Joule's second apparatus is given in Fig. 60. The paddles churning the water in the calorimeter A were driven by two

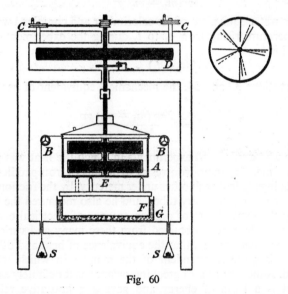

Fig. 60

handwheels CC, the motion being steadied by means of a heavy flywheel D attached to the driving shaft. To prevent rotation of the calorimeter silk cords were applied to the surface of the calorimeter and passed over two pulleys BB to support loaded scale pans SS. The weights in these pans were adjusted so that they were supported in a steady position clear of the floor, thus keeping the calorimeter in equilibrium against the couple due to the rotation of the paddles. In order to steady the driving torque and bring about a more easily maintained equilibrium position, Joule used two sets of paddles each having five arms which moved through four stationary vanes. The paddle arms were attached to the axis in such a way that only one arm was in transit through a vane at any particular instant, the transits following each other at regular intervals. The rotation of the paddle was thus more uniform and the apparatus was subject to less vibration. The paddle shaft was supported on a conical bearing E, and to free the calorimeter from all constraint other than that provided by the cords, the calorimeter was carried by a stand supporting a vessel F floating in water in a larger vessel G. This hydraulic supporting system relieved the thrust on the bearing, and so eliminated the friction here which without this arrangement had been found to be both too great and variable for the maintenance of equilibrium.

In making an experiment the paddles were rotated for 35 minutes, the number of revolutions n being counted mechanically. If the masses of the supporting weights are m_1 and m_2, and r is the radius of the calorimeter, the work done in n revolutions is

$$2\pi n\,(m_1 + m_2)\,gr \quad \bullet \; \bullet \; \bullet \; \bullet \; \bullet \; \bullet \quad [33]$$

and if the water equivalent of the calorimeter and contained water is M, the temperature rising from t_1 to t_2 during the experiment, the heat produced is

$$M\,(t_2 - t_1) \quad \bullet \; \bullet \; \bullet \; \bullet \; \bullet \; \bullet \; \bullet \; \bullet \quad [34]$$

From equations 33 and 34, the mechanical equivalent of heat is given by

$$J \; = \; \frac{2\pi n\,(m_1 + m_2)\,gr}{M\,(t_2 - t_1)}$$

Joule also carried out experiments in which the heat was produced by stirring mercury in an iron vessel by means of iron paddles, by iron rings rubbing against each other under mercury, by the friction of water being forced along capillary tubes, and he also measured the heat produced by the expenditure of a given amount of electrical energy. The constancy of the results obtained from these many and varied experiments established the fact that the equivalence of heat and other forms of energy is quite independent of the manner in which the heat is produced. Joule's work is of great importance as it definitely established that heat is a form of energy and gave a quantitative relationship between heat and work which, considering the resources at his disposal, was of an extraordinary degree of accuracy.

6.06   Hirn's Experiments

During the period that Joule was occupied with his researches on the equivalence of heat and work, Hirn obtained values of the mechanical equivalent of heat in two interesting experiments. His first determination (1857) was made by an impact experiment in which a heavy cylinder of iron, suspended with its axis horizontal, was capable of vertical displacement so that when released it struck against a mass of lead. Since lead is highly inelastic and produces little sound on impact, most of the energy of the blow is converted into heat. The lead mass was supported against a heavy stone "anvil" the recoil of which was measured after the lead had been struck by the released "hammer" of iron. The heat developed in the mass of lead was found from its thermal capacity and temperature rise and, by subtracting the work W_2 done in raising the anvil from the potential energy W_1 lost by the hammer, the value of J could be found from $W_1 - W_2 = JH$. The value obtained by Hirn by this means was 425gm. metres per calorie which, considering the nature of the experiment, is a remarkably good result.

In his second experiment performed in 1861 Hirn obtained a value for J by the converse process of converting heat into work using an ordinary steam engine. From the quantity of steam entering the cylinder at observed conditions of temperature and pressure, he was able to calculate the amount of heat supplied to the engine. The amount of heat remaining in the steam on issuing to the condenser was obtained by passing the steam into a measured mass of cold water and noting the rise in temperature. In this way the quantity of heat used by the engine in a given time was determined, corrections being applied for conduction, convection, and radiation losses, and Hirn found that this quantity was very much larger when the engine was applied to do external work than when not so employed. The amount of work done by the engine during this time was obtained from the area of the indicator diagram and the number of strokes made by the piston. Thus the ratio of the work done to the heat supplied could be calculated. Hirn's value of 420gm. metres per calorie (equal to 766ft.-lb. per B.Th.U.) is, taking into account the difficulties of the investigation, in good accord with the values of J obtained by other methods. The method had the additional importance at the time of being the first direct determination of the amount of work produced by one unit of heat and served to emphasise the fact that when work is converted into heat *all* the work appears as heat.

6.07 Rowland's Experiments

The method of Joule's later experiments on fluid friction was extended and modified by Rowland in 1879 to obtain increased accuracy in the determination of the mechanical equivalent of heat. In applying the method Rowland's object was to reduce to a minimum all the possible sources of error, and to apply corrections with maximum accuracy for all unavoidable sources of error. The two main objections to Joule's experiment were:

(a) The temperatures recorded were not corrected to the standard gas scale, and

(b) The small rate of rise in temperature (about 1°F. per hour) in consequence of which the cooling correction is unduly emphasised in making the final estimate of the temperature rise, which is thus not obtained with any high degree of accuracy (Joule himself was aware of this objection and had in fact designed a further apparatus in which the rate of temperature rise was greatly increased).

In designing his apparatus Rowland modified Joule's method by arranging for the expenditure of a greatly increased amount of energy in churning the water. This was effected by using a steam engine to drive the system of paddles which was very much more elaborate than in Joule's experiment. Rowland thus obtained a rate of rise of temperature of just over $\frac{1}{2}$°C. per minute, which thus materially reduced any errors in applying the cooling correction. The calorimeter in Rowland's apparatus was suspended by a torsion wire, the paddle shaft entering the calorimeter from below. The torque on the calorimeter resulting from the movement of the paddles through the water was balanced by weights attached to cords passing round a wheel rigidly fixed to the top of the calorimeter. A constant temperature jacket surrounded the calorimeter to ensure accurate estimation of the cooling correction, and his mercury thermometers were corrected to the standard gas scale of temperature.

The results of his work on the value of the mechanical equivalent of heat showed a regular variation in its value with temperature. Rowland found that the value of J decreased as the temperature increased from 5°C. to 30°C., after which a steady increase was shown. These results led Rowland to the conclusion that the specific heat of water varied with temperature, reaching a minimum value at a temperature of about 30°C. It was thus necessary to specify the exact range of temperature in defining the calorie, the value of J being given at this temperature in terms of the calorie so defined. Rowland's work on the variation of the specific heat of water with temperature has been confirmed by later workers (see section 2.05) whose results show that the temperature for the minimum value is 37·5°C.

6.08 Mayer's Evaluation of J

As a matter of historical interest a determination of the value of the mechanical equivalent of heat made by Mayer in 1842 from the values of the specific heats of a gas at constant volume and constant pressure is worth noting. It had been found experimentally that the value of the specific heat at constant pressure was always greater than the specific heat at constant volume. Mayer suggested that this was due to the additional heat energy required to expand the gas against the external pressure, that is, to enable it to do external work. Reasoning in this way Mayer proceeded to obtain a relation between the two specific heats of a gas (see section 7·02) from which he obtained a value for the mechanical equivalent of heat. The value so obtained showed only approximate agreement with Joule's value as only unreliable data concerning specific heats was available to Mayer at the time.

6.09 Later Determinations of J—Callendar and Barnes' Experiment

Since Rowland's experiments the mechanical equivalent of heat has been obtained in a variety of ways, both direct and indirect, by several workers. Many of these experiments are of great accuracy, and the value of J has been computed to a high degree of precision. As an example of the indirect methods we shall describe the experiment of Callendar and Barnes (1902) who obtained a very accurate value of J by an electrical method. The apparatus, which was based on the method of continuous flow, was devised by Callendar, the actual experimental work being done by Barnes. The principle underlying the electrical methods of determining the mechanical equivalent of heat is to measure the heat produced when a given amount of electrical energy is dissipated. If the electrical energy is obtained by passing a current of I amperes through a conductor across which a P.D. of E volts is maintained, the electrical energy transformed in t seconds is EIt joules, and if the amount of heat generated is H calories, the mechanical equivalent of heat is obtained from the relation:

$$J = \frac{EIt}{H} \text{ joules per calorie}$$

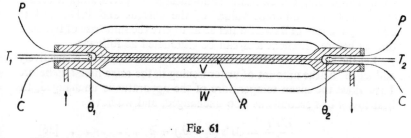

Fig. 61

The scheme of Callendar and Barnes' apparatus is shown in Fig. 61. A steady flow of air-free water was passed along a narrow glass tube, and was heated by an electric current carried by a fine platinum wire R secured at each end to a thick copper tube. The platinum wire was stranded to ensure thorough mixing of the water so as to establish a uniform temperature at any section across the flow tube. The copper tubes were mounted at the inflow and outflow ends of the apparatus

H.T.–5+

as indicated, and were supplied with two sets of leads, one pair CC serving to introduce the current to the platinum wire, and the other pair PP being used for measuring the P.D. across it. The temperatures of the water as it entered and left the apparatus were taken by a pair of differential platinum resistance thermometers T_1, T_2 inserted in the copper tubes whose high thermal conductivity served to equalise the temperatures in the immediate vicinity of the thermometer spirals. The current through the apparatus was accurately measured by inserting a standard resistance in series with the platinum wire and by comparing the P.D. across it against a standard Clark cell using a potentiometer. The P.D. across R was also measured by a potentiometer in terms of the E.M.F. of the standard Clark cell. All measurements were taken after the apparatus had attained a steady state, and the amount of water issuing from the apparatus in a given time was carefully weighed, the time of flow being recorded automatically by an electric chronograph reading to 0·01 second. In a typical experiment carried out by Callendar and Barnes, the amount of water passing through the apparatus was 500gm. in 15 minutes, the temperature rise being from 8 to 10°C.

If a current of I amperes passes through the platinum wire for an applied P.D. of E volts, and the temperature of the M gm. of water issuing in t seconds is raised from θ_1 to θ_2°C., then

$$\frac{EIt}{J} = Ms(\theta_2 - \theta_1) + h \quad \cdots \cdots \quad [35]$$

where s is the mean specific heat of water between the temperatures θ_1 and θ_2 and h is the heat lost in calories from the apparatus during the time t of the experiment. In order to make the heat loss small and regular, the flow tube was surrounded by a vacuum jacket V which was in turn enclosed in a copper jacket W through which was rapidly circulated water at a constant temperature. By performing a second experiment with different values of the current and P.D., and by adjusting the rate of flow of water so as to give the same rise of temperature as before, the heat loss can be eliminated as follows:

Let the new values of the current and P.D. be I' and E' respectively, and let M' be the amount of water issuing in the *same time* t as before. Then since the mean temperature of the apparatus is unchanged, the heat loss h in t seconds as also unchanged, and we have

$$\frac{E'I't}{J} = M's(\theta_2 - \theta_1) + h \quad \cdots \cdots \quad [36]$$

Hence by subtracting the two equations 35 and 36,

$$\frac{(EI - E'I')t}{J} = (M - M')s(\theta_2 - \theta_1)$$

from which $\quad J = \dfrac{(EI - E'I')t}{(M - M')s(\theta_2 - \theta_1)} \quad \cdots \cdots \cdots \quad [37]$

Each individual measurement in this experiment is capable of being obtained with a high degree of accuracy. The temperatures θ_1 and θ_2 can be obtained to one ten-thousandth part of a degree with the platinum resistance thermometers. Since these temperatures are stationary, there is no question of thermometric lag. The values of the currents and P.D.'s can be obtained to a similar degree of accuracy. The final accuracy of the experiment however depends on the degree of accuracy with which the E.M.F. of the standard Clark cell is known in true volts. The value accepted by Callendar and Barnes was probably not within the accuracy limit of their other readings, the final result for J thus being somewhat less accurate than one part in 10,000. Their value for J in terms of the 15°C. calorie is 4·183 joules per calorie.

By using this value for J it is possible to obtain the mean specific heat s of water from equation 37 at various temperatures by carrying out experiments over a succession of small temperature rises. Callendar and Barnes did in fact make such observations and obtained results showing the variation of the specific heat of water with temperature. Some of the values they obtained are given in table 4.

In conclusion the special advantages of continuous flow calorimetry (of which Callendar and Barnes' experiment is a classical example) will be listed. They are:

(1) The conditions of the experiment are under the full control of the investigator, who is thus enabled to use the apparatus to the limit of its sensitiveness.

(2) Since readings are taken in the steady state, the highest degree of instrumental accuracy is possible. Thus the temperatures θ_1 and θ_2 in Callendar and Barnes' experiment could be taken with slow reading platinum thermometers (since there is no question of thermometric lag) to an accuracy much higher than is possible in ordinary "mixture" calorimetry in which the temperature is continuously changing.

(3) The thermal capacity of the apparatus is not involved in the calculations as there is no change of temperature at any point when the final readings are taken.

(4) Heat losses are more regular and certain, and by suitably adjusting the conditions of the experiment, they can be completely eliminated.

6.10   Other Determinations of J

As examples of the many more recent determinations of J the very accurate experiments of Jaegar and Steinwehr (1921) and of Laby and Hercus (1927) will be briefly described.

(a)   **Jaegar and Steinwehr's Experiment.** These workers used an ordinary calorimetric method in which the rise in temperature of a known mass of water was found as a result of being heated by means of an electric current passing through a heating coil of constantan immersed in the water. If the P.D. applied to the coil is E volts, and a current of I amperes passes through the coil for t seconds, the heat supplied is $\dfrac{EIt}{J}$ calories. Hence if the mass of water is M and the water equivalent of the apparatus is m and the temperature rises from θ_1 to θ_2°C.

$$\frac{EIt}{J} = (M + m)\, s\, (\theta_2 - \theta_1)$$

where s is the mean specific heat of water between the temperatures θ_1 and θ_2. Thus:

$$J = \frac{EIt}{(M + m)\, s\, (\theta_2 - \theta_1)}$$

The current through the heating coil was measured by finding the potential drop across a standard 0·1 ohm resistance placed in series with it, and the P.D. across the heater was also found by a potentiometer against a standard Weston cell. The time of switching the current on and off was automatically recorded on a chronograph. In order to reduce the relative importance of uncertainties in the evaluation of m, Jaegar and Steinwehr used a large mass of water (50 kilograms) in a thin-walled calorimeter. Thus the thermal capacity of the apparatus was made relatively small, so that with the dimensions of the apparatus actually used, it was only necessary to obtain a value for m to 1 per cent without affecting the final accuracy sought (1 in 10,000).

The calorimeter was placed on its side and supported on porcelain blocks inside a double-walled vessel filled with water at a regulated temperature near that of the calorimeter. The water was thoroughly stirred by a system of electrically driven paddles, a small correction being applied for the heat so generated. The cooling correction was made small and regular under the conditions of the experiment, and in order that Newton's law of cooling should be strictly applicable in making the correction, the experiments were carried out using only small rises of temperature (of from 1 to 4°). To calculate the exact correction in any experiment observations were taken of the small temperature changes before the beginning and after the end of the application of the heat. The temperature rise of the water was very accurately taken using a standard platinum resistance thermometer. The value obtained by Jaegar and Steinwehr was 4·186 international joules per 15°C. calorie.

(b) **Experiment of Laby and Hercus.** In their determination of J, Laby and Hercus used a combination of the continuous flow method of Callendar and Barnes and the friction balance principle used in the experiments of Joule and Rowland. Their calorimeter consisted of a set of 14 copper tubes let into vertical slots cut in a pile of stalloy stampings contained in a vacuum vessel. A steady stream of distilled water was passed through these copper tubes, the inflow and outflow temperatures being recorded by a pair of platinum resistance thermometers connected differentially. The calorimeter was attached at its upper end to a torsion wheel which was mounted on a ball race and also suspended by a torsion wire. This arrangement comprised the "stator" of an induction dynamometer, the "rotor" being an electro-magnet similar to that of a direct current motor. The electro-magnet was energised by a steady current supplied from accumulators, and was supported on ball bearings and rotated about the vertical axis of the stator by a small electric motor. The eddy currents thus produced in the copper tubes in the calorimeter caused a couple to be applied to the stator, and the stator was prevented from rotating by the application of an opposing couple provided by weights attached to the torsion wheel by tungsten wires passing over pulleys supported on agate knife edges. In this way the energy of the eddy currents was dissipated into heat which was carried off by the stream of water flowing through the copper tube. The energy supplied in a given time was measured in mechanical units by the work done by the couple applied to the torsion wheel as the rotor made a given number of revolutions. The heat produced was calculated as in Callendar and Barnes' experiment. Very great care was taken to reduce heat losses to a minimum (they were actually less than 0·04 per cent of the total heat developed), and special attention was paid to secure steadiness in the rate of rotation, magnetic field, and water flow. The value finally given by Laby and Hercus was

$$J = 4{\cdot}1852 \pm 0{\cdot}0008 \text{ joules per } 15° \text{ calorie}$$

a result which is considered to be of the highest accuracy yet achieved.

6.11 Laboratory Methods of Finding J

(a) **Simple Electrical Method.**
A simple electrical method for determining the value of J can be carried out using the circuit shown in Fig. 62. A heating coil is supplied with a steady current of I amps. from a battery of accumulators, the current being measured by the ammeter A included in the circuit. A high resistance voltmeter V records the P.D. of E volts across the coil, and thus from the meter readings the amount of electrical energy dissipated in t seconds in the coil can be found. This is equal to EIt joules. To find the heat equivalent of this energy, the heating coil is immersed in distilled water contained in a copper

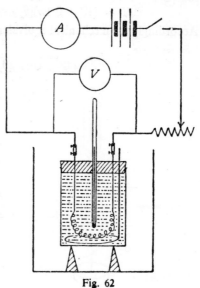

Fig. 62

calorimeter placed on heat insulating supports inside a larger protecting calorimeter. Keeping the ammeter reading constant by means of the control rheostat, the rise in temperature of the water, thoroughly mixed by a stirrer of stout copper wire, is recorded at the end of the time period t seconds. To allow for the radiation losses during this period the thermometer is again read at the end of a further period of $\dfrac{t}{2}$ seconds after the current has been switched off. Then, assuming Newton's law to hold under the conditions of the experiment, it follows that since the rate of cooling is proportional to the temperature excess, the cooling correction for t seconds is equal to the average rate of cooling × the time (t), i.e., to half the final rate of cooling × time. This is clearly equal to final rate of cooling × half the time. Thus if the original temperature of the calorimeter and contents is $\theta_1°$C., and the temperature rises to $\theta_2°$ C. in t seconds by the electrical heating and subsequently falls to $\theta_3°$C. in $\dfrac{t}{2}$ seconds after switching off the current, the final rate of cooling may be taken to be $\dfrac{\theta_2 - \theta_3}{t/2}°$C. per second, and the cooling correction is thus $\dfrac{\theta_2 - \theta_3}{t/2} \times \dfrac{t}{2} = \theta_2 - \theta_3$. Hence the final temperature corrected for radiation losses is $[\theta_2 + (\theta_2 - \theta_3)]°$C. If then the weight of the calorimeter (and stirrer) is $w$gm. and the contained water weighs Wgm., the amount of heat H developed is, taking the specific heat of copper as 0.10,

$$H = (W + 0.1w)[\theta_2 + (\theta_2 - \theta_3) - \theta_1] \text{ calories.}$$

Accordingly the mechanical equivalent J in joules per calorie is given by

$$J = \frac{EIt}{(W + 0.1w)[\theta_2 + (\theta_2 - \theta_3) - \theta_1]}$$

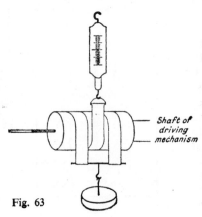

Fig. 63

(b) **Callendar's Brake Band Method.** Callendar's method is based on the brake band principle, and a laboratory form of it supplied with a solid calorimeter of known mass is shown in Fig. 63. A silk friction band encircles the copper calorimeter, its upper end being attached to a spring balance, and its lower end supporting a large weight. The calorimeter is attached by a block of heat insulating material to

Shaft of driving mechanism

the shaft of the driving mechanism which can be rotated either manually or by an electric motor, the number of revolutions being counted by a revolution indicator attached to the apparatus. A hole is bored in the calorimeter into which fits a $\frac{1}{10}$°C. thermometer whose bulb is dipped in oil to ensure good thermal contact. In performing the experiment the position of the spring balance is adjusted until the upward tension is just sufficient to float the attached load when the calorimeter is rotated at a steady rate. If the upward tension varies during the experiment, the readings of the spring balance should be taken every 50 revolutions, say, and the mean taken. The work done against the friction of the silk brake-band is converted into heat which raises the temperature of the copper calorimeter.

If T_1 is the upward tension and T_2 the downward tension in gm. wt., the force of friction $= (T_2 - T_1) g$ dynes (T_2 = mass of floated load), and if r is the radius of the copper cylinder, the frictional couple is $(T_2 - T_1) gr$ dyne-cm. Hence for n revolutions the work done is

$$(T_2 - T_1) gr\ 2\pi n \text{ ergs.}$$

If θ_1 is the initial temperature of the copper calorimeter (of mass M and specific heat s) and θ_2 is the final temperature, the heat developed is

$$Ms\ (\theta_2 - \theta_1 + \delta\theta) \text{ calories,}$$

where $\delta\theta$ is a correction for radiation losses (see below) during the experiment. Hence

$$J = \frac{(T_2 - T_1) gr\ 2\pi n}{Ms\ (\theta_2 - \theta_1 + \delta\theta)} \text{ ergs per calorie.}$$

The radiation correction can be obtained by taking the time t for the n revolutions and continuing the rotation of the calorimeter after this time until the temperature rises by a further 1°C. The calorimeter is now allowed to cool, and the time t' is taken for the temperature to fall 1°C. below the final temperature θ_2 of the experiment. The rate of cooling at θ_2 is thus $\frac{2}{t'}$°C. per minute, and since the rise in temperature of the calorimeter is virtually constant during the experiment, the mean rate of cooling during the experiment is half the rate of cooling at the final temperature. Thus for the t minutes of the experiment the cooling correction $\delta\theta$ is $t \times \frac{1}{2}$ of $\frac{2}{t'} = \frac{t}{t'}$°C. This represents an alternative method of dealing with the radiation correction to that given in the previous experiment.

(c) **Searle's Friction Cone Method.** Another form of simple friction balance (due to G. F. C. Searle) is shown in Fig. 64. In this

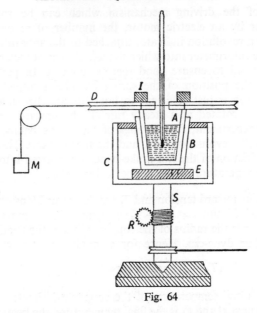

Fig. 64

experiment the heat generated by friction between two truncated brass cones A and B, one fitted inside the other, is measured by recording the rise of temperature of a known mass of water contained in the inner cone which serves as calorimeter. The outer cone B is fitted into an ebonite disc E fixed to the base of a brass cylinder C, and is held in position by a ring of ebonite as shown. The brass cylinder is carried by a vertical spindle S which can be driven manually or by an electric motor, the number of revolutions being recorded by a revolution counter R. The inner cone is securely attached to a grooved wooden disc D which is weighted by an iron ring I to produce a suitable pressure between the cones. On rotating the outer cone the inner cone tends to rotate with it, but is prevented from so doing by the application of an opposing couple provided by a weight M attached to a cord passing round the circumference of the disc and over a pulley as indicated. By suitably adjusting the speed of rotation the weight M can be kept steadily floating, when the rise in temperature is taken for a given number of revolutions.

Let m be the mass of water in the inner cone.

w the mass of both brass cones of specific heat s.

θ_1°C. the initial temperature.

θ_2°C. the final temperature.

Then the heat developed is

$$(m + ws)(\theta_2 - \theta_1 + \delta\theta) \quad . \quad . \quad . \quad . \quad . \quad [38]$$

where $\delta\theta$ is a correction for radiation losses which can be evaluated in a manner similar to that of the preceeding experiment.

Let M be the mass of the attached load.

   r the radius of the disc.

   n the number of revolutions recorded by the revolution counter.

Then the frictional couple is

$$Mgr \text{ dyne-cm.}$$

and the work done in n revolutions is

$$2\pi n \, Mgr \text{ ergs} \quad . \quad . \quad . \quad . \quad . \quad . \quad . \quad . \quad [39]$$

Hence from equations 38 and 39 the mechanical equivalent of heat is

$$J = \frac{2\pi n \, Mgr}{(m + ws)(\theta_2 - \theta_1 + \delta\theta)} \text{ ergs per calorie.}$$

6.12   Law of Conservation of Energy and the First Law of Thermodynamics

The work of Joule and other workers described in this chapter establishes the fact that heat is a form of energy, and further that there is a fixed equivalent between heat and mechanical work which may be summarised by the following statement:

When a given quantity of work is completely converted into heat an equivalent amount of heat is produced, and conversely when heat is transformed into work a definite quantity of work is produced. This statement is known as the *First Law of Thermodynamics,* which may be expressed as follows: $W = JH$

where W represents the number of work units equivalent to the H units of heat, J being the constant factor which we have referred to in the foregoing sections of this chapter and known as Joule's equivalent.

The First Law of Thermodynamics is but a particular form of a wider generalisation based upon experience which has become known as the *Law of Conservation of Energy.* The implication of this law, upon which the whole of modern physical science is based, is that although energy can be converted from one form to another, it can never be created or destroyed, or that the total energy of an isolated system is always constant. The work of Joule and others thus supported the law and served to establish it on a firm basis. The validity of the First Law of Thermodynamics is thus based on experience and on the assumption that this experience is universal.

QUESTIONS.   CHAPTER 6

1. Discuss the equivalence of heat and energy.

Describe an accurate experiment to determine the mechanical equivalent of heat and show how the result is derived from the observations. Discuss the principal sources of error in the experiment you describe.   (Camb. Schol.)

2. Give an account of an accurate method of finding the mechanical equivalent of heat.

A bullet of mass ¼oz. and specific heat $\frac{3}{50}$ strikes a target when travelling at 1,000ft. per second and is brought to rest. If 25 per cent of its kinetic energy is converted into heat which remains in the bullet, what is the rise in temperature of the bullet?

(J = 778ft.–lb. per B.Th.U.)   (O.)

3. What are the chief reasons for believing that heat is a form of energy? Explain the statement that $4 \cdot 2 \times 10^7$ ergs are equivalent to 1 calorie and describe an experiment involving the conversion of mechanical energy into heat by which it may be checked.

Calculate the number of foot pounds weight which are equivalent to the quantity of heat required to raise the temperature of 1lb. of water 1°F. Assume that 1ft. = 30·5cm. and that the acceleration due to gravity = 980 cm. sec.$^{-2}$ (N.)

4. Outline the development of the ideas of heat. What experiments would you carry out to illustrate this development? Mention especially experiments to show the conversion of heat into mechanical energy.   (S.)

5. Give a short account of the observations of Rumford and Joule on the " production of heat by mechanical means " and their significance in the abandonment of the caloric theory.

Describe, in terms of the simple kinetic theory of matter, what becomes of the energy supplied to raise the temperature of a block of copper from room temperature to 100°C.

A steam engine which develops 80h.p. uses 120lb. of fuel per hour. If the calorific value of the fuel is $1 \cdot 13 \times 10^4$ B.Th.U. per pound, estimate the efficiency of the engine.

(Assume that 1 B.Th.U. = 780 ft.-lb.-wt., 1h.p. = 550 ft.-lb.-wt. sec.$^{-1}$)   (L.)

6. What do you understand by the equivalence of heat and energy? State the units in which each is usually measured in the metric system, and describe **one** method of finding the relation between them.

How long will it take a 4kW. electric immersion heater to raise the temperature of a perfectly lagged tank containing 150Kg. of water from 20°C. to 70°C., if the water equivalent of the tank and heater is 10Kg.?

(Mean specific heat of water between 20°C. and 70°C. = 4·18 joules gm.$^{-1}$ deg.$^{-1}$)   (O. and C.)

7. Give *two* examples of the transformation of heat into mechanical energy and *two* of the reverse transformation.

Describe an experiment to determine the mechanical equivalent of heat.

A machine punches out metal discs from a sheet $\frac{1}{16}$in. thick at the rate of 100 per minute, the average resisting force being 50lb. wt. Assuming that all the work done by the machine is converted into heat, find the heat generated per minute.

(778ft.-lb. wt. are equivalent to 1B.Th.U.)   (N.)

8. Explain the ideas underlying the use of the term *mechanical equivalent of heat* and describe a non-electrical *laboratory* method of determining the value of this equivalent.

A steam engine with an output of 10h.p. has an efficiency of 15 per cent. How many lb. of coal of calorific value 8,000lb.°C. heat units per lb. are consumed per hour? (Assume that 1,400ft.-lb. wt. is equivalent to 1lb.°C. heat units.)   (N.)

9. Give a critical account of the methods available for determining the mechanical equivalent of heat.

In the absence of bearing friction a winding engine could raise a cage weighing

1,000kg. at 10 metres per sec., but this is reduced by friction to 9 metres per sec. How much oil initially at 20°C. is required per sec. to keep the temperature of the bearings down to 70°C.?

(Sp. ht. of oil = 0·5; g = 981 cm. per sec. per sec.; J = 4·2 × 10⁷ ergs per cal.)

(O. and C.)

10. Give a *short* account of Joule's classical experiment to determine the mechanical equivalent of heat by the water-churning method.

A river, flowing horizontally, reaches a vertical fall of 100 metres. If the temperature of the water at the top is 20°C. what will be the water temperature (*a*) half way down the fall, (*b*) in the pool at the bottom of the fall? (W.)

11. Describe a method of determining the mechanical equivalent of heat, stating the precautions that have to be taken to obtain an accurate result.

A copper calorimeter weighing 50gm. contains 400gm. of water at a temperature of 18°C. A stirrer provided with a pulley, round which a string passes, is set rotating about a vertical axis at a uniform rate by means of a small motor, and stirs the water. It is found that when the stirrer has made 3,000 revolutions the temperature of the water has reached 21°C. Assuming that ⅔ of the energy supplied to the stirrer is used up in heating the calorimeter and water, determine (*a*) the total energy supplied to the stirrer, and (*b*) the moment of the applied forces about the axis of the stirrer.

(Specific heat of copper = 0·1; mechanical equivalent of heat = 4·2 × 10⁷ ergs per gm. cal.) (W.)

12. Draw a diagram of an apparatus for finding the mechanical equivalent of heat in which the heat is produced by friction. Describe the experimental procedure very briefly, and show how the result is calculated from the observations.

In one of Rumford's experiments the work done by one horse raised the temperature of 26·6lb. of water from 32°F. to 212°F. in 2·5 hours. If 25 per cent of the heat generated was lost find, in ft.-lb. wt. per min., the rate at which the horse worked.

(1lb. degF. = 778ft.-lb. wt.) (N.)

13. Define *joule, calorie.*

Describe a friction experiment to show that 4·2 joules are equivalent to 1 calorie.

An engine working a cement mixer raises the temperature of the mixer and contents, of total water equivalent 220lb., by 15°F. in 15 minutes. Assuming that 50 per cent of the heat generated is lost, calculate the effective horse power of the engine.

(J = 780ft.-lb. wt. per B.Th.U.; 1 h.p. = 550ft.-lb. wt. per sec.) (N.)

14. Describe a mechanical method of finding the number of units of mechanical energy which are equivalent to 1 unit of thermal energy. Point out two limitations of the method.

A beam of 10¹⁴ protons per second strikes a metal target which is cooled by a steady stream of oil flowing through it. When steady conditions have been reached it is found that the difference in temperature between inflowing and outflowing oil is 2·5 degC. Find the mass of oil flowing through the target per second. Indicate any assumptions you make.

Use the following data:

velocity of protons = 10⁹ cm. sec.⁻¹
mass of proton   = 1·67 × 10⁻²⁴gm.
specific heat of oil = 0·55 cal. gm.⁻¹ degC.⁻¹ (N.)

15. Give an account of an electrical method of determining the mechanical equivalent of heat. Justify the use of an electrical method.

A temperature difference of 0·2°C. is observed between water at the top and the bottom of a certain waterfall. What is the height of the waterfall? Discuss the validity of your calculation.

($J = 4·2$ joule calorie^{-1}.)   (Oxford Schol.)

16. Describe in detail an experiment to find the value of the mechanical equivalent of heat by a mechanical method.

Briefly compare the physical principles involved on the conversion of mechanical work into heat and in the conversion of heat into mechanical work.

A small electrical heating coil is immersed in 100cm.^{3} of oil (density 0·80 gm. cm.$^{-3}$) contained in a calorimeter of thermal capacity 8·0 cal. degC.$^{-1}$. When electrical energy is supplied to the heating coil at a rate of 4·3watts, the temperature of the calorimeter and oil rises to a steady value of about 50°C. When the current is switched off this temperature commences to fall at a rate of 1·20 degC. min.$^{-1}$ Calculate the specific heat of oil. If the experiment is repeated with all the data and conditions the same except that the oil is replaced by an equal volume of water, at what rate will the temperature fall?   (N.)

17. What is meant by the mechanical equivalent of heat, J? Describe an experiment to measure this quantity.

The temperature of 1gm. of a gas is raised 1°A. first with the volume kept constant, and then with the pressure constant. The quantities of heat required for these two processes differ by 1cal. Explain this, and deduce a value of J given that the gas constant R is $8·3 \times 10^7$ erg° A.$^{-1}$ (gm. molecule)$^{-1}$, and the molecular weight of the gas is 2.   (Oxford Schol.—subsid.)

18. Summarize the chief methods used to determine the mechanical equivalent of heat and give a full account of one of them.

The specific heat of carbon dioxide at constant pressure is 0·202 cal. gm.$^{-1}$ degC.$^{-1}$, its density at S.T.P. is 1·977 gm. per litre, and the velocity of sound through it at 0°C. is 259 metres per sec. Calculate from the data a value for the mechanical equivalent of heat assuming that standard pressure is $1·013 \times 10^6$ dynes. cm.$^{-2}$ What explanation can be given to account for the low value obtained?   (N.)

(Refer to sections 7.02, 7.08, and 7.15 for the calculation of this question.)

19. Describe fully a non-electrical method for determining the number of joules equivalent to one calorie.

A continuous flow calorimeter consists of a long horizontal glass tube containing an electrically heated constantan wire. Oil at 16°C. enters the tube and leaves at 18°C. when the rate of flow is 20c.c. per minute, the potential difference across the heater is 6 volts and the current through the heater is 0·25 amp. If the rate of flow is then changed to 30c.c. per minute, the steady temperatures at the inlet and outlet are still the same provided that the p.d. is increased to 7 volt and the corresponding current is 0·29 amp.

Calculate the specific heat of the oil in calories per gram, given that its density is 0·9 g. per c.c.   (A.)

20. Describe the continuous flow method of Callendar and Barnes for the measurement of the mechanical equivalent of heat and point out the advantages of the method.

In an experiment using this method when the rate of flow of water through the tube was 60 gm. per min., the current flowing through the wire 2 amps. and the potential difference between its ends 5 volts, the difference between the temperatures of the water entering and leaving the tube was 2·3°C. On increasing the rate of flow to 87·6 gm. per min. and the current to 2·4 amps. the rise of temperature of the water on passing through the tube was again 2·3°C. Deduce a value for the mechanical equivalent of heat.   (Camb. Schol.)

21. Describe in detail one method for measuring the mechanical equivalent of heat. Mention the principal sources of error and outline the precautions necessary to overcome them.

A jet aircraft weighing 20,000kg. uses paraffin as fuel. Estimate the fuel consumption required to maintain a rate of climb of 50 m. sec.$^{-1}$ Energy losses due to imperfections in the engine, or to wind resistance, may be neglected.

(Heat of combustion of paraffin = 10,000 cal. g.$^{-1}$ Assume g = 1,000 cm. sec.$^{-2}$; $J = 4 \times 10^7$ erg cal.$^{-1}$) (Camb. Schol.)

22. Give an account of a continuous flow method of determining the mechanical equivalent of heat. Indicate the advantages of this type of calorimetry.

A lump of copper of mass 400gm. is cooled to $-180°C$. and immersed in 200gm. of water at 15°C. in a copper calorimeter of mass 100gm. Calculate the mass of ice formed on the copper assuming no heat exchange with the surroundings.

(Specific heat of copper = 0·1 cal. per gm. per °C. Latent heat of fusion of ice = 80 cal. per gm.) (W.)

23. Describe the Callendar and Barnes continuous-flow method of determining the mechanical equivalent of heat, pointing out the precautions which must be taken to ensure an accurate result.

How did the experimental results vary with the mean temperature of the water during the experiment? Explain this variation. (N.)

24. Describe an electrical method for the determination of the mechanical equivalent of heat.

An X-ray tube is run at 200,000 volts, 30 milliamp. Assuming that practically all the energy of the electron beam is converted into heat at the target, which is water cooled, find the least rate at which the water, initially at 20°C., must flow in order to ensure that it does not boil. Indicate why this rate leaves a margin of safety. (J = 4·18 joules per calorie.) (N.)

25. Explain what is meant by the mechanical equivalent of heat, and describe briefly some method by which it can be measured, giving the theory of the method.

A block of iron weighing 1 kilogram is struck 1,000 times by a hammer weighing 3·3 kilograms falling from the height of 1·2 metres. Assuming a quarter of the heat generated remains in the iron, find its rise in temperature.

(Mechanical equivalent of heat = $4·2 \times 10^7$ ergs./cal., g = 980 cm./sec.^{2}, Specific heat of iron = 0·11 cal./gm. °C.) (Cape Town)

26. (i) Discuss the significance of the expression " heat is a form of energy ".

(ii) Water in a calorimeter is stirred by a paddle rotated by a constant couple. Calculate the mechanical equivalent of heat, if the mass of water is 4,200gm., the water equivalent of calorimeter and stirrer is 700gm., the couple is produced by a force of 5 kg. wt. acting at a distance of 20cm., and temperature of the water rises from 15·0°C. to 26·0°C. in 27min., while the paddle is turned at the rate of 135 revolutions per min. (Melbourne)

CHAPTER VII

SPECIFIC HEATS OF GASES

7.01 Work Done by a Gas during Expansion

Consider a mass of gas at a pressure p dynes per sq. cm. contained in a cylinder fitted with an air-tight frictionless piston (Fig. 65). On

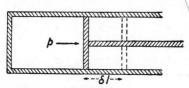

Fig. 65

supplying heat to the gas the resulting expansion will cause the piston to be pushed back along the cylinder so as to maintain the pressure of the gas constant. In other words the gas will perform external work against the pressure of the atmosphere. If the area of the piston head is A sq. cm., the thrust exerted by the gas on it will be pA dynes, and if the piston is moved a distance δl cm. along the cylinder, the work done by the gas will be $pA\delta l$ ergs, or since $A\delta l$ represents the increase in volume δv of the gas, then the work done by the gas is $p\delta v$. Hence if the pressure of the gas remains constant, the work done in expanding the volume from v_1 to v_2 is $p (v_2 - v_1)$ ergs.

If the pressure of the gas does not remain constant during an expansion, the work done will be given by the expression $\int pdv$. Thus for a gas expanding from volume v_1 to v_2 under the conditions shown in Fig. 66, the work done will be

$$\int_{v_1}^{v_2} pdv \text{ ergs.}$$

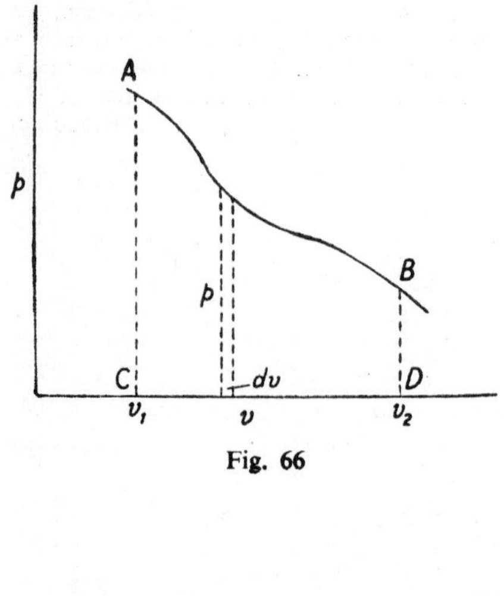

Such a diagram is called an *indicator diagram*, the work done being obtained from the area between the graph and the volume axis bounded by the line, AC and BD drawn through the values v_1 and v_2 respectively.

In the case of an *isothermal* change from initial conditions

Fig. 66

p_1v_1 to final conditions p_2v_2, the work done on expanding will be

$$RT \int_{v_1}^{v_2} \frac{dv}{v} \text{ (since } pv = RT)$$

$$= RT (\log_e v_2 - \log_e v_1) \text{ ergs.}$$

$$= 2 \cdot 3026 \, RT \, (\log v_2 - \log v_1)$$

using logs. to the base 10.

7.02   The Two Principal Specific Heats of a Gas

The specific heat of a substance has been defined as the amount of heat energy required to raise the temperature of one gram of the substance through one degree. The heat so supplied is used up in increasing the intrinsic energy of the substance, *i.e.*, to increase the average kinetic energy of the molecules; but it is clear that if the substance is allowed to expand during the heating process, work will be done against the external pressure, and accordingly an equivalent amount of heat energy must be supplied for this expansion to take place. It is thus evident that the value of the specific heat depends on the conditions under which the change of temperature takes place. In the case of solids and liquids the attendant volume changes are very small, and the thermal equivalent of the external work done on expanding is thus an insignificant proportion of the total heat energy supplied, and consequently there is little variation in the values of the specific heat with the conditions of the experiment. With gases however, where the volume changes on heating can be considerable, resulting in a large part of the heat absorbed being used up in doing external work, the value of the specific heat shows marked dependence on the conditions under which heat is supplied.

The heat equivalent of the work done when a gas expands from a volume v_1 to a volume v_2 is $\int_{v_1}^{v_2} \frac{pdv}{J}$, the value of which is subject to great variation according to the amount of expansion which takes place when the gas is heated under the particular conditions. In order to secure definite results it is customary to speak of two specific heats in connection with gases. These are the *specific heat at constant volume* (c_v) and the *specific heat at constant pressure* (c_p), being the amount of heat required to raise the temperature of 1gm. of the gas through 1°C. under conditions of fixed volume and fixed pressure respectively.

The numerical value of the specific heat of a gas at constant pressure will be greater than its specific heat when measured under conditions of constant volume, the difference between the two specific heats being the thermal equivalent of the work done by the gas in expanding against the external pressure. Thus if a mass m of gas has its temperature raised by an amount δT, then the heat supplied when heated at constant volume will be $m.c_v.\delta T$. If heated at constant pressure however a further amount of heat equal to $\dfrac{p\delta v}{J}$ will be required where δv is the increase in volume at the constant pressure p due to the rise δT in temperature. It therefore follows that

$$mc_p\delta T = mc_v\delta T + \frac{p\delta v}{J}$$

For infinitesimal changes, using the notation of the calculus, this becomes

$$mc_p dT = mc_v dT + \frac{pdv}{J}$$

$$\text{or } c_p - c_v = \frac{p}{mJ} \cdot \frac{dv}{dT}$$

Now assuming that the equation of state of the gas is $pv = RT$, we have for m gm. of the gas

$$pv = mR'T$$

where R' is the value of the gas constant for 1gm. of the gas (see section 5.07). Accordingly

$$\frac{dv}{dT} = \frac{mR'}{p}$$

and therefore

$$c_p - c_v = \frac{R'}{J} \quad \ldots \ldots \ldots \text{[40]}$$

If the molecular heats, i.e., the product of the specific heat and molecular weight, are denoted by C_p and C_v, equation 40 becomes

$$C_p - C_v = \frac{R}{J} \quad \ldots \ldots \ldots \text{[41]}$$

Where R is the gas constant for one gram molecule.

7.03 Calculation of the Value of J from the Specific Heats of a Gas

Using the calculated value of the gas constant R ($8\cdot31 \times 10^7$ ergs per gm. molecule per °C.) it is possible to obtain a value for the mechanical equivalent of heat from the specific heat data of a gas. Thus in the case of hydrogen of molecular weight (M) $2\cdot016$gm., the experimental values of the specific heats are $c_p = 3\cdot40$ and $c_v = 2\cdot42$ calories per gm. per °C., and hence using

$$c_p - c_v = \frac{R'}{J} = \frac{R}{MJ}$$

we have

$$J = \frac{8\cdot31 \times 10^7}{2\cdot016 (3\cdot40 - 2\cdot42)}$$

$$= 4\cdot21 \times 10^7 \text{ ergs per calorie.}$$

A result in good agreement with that obtained by direct methods.

7.04 Internal Work when a Gas Expands—Joule's Experiment

In deriving equation 41 the assumption was made that the gas conformed strictly to the equation of state of a perfect gas, and also that no internal work was done when the gas expanded on being heated under conditions of constant pressure.

This latter assumption is equivalent to stating that no molecular attraction exists as otherwise an additional amount of energy would be required to separate the molecules, that is, to increase their potential energy as the gas expands. To test this point Joule performed an experiment with the apparatus shown in Fig. 67 in which dry air at a pressure of 22 atmospheres, contained in a copper vessel A, was allowed to expand into a similar copper vessel B which was evacuated. Under these conditions the gas in A can expand without doing external work, and if internal work is done against the mutual attractions of the molecules, it must come at the expense of their kinetic energy, which means that the gas will be cooled. The two vessels were immersed in a large tank of water, and on opening the stop-cock C Joule could detect no change in the temperature of the water although his thermometers could read directly to $\frac{1}{200}°$F.

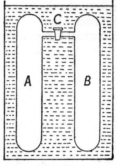

Fig. 67

Strictly speaking it is only the first traces of gas that expand into B which do so without doing external work, the rest of the gas entering B will compress the gas already arrived there. Nevertheless there should be a cooling effect if the potential energy of the molecules is increased as a result of the expansion.

A second experiment performed by Joule with the vessels in separate water tanks showed that the heat produced as the gas in B was compressed was equal to the heat lost by the gas in A in doing the work of compression. The gas as a *whole* however was again not cooled. It thus appears that the internal energy of a gas is not changed by alterations in its pressure and volume at constant temperature, or alternatively that *the internal energy of a gas depends only on its temperature.* This latter statement is sometimes referred to as *Joule's Law.*

It should be noted however that the thermal capacity of the apparatus used in Joule's experiment was very much greater than that of the gas, and hence it was unlikely that the effect of cooling due to the internal work would be detected unless it was comparatively large. All that was shown in Joule's experiment was that there was no such large effect, and thus for general purposes the work done against the inter-molecular forces can be neglected. This matter will be further discussed in section 10·09.

7.05   Measurement of C_v—Joly's Differential Steam Calorimeter

In measuring the specific heat of a gas at constant volume the necessity of containing it in some vessel introduced great difficulties on account of the relatively large thermal capacity of the containing vessel. Any small error in estimating this thermal capacity caused a relatively large error in the specific heat of the gas as deduced from the difference between the thermal capacity of the gas and container and that of the container alone. This difficulty was overcome by Joly who used a differential form of his steam calorimeter (Fig. 68) whereby the thermal

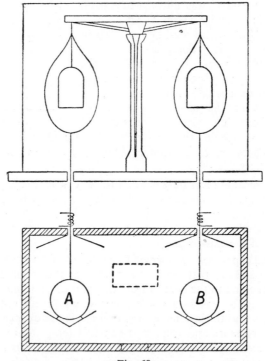

Fig. 68

capacity of a given mass of a gas could be determined directly. Two copper spheres, A and B, 6 to 7cm. in diameter and as nearly alike as possible, were hung by fine platinum wire from opposite arms of a balance inside the same steam chamber. The two spheres were exhausted and the balance counterpoised. The gas under investigation was then introduced into one of the spheres, A say, under high pressure (22 atmospheres), and the balance was once again counterpoised to determine the mass M of the gas in A. When the spheres had attained the temperature t_1 of the enclosure, steam was allowed to enter the enclosure via a wide duct at the back as indicated. Steam condensed on both spheres to raise their temperatures to that of the steam (t_2), but a larger amount condensed on the sphere A to supply the necessary heat to raise the temperature of the mass of gas in it to the final temperature t_2. This extra amount (m) was found by further counterpoise. Then if L is the latent heat of steam, the specific heat c_v is obtained from the heat equation:

$$mL = Mc_v (t_2 - t_1)$$

that is
$$c_v = \frac{mL}{M (t_2 - t_1)}$$

Small catchwater pans were attached to the spheres to prevent water from dripping off them, and they were protected from the water condensed on the roof by the shields as shown. In order to permit free play for the wires at the points where they left the enclosure, and to prevent the escaping steam from vitiating the readings by condensing on the suspension wires outside the steam chamber, the arrangements described in section 2.23 were adopted.

To correct for any slight difference between the thermal capacities of the spheres and their catchwaters, the experiment was repeated with sphere A evacuated and sphere B containing the gas. The error due to the unequal thermal capacities was then eliminated on taking the mean of the values of c_v obtained from the two experiments. Further corrections were necessary for the additional buoyancy of the gas-filled sphere due to the distention resulting from the pressure of the gas inside, and also for the work done by the gas as a result of the thermal expansion of the containing sphere.

The specific heats of gases at low temperatures have been found by Eucken using a Nernst calorimeter (see section 2.13). This method can be applied at sufficiently low temperatures since the thermal capacity of the metal calorimeter falls off, so that the thermal capacity of the gas contained in it is the greater, and accordingly errors arising from the determination of the thermal capacity of the calorimeter are not serious.

7.06 Measurement of C_p—Regnault's Method

In a series of experiments carried out in 1862 Regnault obtained the values of the specific heat at constant pressure of a number of gases by the method of mixtures using the apparatus shown in Fig. 69. The gas was contained under pressure in a large reservoir A which was kept at a constant temperature by surrounding it with a water bath. The pressure of the gas in A was given by the manometer B, readings of which were taken at the beginning and at the end of the experiment.

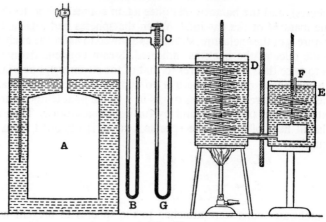

Fig. 69

From A, the gas passed via a regulating valve C to a long copper spiral immersed in an oil bath D where it was heated up to the temperature of the bath before entering the calorimeter E. In the calorimeter the gas passed through a thin brass vessel whose surface area was increased by being divided into a number of chambers to ensure a rapid transfer of heat from the gas to the water in the calorimeter. The gas finally escaped through the spiral tube F to the atmosphere. The flow of gas throughout the experiment was maintained steady by constant adjustment of the regulating valve C so as to obtain a steady difference between the levels of the water manometer G. Thus although the pressure of the gas was falling in A, it was possible to ensure that the pressure of the gas as it entered the oil bath (and calorimeter) was always constant. The mass (m) of the gas passing through the apparatus was obtained from the initial and final readings (p_1 and p_2) of the manometer B, the volume (V) of the container and a knowledge of the density of the gas under given conditions. Thus $m = V(\rho_1 - \rho_2)$ where ρ_1 and ρ_2 are the densities of the gas at the pressures p_1 and p_2 respectively. If then

c_p is the specific heat of the gas,

t_1 the temperature of the oil bath,

t_2, t_3 the initial and final temperatures of the water in the calorimeter,

m_1 the mass of water in the calorimeter, and

w the water equivalent of the calorimeter, brass vessel and spiral F,

we have, remembering that the average fall of temperature of the gas as it passes through the calorimeter is $t_1 - \left(\dfrac{t_2 + t_3}{2} \right)$,

$$mc_p \left[t_1 - \frac{t_2 + t_3}{2} \right] = (m_1 + w)(t_3 - t_2)$$

from which c_p can be calculated.

During the experiment the calorimeter will receive heat from the heater by radiation and also by conduction along the connecting tube, and although the calorimeter is screened from the oil bath by enclosing the former in a wooden box, it will be necessary to apply corrections for this heat transfer. There will in addition be the usual "radiation correction" for the calorimeter itself, and since the gas passes slowly through the apparatus over a considerable time, the various corrections acquire added importance, and thus need to be carefully evaluated to ensure reliable results. Regnault allowed for these corrections by adopting the following procedure. Before the stream of gas was allowed to pass through the apparatus, observations were made of the temperature of the water in the calorimeter for a period of 10 minutes with the heater at the temperature t_1; this determined the rate at which the calorimeter received heat by conduction and radiation from the heater. The gas was then allowed to flow for 10 minutes, the rise in temperature of the calorimeter and contents being noted. The gas flow was now shut off and readings of the temperature of the calorimeter were taken over a further period of 10 minutes, when the effect of radiation and convection losses from the calorimeter against the heat supplied by conduction and radiation from the heater was measured. In this manner Regnault was able to estimate the total correction to be applied to the final temperature of the calorimeter, and it is a great testimony to his skill that the results he obtained were within 2 to 3 per cent of those obtained by more recent methods.

7.07 The Continuous Flow Method for C_p

The method of continuous flow developed by Callendar for the determination of the mechanical equivalent of heat (see section 6.09), has been applied by Swan to the determination of the specific heats of gases at constant pressure. The method was improved subsequently by Scheel and Heuse who used the apparatus shown in Fig. 70 in an accurate series of experiments to obtain values of c_p for gases over a range of temperature between $-180°C$. and room temperature.

The calorimeter is made of glass and is contained in a Dewar vacuum vessel V silvered on the inside to reduce heat losses, the whole being immersed in a constant temperature water bath. Before entering the apparatus at O the gas attains the temperature of the water bath by being passed through a long copper tube immersed in the bath. The temperature of the gas as it enters is taken by a platinum resistance thermometer P_1, and the gas then passes through a spiral to the double-walled jacket, as shown, where it flows up between the walls C and B and down between the walls B and A. The gas now passes over the constantan heating coil H which is situated inside the double-walled

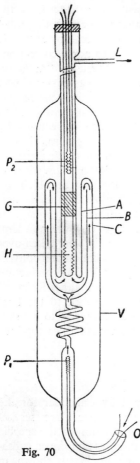

Fig. 70

jacket. By this arrangement any heat escaping from the heater is absorbed by the flow of gas before it enters the central tube, and so is brought back to the main stream of gas. On leaving the heater the gas passes through a packing of copper gauze G which serves to mix the gas thoroughly before its temperature is taken by a second platinum resistance thermometer P_2, after which it leaves the apparatus at L. If then

E is the $P.D.$ across the heating coil in volts,

I the current through it in amperes,

m the mass of gas flowing through the apparatus per second,

$\delta\theta$ the rise in temperature of the gas,

we have $\dfrac{EI}{J} = mc_p\delta\theta + h$

where h represents the small amount of heat lost per second during the experiment. This "radiation" correction can be eliminated by repeating the experiment with different rates of flow, and suitably adjusting the conditions in a manner similar to that described in Callendar and Barnes' experiment (see section 6.09). Some of the values obtained by Scheel and Heuse are given below:

TABLE 17

Specific Heats at Constant Pressure

Air		Hydrogen		Oxygen	
Temp. °C.	Sp. Ht.	Temp. °C.	Sp. Ht.	Temp. °C.	Sp. Ht.
20	0·2408	16	3·403	20	0·218
−78	0·2432	−76	3·157	−76	0·214
−183	0·2525	−181	2·644	−181	0·228

7.08 Isothermal and Adiabatic Changes

Whatever change a perfect gas may undergo the general gas equation $pv = RT$ will always represent the relation between the pressure, volume, and temperature of the gas. There are however certain extreme

changes of an ideal kind of great theoretical interest where special relationships hold good in addition to the general gas equation. These
changes may be understood by considering the behaviour of a gas
subject to compression and expansion by a frictionless piston moving
in a cylinder as in section 7.01. In the first case we will consider the walls
of the cylinder to be made of a perfect heat-conducting substance and
the movement of the piston head to take place *slowly*. On compressing
the gas, the extra work done will appear as heat in the gas and thus tend
to raise the kinetic energy of the molecules, that is, the temperature of
the gas. The heat however is dissipated to the surroundings through the
walls of the cylinder, and the temperature of the gas remains constant
throughout the process. Similarly on expanding the gas by withdrawing
the piston (slowly) heat will be gained from the surroundings to offset
the loss of heat due to the expansion, and again the temperature will
remain constant. Such changes in which compressions and expansions
take place slowly thus giving the gas time to lose or gain heat until the
temperature returns to that of the surroundings, are called *isothermal*
changes, and the relation between the pressure and volume under these
conditions is clearly given by Boyle's law, that is, $pv = $ const.

If now we imagine the walls of the cylinder and the piston head to
be made of a perfect heat insulating substance, there will be no possibility of any heat exchange between the gas and its surroundings. On
compressing a gas under such conditions the work done on the gas will
cause a rise in its internal energy and the temperature will rise. Similarly,
the work done on expansion must be provided at the expense of the
internal energy, and hence there will be a fall in temperature. A change
which occurs when the gas is thermally insulated from its surroundings
in this way is known as an *adiabatic* change, and as will be shown in
the subsequent section, the relation between pressure and volume for
such a change is given by $pv^\gamma = $ constant, where, for a given gas,
γ is a constant. It must be emphasised that this equation represents the
conditions in the ideal adiabatic change as described above. In practice
the material of the cylinder and piston will not be a perfect heat insulator, and further, it cannot be considered to have zero thermal capacity,
so that some heat exchange will occur between the gas and its surroundings. Perhaps the most perfect adiabatic change realised in practice
is provided by the passage of sound waves through a gas, and another
good approximation to the ideal conditions is provided by the bursting of
a tyre. In both these instances the changes take place so suddenly that
there is very little time for heat exchanges between the gas and the
surroundings to occur, and the conditions are thus very nearly adiabatic.

7.09 Equation of an Adiabatic for a Perfect Gas

In the general case when a quantity of heat dQ is given to a gas the
thermal energy supplied will be used up (*a*) in increasing the internal

energy of the gas by an amount dU, and (b) in doing external work of amount dW. This is a particular instance of the law of conservation of energy as expressed in the first law of thermodynamics (section 6.12), and may be stated,

$$dQ = dU + dW$$

In the case of an adiabatic change $dQ = 0$, and hence

$$dU + dW = 0$$

If now we consider the case of one gram of a gas of specific heat c_v undergoing an adiabatic change of volume dv at a pressure p with a corresponding temperature change of dT, we have

$$dW = \frac{pdv}{J} \text{ and } dU = c_v dT \text{ in heat units}$$

and hence $\dfrac{pdv}{J} + c_v dT = 0$ . . . . [42]

But for any change in the state of the gas the gas equation must be satisfied, that is

$$pv = R'T \, (R' \text{ being the gas constant for one gram of the gas})$$

and on differentiation this gives

$$pdv + vdp = R'dT$$ . . . . [43]

Eliminating dT from equations 42 and 43 we have

$$\frac{pdv}{J} + c_v \left(\frac{pdv + vdp}{R'} \right) = 0$$

or $\qquad pdv \left(\dfrac{c_v}{R'} + \dfrac{1}{J} \right) + \dfrac{c_v}{R'} \cdot vdp = 0$

Now from equation 40

$$c_p - c_v = \frac{R'}{J}$$

and on substituting for J in the above equation, and subsequently multiplying throughout by R' we have

$$c_p \cdot pdv + c_v \cdot vdp = 0$$

or, dividing throughout by c_v

$$\frac{c_p}{c_v} \cdot pdv + vdp = 0$$

putting $\gamma = \dfrac{c_p}{c_v}$ and rearranging, this becomes

$$\gamma \frac{dv}{v} + \frac{dp}{p} = 0$$

Hence on integrating,

$$\gamma \log_e v + \log_e p = \text{const.}$$

or $pv^\gamma = \text{const.}$ . . . . [44]

7.10   Temperature Variations in an Adiabatic Change

Equation 44 gives the relation between p and v for an adiabatic change. To obtain the temperature relationships for such a change equation 44 is used in conjunction with the general gas equation $pv = R'T$. Thus:

Substituting $v = \dfrac{R'T}{p}$ in equation 44 we have

$$p\left(\frac{R'T}{p}\right)^{\gamma} = \text{constant}$$

or since R' is constant,

$$\frac{T^{\gamma}}{p^{\gamma-1}} = \text{constant} \quad . \ . \ . \ . \ . \ . \ . \ . \quad [45]$$

which gives the pressure, temperature relationship in an adiabatic.

The volume, temperature relationship is obtained in a similar way by substituting $p = \dfrac{R'T}{v}$ in equation 44, when we obtain

$$Tv^{\gamma-1} = \text{constant} \quad . \ . \ . \ . \ . \ . \ . \ . \quad [46]$$

The following calculation illustrates the use of these relationships.

The pressure of a mass of air at 0°C. is suddenly halved. What is the resulting temperature? $\gamma = 1 \cdot 41$.

Using equation 45, we have

$$\left(\frac{T_1}{T_2}\right)^{\gamma} = \left(\frac{p_1}{p_2}\right)^{\gamma-1}$$

that is

$$\left(\frac{273}{T_2}\right)^{1\cdot41} = 2^{\cdot41}$$

from which, by logs,
$$T_2 = 225°A \text{ or } -48°C.$$

7.11   Work Done in an Adiabatic Change

For a small change we have
$$dW = pdv$$
Hence for a change in volume from v_1 to v_2

$$W = \int_{v_1}^{v_2} pdv = k\int_{v_1}^{v_2}\frac{dv}{v^{\gamma}}, \text{ since } pv^{\gamma} = k$$

$$= \frac{k}{1-\gamma}\left[\frac{1}{v_2^{\gamma-1}} - \frac{1}{v_1^{\gamma-1}}\right]$$

$$= \frac{1}{1-\gamma}\left[\frac{p_2 v_2^{\gamma}}{v_2^{\gamma-1}} - \frac{p_1 v_1^{\gamma}}{v_1^{\gamma-1}}\right]$$

$$= \frac{1}{1-\gamma}\left[p_2 v_2 - p_1 v_1\right] \text{ ergs,}$$

or since $p_1 v_1 = RT_1$ and $p_2 v_2 = RT_2$

$$W = \frac{R}{\gamma-1}\left[T_1 - T_2\right] \text{ ergs.}$$

These expressions also give the change in internal energy since $dU = -dW$.

7.12   Adiabatic and Isothermal Curves

The plot of p against v according to the equation $pv^\gamma = $ const. is known as an adiabatic curve. The value of the constant for any particular gas depends on its heat content, and is in the nature of a parameter. Accordingly for a fixed mass of gas there is a whole family of adiabatic curves, one of which is represented by the curve AB in Fig. 71. As the gas is indefinitely expanded the curve approaches asymptotically to the volume axis to the right. If the state of a gas is represented on this adiabatic by the point P_1, then on transforming the gas adiabatically to the state represented by the point P_2 an amount of work is done equal to the area intercepted between the curve and the volume axis by the ordinates through P_1 and P_2. In other words this shaded area represents the excess of the energy of the gas at P_1 over that at P_2, since the work done in the adiabatic change from P_1 to P_2 must be obtained from the gas itself. Proceeding in this way along the adiabatic curve as it approaches the volume axis at infinity, it is seen that the area between the curve and the volume axis bounded by the ordinate through P_1 is a measure of the total energy of the gas taking the energy zero as the energy possessed by the gas when at infinite expansion.

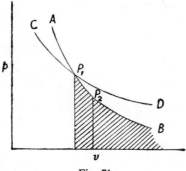

Fig. 71

On the same diagram CD represents the isothermal passing through the point P_1. This curve is one of a family of Boyle's law rectangular hyperbolæ which extend upwards as the temperature rises. It will be seen that the isothermal curve has been drawn less steep than the adiabatic curve through P_1. This is because for a given decrease in pressure the volume change under adiabatic conditions will be less than that under isothermal conditions on account of the fall in temperature accompanying the adiabatic expansion. The relative slopes of the isothermal and adiabatic curves passing through the point P_1 at which the pressure and volume of the gas are p_1 and v_1, may be obtained as follows. The slope of each curve is given by $\dfrac{dp}{dv}$ and is derived in the two cases by differentiating the p, v relationship.

For the isothermal, $pv = $ const.

$\therefore$ on differentiating

$$pdv + vdp = 0$$

$$\therefore \frac{dp}{dv} = -\frac{p}{v}$$

and hence the slope of the isothermal CD at P_1 is $-\dfrac{p_1}{v_1}$

For the adiabatic, $pv^\gamma = \text{const.}$

$\therefore \gamma pv^{\gamma-1}dv + v^\gamma dp = 0$

or $\gamma pdv + vdp = 0$

$\therefore \dfrac{dp}{dv} = -\gamma\dfrac{p}{v}$

and hence the slope of the adiabatic AB at P_1 is $-\gamma\dfrac{p_1}{v_1}$. Thus the slope of the adiabatic at the point where it intersects the isothermal is γ (which is always greater than 1) times as great as the slope of the isothermal at that point.

7.13 Experimental Determination of the Ratio of the Specific Heats of a Gas (γ)

Clement and Désormes' Experiment.—This is a direct method and is of interest historically as being one of the first methods (1819) by which the value of γ has been measured. The principle of the method is to produce an adiabatic cooling by sudden expansion of a gas, and then to allow the gas to absorb heat from the surroundings, the pressure of the gas rising to the value it would have had if the original expansion had taken place under isothermal conditions. The experiment can be carried out in the ordinary laboratory using the apparatus shown in Fig. 72. This consists of a large vessel (of volume 10 litres or more)

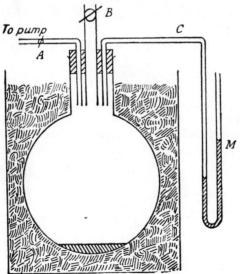

Fig. 72

containing a little concentrated sulphuric acid to keep the gas inside dry. The mouth of the vessel is closed by a rubber bung through which passes a narrow tube A connecting to a pump, a wide tube B fitted with a stop-cock, and the connecting tube to an oil or sulphuric acid mano-meter. The vessel is placed in a box and lightly packed with cotton wool to exclude draughts. Some of the gas under investigation is pumped into the vessel, and when it has taken up the temperature of its surround-ings, the manometer reading is taken (a pressure increase of about 30cm. of oil is convenient). The stop-cock B is now opened and closed again after an interval of about one second. After the adiabatically cooled gas has once again acquired the temperature of its surroundings, the manometer is again read and the atmospheric pressure is taken from a standard barometer. The experiment can be repeated with different initial values of the pressure to obtain an average value for γ which is calculated from the observations as follows.

Let the gas that remains in the vessel initially occupy a volume v which is less than the volume V of the vessel. If then p_1 and p_2 are respectively the initial and final pressures of the gas, and P is the atmospheric pressure, we have for the adiabatic change

$$p_1 v^\gamma = PV^\gamma \quad . \quad . \quad . \quad . \quad . \quad . \quad . \quad . \quad [47]$$

and for the isothermal change

$$p_1 v = p_2 V \quad . \quad . \quad . \quad . \quad . \quad . \quad . \quad . \quad [48]$$

From equations 47 and 48,

$$\frac{p_1}{P} = \left(\frac{p_1}{p_2}\right)^\gamma \quad . \quad . \quad . \quad . \quad . \quad . \quad . \quad [49]$$

or taking logs:

$$\log p_1 - \log P = \gamma \,(\log p_1 - \log p_2)$$

$$\therefore \gamma = \frac{\log p_1 - \log P}{\log p_1 - \log p_2}$$

For small pressure differences the expression for γ can be given in a simpler form as follows. Let $p_1 = P + h_1$ and $p_2 = P + h_2$ where h_1, h_2 are small compared with P. Equation 49 above then becomes

$$\frac{P + h_1}{P} = \left(\frac{P + h_1}{P + h_2}\right)^\gamma$$

that is

$$1 + \frac{h_1}{P} = \frac{\left(1 + \dfrac{h_1}{P}\right)^\gamma}{\left(1 + \dfrac{h_2}{P}\right)^\gamma}$$

or

$$\left(1 + \frac{h_1}{P}\right)^{\gamma-1} = \left(1 + \frac{h_2}{P}\right)^\gamma$$

Expanding by the binomial theorem and neglecting the second and higher powers of $\dfrac{h_1}{P}$ and $\dfrac{h_2}{P}$ we have

$$1 + (\gamma - 1)\frac{h_1}{P} + \quad . \quad . \quad . \quad = 1 + \gamma\frac{h_2}{P} + \quad . \quad . \quad .$$

from which

$$\gamma = \frac{h_1}{h_1 - h_2}$$

Clement and Désormes' method yields only approximate values for γ as it is not possible to obtain a truly adiabatic expansion since *some* heat will pass into the gas from the walls of the vessel however short the time during which the stop-cock is open. This error can be reduced by employing vessels of large capacity since the volume of the gas increases with the cube of the linear dimensions, whereas the surface area varies as the square of the linear dimensions. Another objection is that if the stop-cock is sufficiently wide to permit of a rapid expansion, oscillations are set up on account of the inertia of the gas, and consequently the final state of the gas depends upon the stage of the oscillation on closing the stop-cock. This difficulty has been overcome by Lummer and Pringsheim and later by Partington by measuring the drop in temperature accompanying the expansion.

7.14 Partington's Method

If the temperature of a mass of gas at an initial pressure p_1 falls from $T_1°A$ to $T_2°A$ as it expands adiabatically to atmospheric pressure P, then by equation 45 we have

$$\left(\frac{T_1}{T_2}\right)^\gamma = \left(\frac{p_1}{p_2}\right)^{\gamma-1}$$

$$\therefore \gamma (\log T_1 - \log T_2) = (\gamma - 1)(\log p_1 - \log p_2)$$

from which

$$\gamma = \frac{\log p_1 - \log P}{(\log p_1 - \log P) - (\log T_1 - \log T_2)}$$

In his method Partington measured the fall in temperature by a sensitive bolometer B (Fig. 73) which consisted of a platinum resistance thermometer of fine wire having a very low thermal capacity. The bolometer was connected by leads to one arm of a Wheatstone's bridge network, the effect of conduction along the leads being compensated by using a set of dummy leads connected as shown. With equal resistances in the ratio arms R_1 and R_2, the resistance of the balometer was given by R_3. To detect immediately the rapid changes of temperature, a galvanometer of very short period is required for use with the bridge. Partington used an Einthoven string galvanometer responsive to changes in 0·01 seconds. The bolometer was situated in the centre of the

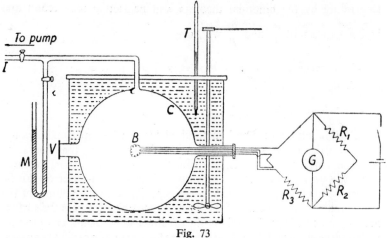

Fig. 73

mass of gas contained in a large spherical copper vessel C of capacity about 130 litres surrounded by a water bath. By using such a large expansion vessel the effects of conduction and convection in the gas were reduced. The gas was introduced in the vessel via the inlet tube I, the difference between the pressure of the gas and that of the atmosphere being measured by the oil manometer M.

Before proceeding with the experiment the aperture of the expansion valve V was adjusted to a size at which no oscillations occurred. This was done by setting the Wheatstone bridge to be slightly off balance and successively stopping down the valve aperture until a steady instantaneous deflection was obtained with the galvanometer. The experiment was then performed by reducing the value of R_3 slightly below the value required for a balance and noticing the reading of the galvanometer as the temperature of the gas fell on opening V. If the galvanometer did not return to zero (balance) position, the initial pressure of the gas was adjusted until this was achieved. To find the temperature corresponding to this reduced value of R_3, iced water was added to the water bath until the bridge was again balanced, indicating that the air in the vessel was at the temperature of the water bath. This temperature, as well as the initial temperature, was taken by a standardised mercury thermometer T placed in the water bath.

7.15 Acoustical Method

It is shown in textbooks on sound that the velocity of longitudinal waves (sound waves) in a body is given by

$$V = \sqrt{\frac{E}{\rho}} \quad \ldots \ldots \ldots \ldots \quad [50]$$

E being the elasticity of the body and ρ its density. For solids E is

Young's modulus, whilst for gases and liquids which will be subject to volume changes as a result of the compressive stresses applied, E will be the bulk modulus. Dealing with the case of gases, the compressions and rarefactions set up by the sound waves in the gas follow each other so rapidly that the changes are adiabatic, and it is thus necessary to evaluate the adiabatic bulk modulus for use in equation 50. Now by definition, the bulk modulus k is

$$k = \frac{\text{compressive stress}}{\text{compressive strain}}$$

$$= \underset{\delta p \to 0}{Lt} - \frac{\delta p}{\dfrac{\delta v}{v}}$$

where δv is the diminution in volume of a volume v of the gas when subject to an increase in pressure δp.

Thus $k = -v\dfrac{dp}{dv}$

For an adiabatic change, $pv^\gamma = \text{const.}$

$$\therefore p\gamma v^{\gamma-1}dv + v^\gamma dp = 0$$

or $\dfrac{dp}{dv} = -\gamma\dfrac{p}{v}$

and $k = \gamma p$

Thus the velocity of sound in a gas is given by

$$V = \sqrt{\frac{\gamma p}{\rho}} \quad \cdots \cdots \cdots \quad [51]$$

Accordingly by measuring the velocity of sound in a gas of known density and at known pressure, it is possible to obtain a value of γ for the gas from equation 51.

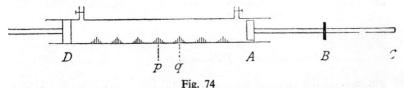

D   p   q   A   B   C

Fig. 74

The measurement of V is carried out using a Kundt's dust tube (Fig. 74). The tube, in which is lightly sprinkled lycopodium powder, is fitted with a tightly fitting piston D at one end and closed at the other by a loosely fitting piston head of cork attached to the end of a long brass rod AC clamped firmly at its mid point B. The brass rod is set in longitudinal vibration by stroking it with a resined cloth, and the piston D is adjusted until the gas column AD resonates to the note in the rod. The lycopodium powder is then agitated and settles down in small heaps at the nodes of the standing vibrations of the gas column. When sharp resonance has been obtained the distance between several of these heaps is measured to obtain the average distance pq between the nodes. The wavelength of the note (of frequency n) in the gas is thus $2pq$, and if V_g is the velocity of sound in the gas, then

$$V_g = n \times 2pq \quad \ldots \ldots \ldots \ldots \quad [52]$$

The experiment is now repeated with air in the resonance tube, and the inter-node distance $p'q'$ obtained for air. Then if V_a is the velocity of sound in air, we have

$$V_a = n \times 2p'q' \quad \ldots \ldots \ldots \ldots \quad [53]$$

From equations 52 and 53,

$$\frac{V_g}{V_a} = \frac{pq}{p'q'}$$

Now the velocity of sound in air at the temperature $t°C$. of the experiment may be obtained from the formula:

$$V_a = 33100 \left[1 + \frac{t°C.}{273} \right]^{\frac{1}{2}} \text{ cm. per sec.}$$

and hence

$$V_g = 33100 \left[1 + \frac{t°C.}{273} \right]^{\frac{1}{2}} \cdot \frac{pq}{p'q'}$$

Thus, having determined the velocity of sound in the gas, and knowing its pressure and density, a value of γ for the gas can be obtained from equation 51.

7.16   Relation between γ and the Atomicity of a Gas

It is shown in section 8.06 that for a perfect monatomic gas

$$pv = \tfrac{1}{3} MC^2$$

where C is the root mean square velocity of the gas molecules.

Now if M is the molecular weight of the gas in grams we have, using the equation of state $pv = RT$

$$\tfrac{1}{3} MC^2 = RT$$

or, for 1gm. of the gas,

$$\tfrac{1}{3}C^2 = \frac{R}{M}T = R'T$$

where R' is the value of the gas constant for 1gm. Hence the kinetic energy of translation of the molecules of 1gm. of the gas (that is, the intrinsic energy per gm.)

$$= \tfrac{1}{2}C^2 = \tfrac{3}{2}R'T \quad \ldots \ldots \ldots \quad [54]$$

If now the temperature of the gas is raised by 1°C., the kinetic energy becomes

$$\tfrac{3}{2}R'(T + 1)$$

and hence the increase in the kinetic energy of the molecules

$$= \tfrac{3}{2}R'(T + 1) - \tfrac{3}{2}R'T = \tfrac{3}{2}R' \cdot$$

By definition, the increase in the intrinsic energy per degree rise in temperature of 1gm. of a gas is the specific heat of the gas at constant volume, hence

$$c_v = \tfrac{3}{2}R' \text{ ergs } = \tfrac{3}{2}\frac{R'}{J} \text{ cals. per gm. per } °C.$$

Now by equation 40, section 7.02,

$$c_p - c_v = \frac{R'}{J}$$

$$\therefore \quad c_p = \frac{3R'}{2J} + \frac{R'}{J} = \frac{5R'}{2J}$$

and

$$\gamma = \frac{c_p}{c_v} = \frac{5}{3}$$

Thus for a perfect monatomic gas the ratio of the specific heats at constant pressure to constant volume should be 1·67.

This result may be obtained by applying the theorem of equipartition of energy. By the application of statistical mechanics to a large collection of particles, the nature of which need not be exactly specified, Boltzmann showed that the total energy of the system of particles is divided equally among the different degrees of freedom. For a collection of molecules in thermal equilibrium the energy associated with each degree of freedom is $\tfrac{1}{2}RT$ if one gram molecule is considered, or $\tfrac{1}{2}kT$ per molecule where k is **Boltzmann's constant,** and is equal to $\dfrac{R}{N}$ (N being **Avagadro's number,** the number of molecules in one gram-molecule). Various definitions have been given of the term "degrees of freedom," but for the purposes of this discussion it may be taken as referring to the number of independent square terms entering into the expression for the kinetic energy of the system. Thus for a monatomic gas in which the energy is purely translational, the kinetic energy of any one molecule may be given by

$$\tfrac{1}{2}mu^2 + \tfrac{1}{2}mv^2 + \tfrac{1}{2}mw^2$$

where u, v, w are the components of its velocity along the three standard directions. Such a molecule thus possesses three degrees of freedom, and the total translational energy will thus be $\tfrac{3}{2}kT$. Therefore for one gram-molecule the total energy is $\tfrac{3}{2}RT$, and for one gram of the gas $\tfrac{3}{2}R'T$ as given by equation 54.

For gases of higher atomicity the number of degrees of freedom is increased as with such gases rotational energy must also be considered. Thus in the case of a diatomic molecule which we may consider to be a sort of dumb-bell structure, there will be two further degrees of freedom on account of rotation about two mutually perpendicular axes at right angles with the line joining the two particles. If further the particles are capable of vibration relative to one another along the line

H.T.-6+

of their common axes, two further degrees of freedom corresponding to the kinetic energy and potential energy of the vibration (see section 2.25) must be added to the total possible degrees of freedom of the molecule. With polyatomic gases similar considerations show that in addition to the three translational degrees of freedom, there will be three degrees of rotational freedom, and a further two degrees of freedom for each mode of vibration possible within the molecule.

In general therefore if there are x degrees of freedom including those due to vibration and rotation, the total internal energy per gram molecule of the gas

$$= \frac{x}{2} \cdot RT$$

or $\qquad = \frac{x}{2} \cdot R'T$ per gram of the gas.

From this it follows that the specific heat at constant volume is

$$c_v = \frac{x}{2} \cdot R' \text{ work units.}$$

$$= \frac{x}{2} \cdot \frac{R'}{J} \text{ cal. per gm. per } °C.,$$

which with equation 40 gives

$$\gamma = \frac{c_p}{c_v} = \frac{\dfrac{x}{2} \cdot \dfrac{R'}{J} + \dfrac{R'}{J}}{\dfrac{x}{2} \cdot \dfrac{R'}{J}}$$

$$= \frac{x + 2}{x} \qquad \cdot \cdot \cdot \cdot \cdot \cdot \cdot \cdot \quad [55]$$

$$= \frac{5 + n}{3 + n}$$

where n is the number of degrees of freedom due to vibration and rotation in addition to the three degrees of translational freedom.

It will also be clear from the above that for one gram-molecule the ratio of the increase in translational energy per degree rise in temperature to the increase of the total energy of the gas per degree rise in temperature will be

$$\frac{\dfrac{3}{2}R}{\dfrac{x}{2} \cdot R} = \frac{3}{x}$$

which, using equation 55, becomes

$$\frac{3(\gamma - 1)}{2} \qquad \cdot \cdot \cdot \cdot \cdot \cdot \cdot \cdot \quad [56]$$

7.17 Theoretical and Experimental Values

The extent of agreement between the theoretical and experimental values of C_v and γ for gases of varying atomicity is shown in tables 18 and 19. The figures in the last column of each of these tables express

TABLE 18
Theoretical values of C_v and γ for various molecules

Type of Molecule	Degrees of freedom	Molecular heat at const. vol. in cals. per °C. $= \dfrac{x}{2} \cdot \dfrac{R}{J}$	$\gamma = \dfrac{x+2}{x}$	$\dfrac{3(\gamma-1)}{2}$
Monatomic........................	3	2·98	1·667	1
Diatomic (without vibration) ...	5	4·97	1·400	0·600
Diatomic (with vibration)	7	6·95	1·286	0·429
Polyatomic (without vibration)..	6	5·96	1·333	0·500
Polyatomic (with vibration)......	8	7·95	1·250	0·375
,, ,, ,, ...	10	9·93	1·200	0·300
,, ,, ,, ...	12 &c.	11·92	1·167	0·250

TABLE 19
Experimental values of C_v and γ for different gases at 15°C.

Gas	Atomicity	C_v	γ	$\dfrac{3(\gamma-1)}{2}$
Argon	1	2·98	1·666	1
Helium	1	2·98	1·666	1
Carbon Monoxide..................	2	4·94	1·404	0·606
Chlorine	2	5·93	1·355	0·537
Hydrogen	2	4·84	1·408	0·612
Nitrogen.............................	2	4·93	1·405	0·607
Oxygen	2	5·04	1·396	0·594
Carbon Dioxide....................	3	6·75	1·302	0·453
Sulphur Dioxide	3	7·49	1·285	0·428
Acetylene	4	6·83	1·280	0·420
Ethylene	6	8·20	1·250	0·375

the proportionate increase in translational energy to the total energy of the gas as given by equation 56. It will be seen that the experimental results for monatomic gases are in very good accord with theory. There is also very close agreement in the case of diatomic gases (with the exception of chlorine) with the theoretical results based on the equipartition of energy assuming no degrees of vibrational freedom. The exceptional case of chlorine with the high C_v value of 5·93 is interesting, as the result is intermediate between the theoretical values of C_v for a diatomic molecule with and without vibrational freedom. This could be explained if it were assumed that in addition to translational and rotational energy a certain proportion only of the molecules possessed vibrational energy. It is interesting to note that the molecular heat at constant volume shows a steady increase with temperature which on the foregoing supposition would imply an increasing proportion of vibrating molecules. For the polyatomic gases the results for C_v in every case quoted are greater than the theoretical value of 5·96 given for the non-vibrating molecule. It would thus appear that the "partial" degrees of freedom of vibrational energy referred to in the case of chlorine are present to varying extents in every case of these gases.

The experimental results for monatomic gases show convincingly that the energy of the molecules of these gases is translational only. The question of spin both for this type of molecule and for a diatomic molecule spinning about the common axis of the molecules is worthy of consideration as there is little or no evidence of energy of spin revealed in the experimental results. The explanation is provided by the quantum theory which permits only of possible values for the angular momentum of a rotating body which are integral multiples of $\dfrac{h}{2\pi}$ (h being Planck's constant). Thus the least value of the angular momentum Iw is $\dfrac{h}{2\pi}$ and hence the minimum energy of spin $\frac{1}{2}Iw^2$ will be $\dfrac{h^2}{8\pi^2 I}$. This quantity is inversely proportional to the moment of inertia of the molecule, which in both the cases of spin mentioned is very small, and hence the energy of spin is correspondingly large. Calculation shows that this energy is very much larger than the mean energy of translation of the molecules, and consequently energy of spin will not be possessed by these molecules. For a fuller account of this point and the general quantum theory of specific heats of gases, the reader is referred to the standard works on the quantum theory.

QUESTIONS. CHAPTER 7

1. Explain what is meant by the term specific heat, and why it is necessary to define two specific heats for a gas. What are they?

A mixture of one volume of oxygen and two volumes of hydrogen at N.T.P. is exploded in a container. What is the maximum pressure, in atmospheres, developed in the container?

(C_v for water vapour 6·48 cal. per gm. mol. per °C. Heat of formation of water vapour 60,000 cal. per gm. mol.) (Camb. Schol.)

2. (a) Explain how it is that the specific heat of a gas can have any value whatsoever.

(b) Explain what is meant by the *two principal specific heats* of a gas, and describe how ONE of these specific heats is determined experimentally. (W.)

3. Define the two *principal specific heats* of a gas, and describe *briefly* an experiment to determine one of them.

State the significance of the *ratio* of the specific heats in the expression for the velocity of sound in a gas.

Derive an expression for the *difference* between the principal molecular specific heats of an ideal gas and calculate its value, given that the volume occupied by a gram molecule of an ideal gas at 0°C. and a pressure of 10^6 dyne. cm.$^{-2}$ is 22·7 litres.

(L.)

4. Distinguish between adiabatic and isothermal changes. Explain how you would attempt to achieve each type of change in experiments with gases.

A fixed mass of an ideal gas whose principal specific heats have a ratio (γ) of 1·67 occupies 200cm.^{3} at 31°C. and at one atmosphere pressure. It is allowed to expand adiabatically until the volume is 300cm.^{3}, after which it is kept at constant pressure until the temperature returns to 31°C. Calculate (a) the pressure and temperature after the adiabatic expansion, (b) the final volume.

Explain why the value γ exceeds unity. (N.)

5. Distinguish between *isothermal* and *adiabatic* expansions. Draw and discuss the isothermal curves relating pressure and volume for a fixed mass of (a) an ideal gas, (b) a real gas (e.g. carbon dioxide) over a wide range of pressures and temperatures.

A fixed mass of an ideal gas occupying 400cm.^{3} at 15°C. is expanded adiabatically until its temperature is 0°C. What is the new volume if the ratio of the principal specific heats of the gas is 1·40? It is then compressed isothermally until the pressure is restored to its original value. Calculate the final volume of the gas and represent the changes on a (p, v) diagram. (N.)

6. What is meant by (a) an isothermal change, (b) an adiabatic change? Find expressions for the work done in both these cases when a gas expands. Calculate in each case the work done when a given mass of gas expands to twice its volume.

(Camb. Schol.)

7. What is meant by an isothermal and an adiabatic change?

Explain how such changes can be made.

A quantity of air ($\gamma = 1·41$) at 20°C. and 76cm. of mercury is suddenly expanded to twice its volume under adiabatic conditions. What will then be its temperature? If its temperature is allowed to rise again to 20°C., what will be the pressure?

(O.)

8. Explain why the barrel of a bicycle pump gets hot when you are blowing up a tyre.

A quantity of air at 10°C. and a pressure of 15lb. per sq. in. is suddenly compressed to one-third of its volume. What is the resulting temperature and pressure? ($\gamma = 1·4$). (O.)

9. Describe how you would measure the change in temperature of a mass of gas which undergoes a small sudden compression. Indicate the principal sources of error.

A mass of gas initially occupying 3,000c.c. at 15°C. and under a pressure of 1 atmosphere is compressed slowly, without gain or loss of heat, to a volume cf 600c.c. Calculate (a) the final temperature and pressure attained, (b) the work expended in compression. You may assume that pressure and volume are related by the condition $pv^{1·4} = constant$ when there is no gain or loss of heat.

(1 atmosphere = $1·013 \times 10^6$ dynes per sq. cm.) [N.]

10. State the conditions under which a perfect gas obeys (a) the relation $pv = RT$, (b) the relation $pv^\gamma = $ constant.

100c.c. of air at 0°C. and 760mm. pressure are compressed adiabatically to

20c.c. Find (a) the new pressure, (b) the new temperature, (c) the work done in compression. (For air $\gamma = 1\cdot4$.)   (Camb. Schol.)

11. Derive an expression for the difference between the specific heat of an ideal gas at constant pressure and the specific heat at constant volume.

Find the specific heat at constant pressure of nitrogen given that its specific heat at constant volume is $0\cdot175$ calorie per g. per degC. and that its density at S.T.P. is $1\cdot25$g. per litre. The density of mercury is $13\cdot6$g. per c.c.   (A.)

12. What is an *adiabatic expansion*? Deduce an expression connecting the volume and pressure of a perfect gas undergoing such an expansion.

Why does the air escaping from a motor car tyre valve feel cold?

The pressure of a gas in a perfectly insulated container is suddenly halved and the temperature falls from 0°C. to -50°C. What is γ for this gas?   (S.)

13. Describe a method of determining the specific heat of a gas at constant volume.

$3\cdot9$gm. of a gas occupy 3 litres at 15°C. and 76cm. of mercury pressure. It is supplied with 60 calories of heat. What is its new pressure and temperature if its volume is kept constant? If it is now allowed to expand adiabatically back to its original pressure what is its new volume and temperature?

(S.H. of the gas at constant volume $= 0\cdot17$ cal. gm.$^{-1}$ degC.$^{-1}$ S.H. of the gas at constant pressure $= 0\cdot24$ cal. gm.$^{-1}$ degC.$^{-1}$)   (S.)

14. Describe and explain how you would use a constant volume air-thermometer to determine the temperature, on the absolute air-scale, of the melting point of ice.

The temperature of the charge in the cylinder of an internal combustion engine is raised from 127°C. to 527°C. on the compression stroke before ignition occurs. Assuming that during compression $pV^{1\cdot35}$ is constant, calculate a value for the ratio of the pressures in the cylinder before and immediately after the stroke.   (L.)

15. Establish the relationship between (i) the volume and the pressure, (ii) the volume and the absolute temperature, of an ideal gas undergoing a reversible adiabatic change.

One mole of an ideal diatomic gas ($\gamma = 1\cdot40$), initially at S.T.P., is compressed to one-tenth of its initial volume by a reversible adiabatic process. Calculate (i) the temperature of the compressed gas, (ii) the work done in compression.

How much work would have been done if the same change in volume had been brought about by a reversible isothermal compression?

($R = 8\cdot32 \times 10^7$ erg. mole^{-1} °C.$^{-1}$, $\log_e 10 = 2\cdot303$.)   [O. and C.]

16. Explain why the specific heat of a gas at constant pressure is not equal to its specific heat at constant volume. Show that the specific heat of an ideal diatomic gas at constant volume is $\frac{5}{2}R$ per mole.

An ideal gas is heated under such conditions that the product of pressure and absolute temperature is kept constant. What is the difference between the specific heat under these conditions and the specific heat at constant volume?

(Oxford Schol.)

17. A flask of volume V containing air at a pressure p and temperature T is connected by a closed stopcock to a deflated rubber balloon. The stopcock is opened for a short time to allow the pressure in the flask and balloon to reach atmospheric pressure p_0 and it is then closed. What is the temperature of the air in the balloon immediately after closing the stopcock, and what is then the diameter of the balloon, assumed to be spherical? Calculate also the mass of gas transferred to the balloon. State any assumptions you have to make to arrive at a result.   (Oxford Schol.)

18. **Distinguish between heat and energy.** Summarize without proof the main results obtained by applying the First Law of Thermodynamics (Law of Conservation of Energy) to ideal gases, and discuss whether these results can be used to test the Law.

A mole of an ideal gas, of specific heat at constant volume C_v, is initially held at a pressure P_1 and absolute temperature T_1. It is then expanded adiabatically so that its temperature is T_2 and then brought isothermally to a pressure P_2. Calculate the heat absorbed by the gas in the isothermal expansion. (Oxford Schol.)

19. Define the specific heat at constant volume C_v and the specific heat at constant pressure C_p, and show that $C_p - C_v = R$ for a perfect gas. Describe briefly a method of measuring the ratio $\gamma = C_p/C_v$ and indicate the way in which γ depends on the nature of the gas.

Show that when a perfect gas expands adiabatically from volume V_1 at pressure P_1 to volume V_2 at pressure P_2 the work done by the gas is $\dfrac{P_1V_1 - P_2V_2}{\gamma - 1}$.

(Camb. Schol.)

20. Define the terms " specific heat at constant volume " and " specific heat at constant pressure " for a gas, and explain carefully why they are different. Show that for a perfect gas $C_p - C_v = R$, where C_p, C_v are the specific heats per gram molecule, and R is the gas constant.

A perfect gas is enclosed in a cylinder with a frictionless piston. The force on the piston is continually adjusted so that the ratio of pressure to volume of the gas is held constant. The specific heat, C, per gram molecule of the gas is measured under these conditions. Show that $C = C_v + \frac{1}{2}R$. (Camb. Schol.)

21. Give reasons for the temperature changes you would expect when a gas expands adiabatically (*a*) against a piston which is slowly withdrawn, (*b*) into another cylinder previously evacuated, (*c*) slowly through a valve into the surrounding atmosphere.

A vessel of low thermal conductivity containing air at 0°C. and at a pressure of 2 atmospheres is opened and then quickly closed when the excess pressure has been released. What will be the final pressure in the vessel if its surroundings are maintained at 0°C.

(C_p/C_v for air = 1·4.) (Camb. Schol.)

22. Show from first principles that for the adiabatic expansion of an ideal gas $pT^{\frac{\gamma}{(1-\gamma)}}$ is constant, where $\gamma = C_p/C_v$ is the ratio of the specific heat of the gas at constant pressure to that at constant volume.

A large localized mass of air at ground level is heated to a temperature of 27°C. while the temperature of the rest of the atmosphere remains at 17°C. at all relevant heights. Estimate how high the mass of hot air rises if it does not mix with the surrounding colder air.

(Molecular weight of air = 29, γ for air = 1·4. Assume $g = 1,000$ cm. sec.$^{-2}$)

(Camb. Schol.)

23. Derive the relation between the pressure and volume of a perfect gas undergoing reversible adiabatic expansion.

A hollow cylinder of internal cross-sectional area 1cm.^{2} is closed at both ends. The cylinder is divided into two equal halves, each of length 100cm., by a frictionless piston of mass 100g., and each half is filled with a gas at a pressure of 10^6 dyne cm.$^{-2}$ The cylinder is held horizontal and the piston is displaced a small distance from the centre and released. Show that the piston performs simple harmonic motion. The period of the motion is 0·37s. Determine γ ($= C_p/C_v$) for the gas. (The cylinder and piston may be assumed to have negligible thermal conductivity.)

(Camb. Schol.)

CHAPTER VIII

KINETIC THEORY OF GASES

8.01   The Molecular Theory of Matter

The kinetic theory of gases is but one branch of the more general molecular theory of matter. According to this theory matter is essentially discontinuous being composed of a very large number of tiny particles, called molecules, each of which exhibits the characteristic properties of the substance of which it is a part. A molecule of a substance is the smallest part of that substance which is capable of a free existence whilst retaining the properties of the substance. It is not asserted that a molecule is not capable of further subdivision, rather that if such subdivision is effected the result will be to produce other particles whose properties are chemically and physically different from those of the original substance. Thus the molecule of carbon dioxide on division gives molecules of carbon and oxygen, and in dealing with the elements, division of the molecule yields atoms of the substance—the atom being the smallest particle of matter capable of taking part in a chemical change. The molecules of certain elements such as the inert gases, are monatomic, *i.e.*, the molecule comprises only one atom of the substance, in which cases the molecule and atom are identical.

Early ideas as to the discontinuity of matter can be traced back to the Greek and Roman philosophers such as Democritus, Leucippus, and Lucretius. Their ideas were purely speculative however, and having no firm foundation, produced no further developments. The knowledge of the compressibility of gases established experimentally by Boyle and Mariotte working independently led Bernouilli in 1730 to a consideration of the properties of gases, on the assumption that they were comprised of particles in rectilinear motion. On this basis he successfully deduced Boyle's law, and may thus be considered as the founder of the modern kinetic theory. It is to Dalton however that credit must be given for laying the firm foundation of the atomistic theory of matter. Dalton's statement of the atomic theory was made in 1801, and appears to have been based on many converging lines of evidence. It may however be considered to have been definitely established by his investigations on the combining weights of elements when uniting to form one or more compounds as embodied in the laws of Constant and Multiple Proportions. The more recent discovery of isotopes and the further division of the atom into protons, electrons, &c., invalidates Dalton's original statements that the atoms of an element are identical in weight and are indivisible, but otherwise there is an overwhelming accumulation of evidence, both direct and indirect, to justify the fundamental principle of Dalton's theory.

Shortly after the announcement of the atomic theory, Gay-Lussac published the results of his researches on the combining volumes of gases. He showed that when gases react together they do so in volumes which bear a simple relation to one another and to the volume of the gaseous product. In view of Dalton's work on the combining weights of elements, Gay-Lussac pointed out that there must be a simple relation between the atomic weights of the elements and the combining volumes, and the assumption was made that equal volumes of gases contain equal numbers of atoms. This however, as Dalton pointed out, led to contradictions, and it was left to Avogadro to remove the fallacy of Gay-Lussac's reasoning by stating that the ultimate particles involved when gases combine are not the atoms themselves but aggregates of them. Avogadro called these aggregates *molecules*, and Avogadro's hypothesis that "equal volumes of all gases at the same temperature and pressure contain equal numbers of molecules," whilst removing Gay-Lussac's difficulty, served also to modify the atomic theory and introduced the concept of the molecule as the smallest particle of matter capable of free existence.

8.02 The Kinetic Theory of Matter

Matter then in its various states is an agglomeration of very tiny particles called molecules. Many of the phenomena exhibited by matter are only explicable if we further assume that these molecules are in constant and rapid motion. Thus it is difficult to see how gases could diffuse one into the other if the molecules were at rest. Further, the fact that the molecules of a gas show no sign of settling under the influence of gravity but maintain a uniform pressure on the walls of the containing vessel suggests strongly that the gaseous molecules are in a state of incessant motion. Similarly the observed effects of diffusion of liquids and the free evaporation from liquid surfaces, together with the vapour tension of solids and the diffusion that takes place at the common surface of two solids pressed together for some considerable period, indicate that molecular movement is not limited to gases. In liquids the movement of the molecules is more restricted than in the case of gases, and it is even more restricted in the solid state, but the evidence strongly suggests that the molecules of matter in all its forms are in ceaseless motion. This is known as the kinetic theory of matter, and from a physical point of view an essential feature of the theory is the identification of the energy of the molecules with the heat content of the substance—in conformity with the dynamical theory of heat discussed in chapter VI.

6*

The kinetic theory has been most fully developed in the case of gases where the almost complete absence of cohesive forces introduces a great simplification in the analytical treatment. The mathematical basis for the kinetic theory of gases was firmly established by Maxwell and Clausius, and later was extended by the work of Boltzmann, Jeans, and others. The laws governing the cohesive forces in the more condensed liquid and solid states are not sufficiently well known as yet to permit of a corresponding development of the mathematical treatment in these cases. Nevertheless the general kinetic theory of matter is useful because it visualises processes, and in the case of gases particularly leads to deductions which are in marked accord with the experimental evidence.

8.03 The States of Matter

The solid state: As mentioned above the molecular movement is most restricted in the solid state. Solids possess a definite size and shape which can only be altered by the application of very large forces. The molecules in this state are very closely packed together, and in consequence they exert considerable forces on one another. It is these cohesive forces which prevent the free movement of the molecules throughout the substance, and give it its rigid structure. X-ray analysis of the solid state indicates that the molecules are arranged in an orderly pattern throughout the substance, and it is assumed that the molecules are capable of vibratory motion about these fixed positions of equilibrium. The energy of vibration of the molecules is determined by the temperature of the body, and as the temperature is raised, the amplitude of vibration is increased causing the body to expand as the molecules move further apart. As the temperature is steadily increased a point is reached at which the vibrations of the molecules become so violent as to overcome the cohesive forces. The solid then loses its rigid structure, the molecules acquiring freedom of movement throughout the bulk of the now liquid substance. We thus see that in order to effect the change of state from solid to liquid an amount of energy must be supplied to the body to overcome the cohesive forces in the solid structure, and this energy is a measure of the latent heat of fusion of the solid.

The liquid state: In the liquid state the motions of the molecules are less restricted than in the solid state. The reduction of the strong cohesive forces between the molecules in the solid state effected at the change of state from solid to liquid, results in the liquid molecules possessing individual freedom of movement. It is due to this translational movement of the molecules that a liquid has the property of mobility whereby it can readily adapt itself to the shape of the containing vessel. Although a liquid can change its shape, its volume remains approximately constant, and this indicates that there still exist appreciable forces of cohesion between the liquid molecules. The great resistance to compression shown by liquids also suggests that the molecules are sufficiently close together to exert considerable forces on one another. It is these forces acting on the molecules near the free surface of a liquid that account for the phenomenon of surface tension. For a molecule situated within the bulk of the liquid the forces exerted on it by the surrounding molecules will, by symmetry, balance out and the molecule is free to pursue its random motion within the liquid. Near the surface, however, the molecule will be acted upon by a mean force of attraction, due to the molecules below it, which will be directed towards the interior of the liquid. This results in the surface behaving as if it was enclosed by an elastic membrane, and accounts for the formation of drops and other well known surface tension effects.

Evaporation of an exposed liquid is accounted for by the more energetic molecules possessing a sufficiently large upward velocity to carry them clear of the range of molecular attraction. This explains the cooling effect observed during evaporation since it is only the molecules possessing high velocities that can thus escape, and the remaining molecules must therefore have a lower average kinetic energy, and the temperature of the liquid mass is accordingly lowered.

The gaseous state: In the gaseous state the average distance between the molecules is very much greater than in either the solid or the liquid states. Thus, to take an example, since 1c.c. of water at 100°C. produces about 1600c.c. of steam at normal atmospheric pressure, it follows that the average distance between the molecules of the steam is $\sqrt[3]{1600}$ or approximately 12 times as great as the average distance between the water molecules. In consequence of the increased inter-molecular distances in the gaseous state, the molecules escape almost completely from their mutual attractions, and with the forces of cohesion thus being reduced to practically negligible dimensions, the motions of the molecules become rectilinear, and are disturbed only by their mutual collisions and by collisions with the walls of the containing vessel. The molecules of a gas are thus in incessant, rapid, haphazard motion, and the freedom of these motions accounts for the rapid diffusion of gases and their property of indefinite expansibility.

The pressure exerted by a gas is explained by the numerous collisions (assumed perfectly elastic) taking place at the walls of the containing vessel. The rate of change of momentum thus produced over unit area of the surface gives a measure of the pressure exerted by the gas at the walls (see section 8.06). If the volume occupied by a gas is halved, the number of collisions at the containing walls will be doubled, and thus the pressure will be doubled as required by Boyle's law. Further, since the temperature of the gas is determined by the average kinetic energy of the molecules, a rise in temperature of a fixed volume of gas will result in more vigorous and more numerous molecular collisions at the walls per second, and hence the pressure will rise. Thus we see in a qualitative way that the kinetic theory of gases accounts for the well known gas laws.

The continuous collisions taking place in a gas will result in alterations in the speeds and directions of movement of the molecules. The average molecular speed, and thus the total kinetic energy of the gas, is not however affected by the collision process. That the average molecular speed is not reduced by collisions is shown by the fact that a given volume of gas remains at the steady temperature of its surroundings, and does not show the fall in temperature to be expected from a continued lowering of the kinetic energy of the gas molecules. This justifies the assumption that the collisions of the molecules are perfectly elastic, resulting in a conservation of the total kinetic energy of the assemblage of molecules, the average value of which thus remains constant.

8.04 Distribution of Molecular Velocities in a Gas

In the simple analytical treatment of the kinetic theory of gases to be given in the subsequent sections of this chapter, the results will be expressed in terms of the average (root mean square value) velocity of the molecules of the gas. Provided it is assumed that the gas is in a state of molecular chaos, that is, the velocities and directions of movement of the molecules are distributed entirely at random, and that the collision processes are perfectly elastic, the general deductions that result from the following simplified treatment are independent of any particular law governing the distribution of velocities amongst the molecules. The mathematical treatment of the laws of distribution of velocities is outside the scope of this book, although reference may be made in general terms to Maxwell's work* (later improved by Boltzmann, Jeans and others) in this connection.

Maxwell showed that whatever the initial distribution of velocities, any measure of uniformity would immediately be destroyed by collisions and a state would be attained in which the number of molecules possessing velocities between any given range is no longer affected by molecular encounters. By an application of statistical mechanics Maxwell obtained a law of distribution† identical in form with Gauss' law of errors, an instance of which is given by the distribution of shots round a target by marksmen of equal skill. As applied to the case of molecular speeds Fig. 75 shows a graphical representation of the distribution law where F is the fractional number of the total number of

---

* For the mathematical treatment of the laws of distribution of molecular velocities the reader is referred to "The Dynamical Theory of Gases" by Jeans.—*Cambridge University Press.*

†Maxwell's law is of the form
$$F = A \, c^2 e^{-Bc^2} dc$$
where F is the fractional number of the total number of molecules having velocities within the range dc of the velocity c, and A and B are (positive) constants.

molecules having speeds between given limits, and c represents the molecular velocity. It is seen from the graph that the majority of the molecules have velocities lying within a limited range of the velocity given by the point A —this velocity is known as the *most probable velocity*. For velocity values increasingly diverging from the most probable velocity, the number of molecules having these velocities falls off rapidly, and only a very small proportion of the molecules have very low or very high velocities.

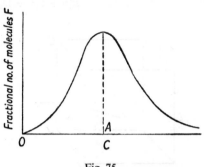

Fig. 75

The effect of increasing the temperature of the gas is to cause a wider distribution of the velocities—the curve becoming more flattened at the maximum, which is now displaced to the right corresponding to an increased value of the most probable velocity.

8.05 Kinetic Theory of Gases—Fundamental Assumptions

Before proceeding with the analytical treatment of the kinetic theory of gases we shall summarise here the assumptions on which the model of our ideal gas will be based. These are:

(a) The gas consists of a very large number of exceedingly small particles —identifiable with chemical molecules—which are in incessant, rapid, haphazard motion.

(b) For a given gas all these particles are identical, and there is a great number of them in the smallest volume of gas we can consider.

(c) The average distance between the particles (called the "mean free path") is so large compared with their linear dimensions that the particles may be considered to be of negligible size.

(d) The particles exert no appreciable force on one another, so that, between collisions, they travel in straight line paths.

(e) The duration of a collision is negligible compared with the time elapsing between collisions.

(f) All collisions are perfectly elastic. There is thus no loss of kinetic energy during a collision process, and a particle rebounds from the containing wall with a velocity equal to that with which it strikes it.

8.06 Calculation of the Pressure of a Gas

Consider a cubical vessel of side l cm. (Fig. 76) containing n gas molecules each of mass m. Let one of these molecules be approaching the side $ABCD$ with a velocity c_1 in the direction shown. We can resolve this velocity into three components u_1, v_1, w_1 along three mutually perpendicular axes parallel to three adjacent edges of the cube, when

$$u_1{}^2 + v_1{}^2 + w_1{}^2 = c_1{}^2$$

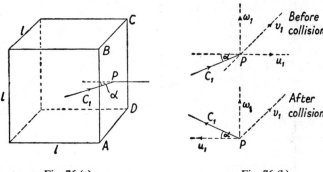

Fig. 76 (a)   Fig. 76 (b)

On collision with the face $ABCD$, only the component velocity u_1 normal to the face will be affected, and since the impact is perfectly elastic the molecule will rebound with its original velocity c_1, the component u_1 being reversed in direction as indicated. Thus the change of momentum normal to the face $ABCD$ due to the impact of this molecule there is $mu_1 - (-mu_1) = 2mu_1$. After recoil the molecule travels to the opposite face, from which it rebounds after collision there, and again travels towards the face $ABCD$. Now the time between successive

collisions at $ABCD$ is $\dfrac{2l}{u_1}$ and hence the change of momentum per

second at this face due to collisions with it by the molecule considered

$= 2mu_1 \times \dfrac{u_1}{2l} = \dfrac{mu_1{}^2}{l}$. We thus have taking all the n molecules having

component velocities of u_1, u_2, u_3 . . . . . u_n perpendicular to the face $ABCD$, a total change of momentum per second at this face of

$$\frac{m}{l}(u_1{}^2 + u_2{}^2 + u_3{}^2 + \quad . \quad . \quad . \quad . \quad . + u_n{}^2) \quad . \quad . \quad . \quad . \quad [57]$$

Now by Newton's second law of motion, the rate of change of momentum is equal to the impressed motive force, and accordingly equation 57 gives the thrust exerted by the molecules on the side $ABCD$ of the cube. This thrust is distributed over an area of l^2 sq. cm., and we thus see that the pressure p_x exerted by the gas on the face $ABCD$ is

$$p_x = \frac{m}{l^3}(u_1{}^2 + u_2{}^2 + u_3{}^2 + \quad . \quad . \quad . + u_n{}^2) = \frac{nm}{l^3}\overline{u^2}$$

Where $\overline{u^2}$ is the mean of the values $u_1{}^2$, $u_2{}^2$, . . . $u_n{}^2$.

Similarly if p_y and p_z represent the pressures on the faces of the cube perpendicular to the directions of the component velocities v_1 and w_1 respectively, we have

$$p_y = \frac{m}{l^3}(v_1^2 + v_2^2 + \ . \ . \ . + v_n^2) = \frac{nm}{l^3}\overline{v^2}$$

and $p_z = \frac{m}{l^3}(w_1^2 + w_2^2 + \ . \ . \ . + w_n^2) = \frac{nm}{l^3}\overline{w^2}$

Now it is obvious from considerations of symmetry that

$$\overline{u^2} = \overline{v^2} = \overline{w^2} = \tfrac{1}{3}\overline{c^2}$$

and hence $p_x = p_y = p_z = p$ (the pressure of the gas). Thus we have

$$p = \tfrac{1}{3}\frac{nm}{l^3}\overline{c^2} = \tfrac{1}{3}\frac{nm}{l^3}C^2 \ . \ . \ . \ . \ . \ . \ . \quad [58]$$

where $C = \sqrt{\overline{c^2}} = \sqrt{\dfrac{c_1^2 + c_2^2 + c_3^2 + \ . \ . \ . + c_n^2}{n}}$

C is thus the square root of the mean square velocity or the root mean square velocity as it is more generally called. It is not the same as the average velocity as can readily be demonstrated by considering simple numerical cases. It can be shown however that the root mean square value of the velocity is simply related to the arithmetical mean value of the molecular speeds.*

Since the volume v of the gas in the cubical vessel is clearly equal to l^3, and the total mass M of the gas $= nm$, we may write equation 58 as

$$pv = \tfrac{1}{3}nm\,C^2 = \tfrac{1}{3}M\,C^2 \ . \ . \ . \ . \ . \ . \ . \quad [59]$$
$$= \tfrac{2}{3} \times \tfrac{1}{2}M\,C^2$$
$$= \tfrac{2}{3} \times \text{total kinetic energy of the gas.}$$

A more rigid statistical treatment of the problem (for which see Jeans *op. cit.*) leads to an expression for the pressure of a gas which is identical with that obtained by the somewhat over-simplified method followed above. Nevertheless the general physical principles on which the method is based are sound, and we shall now proceed to a further consideration of the equations obtained and to certain deductions and consequences emerging from them.

8.07 Boyle's Law

In the ideal type of gas considered by the kinetic theory the energy of the molecules is purely translational, that is, it is kinetic energy.

---

* Thus according to Maxwell's law of distribution of molecular velocities, the average velocity $\bar{c}$ and the root mean square velocity C are related by the equation

$$\bar{c} = \sqrt{\frac{8}{3\pi}}\,C = 0\text{·}921\,C.$$

Thus it is reasonable to suppose that when heat is supplied to the gas there is an increase in the total kinetic energy of the gas which is proportional to the temperature rise observed. Thus the temperature of the gas is related to its kinetic energy, and since

$$pv = \tfrac{2}{3} \times \text{total kinetic energy of the gas}$$

it follows that at constant temperature

$$pv = \text{const.}$$

which is Boyle's law.

8.08   Temperature and Kinetic Energy

We shall here examine in more detail the relationship between kinetic energy and temperature to which reference has been made in the previous section. If we consider one gram-molecule of a gas we may write equation 59 as

$$pv = \tfrac{1}{3} MC^2 = \tfrac{1}{3} Nm\ C^2 \quad \ldots \quad [60]$$

Where M is the molecular weight of the gas and N is Avogadro's number (the number of molecules in one gram-molecule of a gas).

Now from experimental evidence we know that for one gram-molecule of a gas the relation between pressure, volume, and temperature is given by

$$pv = RT \quad \ldots \ldots \ldots \ldots \quad [61]$$

R being the gas constant.

From 60 and 61 we thus have

$$\tfrac{1}{3} MC^2 = \tfrac{1}{3} NmC^2 = RT$$

$$\text{or } \tfrac{2}{3} \times \tfrac{1}{2} MC^2 = RT \quad \ldots \ldots \ldots \quad [62]$$

From which we see that the kinetic energy of one gram-molecule of the gas is equal to $\tfrac{3}{2} RT$.

Thus the perfect gas equation in conjunction with the kinetic theory of gases directly relates the kinetic energy of a gas to its absolute temperature, and accordingly the velocity of the molecules will be zero at the absolute zero of temperature.

For one molecule, the kinetic energy may be written

$$\tfrac{1}{2} mC^2 = \tfrac{3}{2} \frac{R}{N} T$$

$$= \tfrac{3}{2} kT \quad \ldots \ldots \ldots \quad [63]$$

Where k is the value of R for one molecule of the gas and is called *Boltzmann's constant*.

The kinetic definition of temperature implicit in equations 62 and 63 requires amplification when the problem of a gaseous mixture is considered. If a vessel contains a mixture of gases whose molecules have masses of m_1, m_2, m_3, ... &c., it is well known that after a sufficient lapse of time a state of constant temperature is attained throughout the vessel, on the walls of which there is exerted a uniform pressure. The interchange of energy between the particles of the gaseous mixture is effected by elastic collisions depending on the mass and speed of the particles but not on their nature. It would thus appear that there would be no kinetic difference between the various sets of molecules which at the same temperature would accordingly possess identical amounts of kinetic energy. The actual proof of this important principle was given by Maxwell by an application of the method of statistical mechanics to a system containing particles of different masses exchanging energy by collisions. Maxwell showed that if the root mean square velocities of the particles of masses stated above are C_1, C_2, C_3 . . &c., in the state of thermal equilibrium, then

$$\tfrac{1}{2}m_1C_1^2 \ = \ \tfrac{1}{2}m_2C_2^2 \ = \ \tfrac{1}{2}m_3C_3^2 \ = \ \&c. \quad . \ . \ . \ . \quad [64]$$

This important result completes the kinetic interpretation of temperature, for from it we see that the condition for no net transfer of energy between two or more gases when brought into contact, is that the mean kinetic energies of their individual molecules should be the same. We have previously seen (equation 63) that for any given gas the absolute temperature is directly proportional to the mean kinetic energy of the molecules of the gas, and by equation 64 it therefore follows that at a given temperature the mean kinetic energy of the individual molecules of all gases are identical. From this it follows that Boltzmann's constant k in equation 63 is a universal constant.

8.09 Avogadro's Hypothesis

From the kinetic interpretation of temperature given in the preceeding section, the hypothesis of Avogadro, namely that equal volumes of all gases at the same temperature and pressure contain equal numbers of molecules, immediately follows. Thus if two equal volumes v of different gases contain respectively n_1 and n_2 molecules of masses m_1 and m_2, the pressure exerted by each gas being p, we have from equation 59

$$pv \ = \ \tfrac{1}{3}n_1m_1C_1^2$$

and $$pv \ = \ \tfrac{1}{3}n_2m_2C_2^2$$

C_1 and C_2 being respectively the root mean square velocities of the molecules of the two gases. Now if further the gases are at the same temperature, we have from equation 64

$$\tfrac{1}{2}m_1C_1^2 \ = \ \tfrac{1}{2}m_2C_2^2$$

and thus it follows that

$$n_1 \ = \ n_2$$

which establishes Avogadro's hypothesis.

8.10 Molecular Velocities

Since $pv = \frac{1}{3}MC^2$
it follows that

$$C = \sqrt{\frac{3pv}{M}} \quad \cdots \cdots \quad [65]$$

$$= \sqrt{\frac{3p}{\rho}} \quad \cdots \cdots \quad [66] \text{ where } \rho = \frac{M}{v}$$

$$= \sqrt{\frac{3RT}{M}} \quad \cdots \cdots \quad [67] \text{ putting } pv = RT$$

Any of these three equations enables us to calculate the magnitude of the molecular velocities (root mean square value) for a given gas under stated conditions. Thus for hydrogen at *N.T.P.* taking ρ as 0.8987×10^{-4} gm. per c.c. we have from equation 66

$$C = \sqrt{\frac{3 \times 76 \times 13.6 \times 981}{0.8987 \times 10^{-4}}}$$

$$= 1.84 \times 10^5 \text{ cm. per sec.}$$

Having thus calculated a value for the average molecular velocity, its value at any other temperature may readily be obtained since by equation 67 the molecular speed for a given gas is proportional to the square root of the absolute temperature. Table 20 below gives the root mean square velocity of the molecules of certain selected gases calculated for 0°C. To obtain the mean velocity the values given must be multiplied by the factor $\sqrt{\dfrac{8}{3\pi}}$ (see footnote, page 167). Thus in the case of hydrogen the mean molecular velocity of 0°C. is 1.69×10^5 cm. per sec.

TABLE 20

Molecular velocities at 0°C.

Gas	Molecular Weight	Root mean square velocity. cm. per sec.
Hydrogen........................	2·016	18·4 × 10⁴
Helium...........................	4	13·1 × 10⁴
Nitrogen	28	4·95 × 10⁴
Oxygen	32	4·61 × 10⁴
Argon	40	4·14 × 10⁴
Carbon dioxide	44	3·95 × 10⁴
Chlorine	71	3·11 × 10⁴

It may be noted that the calculated values of the molecular velocities are, as should be the case, of the same order of magnitude as the velocities of flow of the gases through small orifices into a vacuum space, and also of the same order of magnitude as the velocity of sound in the gases (see section 7.15).

8.11 Graham's Law of Diffusion

The effects observed when gases pass through extremely fine channels, such as those provided by the walls of unglazed porcelain vessels, is called diffusion. Graham (1832) investigated the diffusion of gases through porous substances, and found that the rate of diffusion of any gas was inversely proportional to the square root of its density—a result which is known as *Graham's law of diffusion*. On the kinetic picture of a gas that has been built up in this chapter it will be clear that the diffusion rate will be determined by the average velocity of its molecules, a quantity which, as previously mentioned, is proportional to the root mean square velocity. Now from equation 66 we see that the root mean square velocities C_1 and C_2 of the molecules of two gases of densities ρ_1 and ρ_2 respectively at a pressure p are given by

$$C_1 = \sqrt{\frac{3p}{\rho_1}} \text{ and } C_2 = \sqrt{\frac{3p}{\rho_2}}$$

thus $\dfrac{\text{rate of diffusion of gas}_1}{\text{rate of diffusion of gas}_2} = \dfrac{C_1}{C_2}$

$$= \sqrt{\frac{\rho_2}{\rho_1}}$$

and it thus follows that the kinetic theory of gases provides a theoretical basis for Graham's empirical law.

It should not be inferred from the above that gaseous diffusion takes place at the speeds of the average velocity of the gas molecules as calculated in section 8.10. The diffusion process is hindered by continual collisions of the molecules with each other with consequent frequent changes occurring in the direction of motion of the molecules. Thus at the end of a given time any particular molecule will only be displaced a very small distance from its original position, the diffusion rate accordingly being very much less than the mean velocity of the molecules of the gas.

8.12 Dalton's Law of Partial Pressures

If a volume v contains n_1 molecules each of mass m_1 of a given gas for which the root mean square velocity is C_1 and n_2 molecules of mass m_2 of another gas with root mean square velocity C_2, &c., it can be shown by reasoning similar to that given for a simple gas in section 8.06 that the pressure p of the mixture is

$$p = \tfrac{1}{3}\frac{n_1 m_1 C_1^2}{v} + \tfrac{1}{3}\frac{n_2 m_2 C_2^2}{v} + \quad . \quad . \quad . \quad . \quad . \quad . \quad [68]$$

Now if the first set of molecules were alone present in the space the pressure p_1 they would exert is given by $p_1 = \tfrac{1}{3}\dfrac{n_1 m_1 C_1^2}{v}$, and the pressure p_2 of the second set if present alone is $p_2 = \tfrac{1}{3}\dfrac{n_2 m_2 C_2^2}{v}$, &c. Thus we see that

$$p = p_1 + p_2 + \quad . \quad . \quad . \quad . \quad .$$

In other words the total pressure exerted by the gaseous mixture is the sum of the individual pressures that would be exerted if the several gases occupied the space in turn alone—which is Dalton's law of partial pressures (see section 9.20).

Using Maxwell's result of the equipartition of energy amongst the particles when thermal equilibrium is established (equation 64), we have from equation 68

$$p = \tfrac{2}{3}\left(\frac{n_1 + n_2 + \quad . \quad . \quad . \quad . \quad .}{v} \right). \tfrac{1}{2}m_1 C_1^2$$

from which it follows that the pressure of a gaseous mixture at a given temperature depends only on the number of molecules per cubic centimetre of the volume.

8.13 Validity of the Deductions from the Kinetic Theory

We thus see that all the well-known gas laws are predicted by the kinetic theory, and are in fact contained in the expression for the pressure given in equation 59 together with the statistical conception of temperature given in equation 64. It should be made clear however that the validity of the deductions made in sections 8.07 and 8.09 to 8.12 are restricted by the limits imposed by the preliminary assumptions given in section 8.05 on which the theory is based. In view of these, particularly in regard to the negligible sizes of the molecules and the non-existence of molecular attractions, it is evident that at best the laws can only represent the state of an ideal gas which is most closely realised in practice by a gas under conditions of extreme rarefaction. For real gases under normal conditions of temperature and pressure there will be varying degrees of departure from the ideal conception of a gas implicit in the foregoing treatment. For gases at low temperature and under high compression it will no longer be permissible to ignore the sizes of the molecules relative to the free volume of the gas, nor will it be possible to disregard the effect of inter-molecular forces. The attempts of van der Waals and others to modify the kinetic theory of gases in regard to these effects and so to obtain an equation of state representative of the actual behaviour of real gases, will be discussed in chapter X (sections 10.04 to 10.08). We shall not further pursue the question of inter-molecular forces here, however, but in the remaining sections of this chapter some extensions of the simple theory will be developed and certain applications of special interest will be discussed.

8.14 The Mean Free Path

Between collisions a molecule will traverse a rectilinear path. The length of this path is of course an extremely variable quantity, but it is possible to determine its average value by making certain assumptions. The average value so obtained is called the *mean free path* of a mole-cule. In dealing with this problem it is usual to extend our conception of the collision process from that of the elastic impact of tiny solid particles implicit in the treatment previously given, and to consider the molecule as surrounded by a sphere of action within which the centre of another molecule cannot enter. It is not easy to give exact definition to the mechanics of the collision process, as this would involve more details of the structure of the molecule and its associated field of force than are yet available. However we may generally assume that the near approach of two molecules results in mutual forces of repulsion being called into play, which cause a reversal or change of direction of the two molecules when their centres are sufficiently near to each other. The conception of a molecular sphere of action surrounding a molecule is convenient for calculation and gives a reasonable if albeit approximate representation of molecular encounters.

If the diameter of the molecular sphere of action is σ, then we may imagine the molecule to trace out a cylinder of diameter 2σ so that all molecules such as B (Fig. 77), with centres outside the cylinder, will not

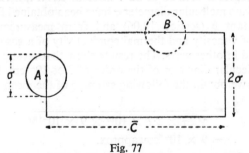

Fig. 77

undergo collision by the molecule A. Let us for a moment imagine all molecules other than A to be stationary, then as A moves with the mean velocity $\bar{c}$, it will collide in a period of one second with all the molecules whose centres lie within the volume of the cylinder of diameter 2σ and length $\bar{c}$. If the number of molecules per c.c. is n, the number of collisions ν made by the molecule A in one second is

$$\nu = \pi\sigma^2 \bar{c} n$$

This is known as the *collision frequency*. If now we represent the average distance travelled between successive collisions by λ it is clear that ν will also be given by $\dfrac{c}{\lambda}$. Thus

$$\frac{\bar{c}}{\lambda} = \pi\sigma^2\bar{c}n$$

or

$$\lambda = \frac{1}{\pi n\sigma^2} \quad \cdots \cdots \cdots \cdots \quad [69]$$

This expression for the mean free path requires modification when the motions of the other molecules are taken into account. Assuming these velocities to be distributed according to Maxwell's law, a result is obtained similar to equation 69 above, but differing by a numerical factor. Thus Maxwell gives

$$\lambda = \frac{1}{\sqrt{2}. \, \pi n\sigma^2} \quad \cdots \cdots \cdots \quad [70]$$

Since n, the number of molecules per c.c., is proportional to the density ρ of the gas, it is clear that λ varies as $\dfrac{1}{\rho}$ if it is assumed that σ is constant at a given temperature. Thus the mean free path varies inversely as the density, and therefore inversely as the pressure of the gas at any given temperature.

The number of molecules (n) per c.c. at *N.T.P.*—*Loschmidt's number*—can be obtained from Avogadro's number N (see section 8.22) the value of which is about 6×10^{23}. Thus, since one gram-molecule of a gas occupies a volume of 22·4 litres at *N.T.P.*, n is 2.7×10^{19}. Estimates of the molecular diameter σ have been obtained from van der Waal's constant b (section 10.06) and from measurements of the coefficients of viscosity and thermal conductivity of a gas (section 8.15 and 8.16) and in other ways. The results show that most gases have an effective molecular diameter of the order of 3×10^{-8}cm. Hence using equation 70, we obtain the following estimate of the mean free path:

$$\lambda = \frac{1}{\sqrt{2\pi} \times 2.7 \times 10^{19} \times 9 \times 10^{-16}}$$

$$= 9 \times 10^{-6} \text{cm. approx.}$$

Thus, although the average distance travelled by the molecules between successive collisions is very small, it is nevertheless about 300 times the molecular diameter—the relative magnitude becoming greater as the pressure of the gas is reduced and the mean free path correspondingly increased. There is thus every justification, particularly in the case of gases at low pressures, for the assumption that the linear dimensions of the molecules are negligible compared with their average distance of separation.

8.15   Viscosity of Gases

If the contiguous layers of a fluid are maintained in relative motion there are called into play viscous forces, due to internal friction, which tend to oppose the movement of one layer over another. The tangential force F per unit area required to establish a velocity gradient of $\dfrac{dv}{dz}$ perpendicular to the direction of flow against these retarding forces is given by

$$F = \eta \cdot \frac{dv}{dz} \quad . \quad . \quad . \quad . \quad . \quad . \quad . \quad . \quad . \quad . \quad [71]$$

Where η is called the *coefficient of viscosity*—being the shearing stress necessary to maintain unit velocity gradient. The force of viscosity in gases is explained on the kinetic theory as being due to the interchange of molecules between two layers of gas across the plane separating them. Thus if we consider the molecules above the plane AB (Fig. 78) to be moving with a velocity of drift (superimposed on the average thermal velocity $\bar{c}$) which is greater than that of the molecules below it, there

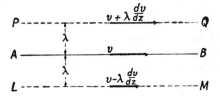

Fig. 78

will be a continual interchange across AB of more energetic molecules from above for less energetic molecules from below. Thus the layer above AB tends to lose momentum in the direction of flow, whereas the layer below AB tends to gain it. This net change of drift momentum taking place per second across AB accounts for the force of viscosity (by Newton's second law of motion, force is equal to the rate of change of momentum).

If there are n molecules per c.c. of the gas their random velocities may be resolved parallel to three mutually perpendicular directions (the co-ordinate axes x, y, z, say), and since no one direction is to be preferred to another, we may thus conceive of the gas to be constituted of three equal groups of molecules travelling parallel to these three axes of reference. Further, in each group of molecules, there will be equal numbers travelling in the positive and negative directions of the corresponding axis, thus giving $\dfrac{n}{6}$ molecules in each of the directions $\pm x$, $\pm y$, $\pm z$. Now consider the molecules crossing unit area of the plane AB from two planes PQ and LM situated a distance from AB equal to the mean free path λ of the molecules. On arriving at AB they will make their first collisions there, and hence their first transfer of momentum.

If the drift velocity of the plane AB is v and there is a velocity gradient of $\dfrac{dv}{dz}$ measured in a direction from PQ to LM perpendicular to AB, the corresponding drift velocities of the molecules in the planes PQ and LM will be $v + \lambda \dfrac{dv}{dz}$ and $v - \lambda \dfrac{dv}{dz}$ respectively. Now in one second there will be $\dfrac{n.\bar{c}}{6}$ molecules travelling in contrary directions across unit area of AB, those moving downwards from PQ transferring a momentum of

$$\tfrac{1}{6}\,n\bar{c}m \left(v + \lambda \frac{dv}{dz} \right)$$

and those moving upwards transferring a momentum of

$$\tfrac{1}{6}\,n\bar{c}m \left(v - \lambda \frac{dv}{dz} \right)$$

where m is the mass of a molecule. Thus the net transfer of momentum per second per unit area of the plane AB, that is, the force of viscosity acting per square centimetre, is

$$\tfrac{1}{3}\,n\bar{c}m\,\lambda\,\frac{dv}{dz} \quad . \quad . \quad . \quad . \quad . \quad . \quad . \quad . \quad . \quad [72]$$

Hence, from the definition of the viscosity coefficient η given in equation 71 we have*

$$\eta = \tfrac{1}{3}\,nm\bar{c}\,\lambda$$
$$= \tfrac{1}{3}\,\rho\bar{c}\,\lambda \quad . \quad . \quad . \quad . \quad . \quad . \quad . \quad . \quad [73]$$

since $nm = \rho$, the density of the gas.

A consideration of the expression for η as given above produces several important consequences. We have seen in the previous section that the mean free path of the molecules is inversely proportional to the density. Thus it would appear that for a given temperature ($\bar{c}$ constant) the viscosity of a gas is independent of its density and therefore of its pressure. This surprising result, first predicted by Maxwell and later experimentally confirmed by him from observations on the decrement of an oscillating disc suspended above a fixed disc in a gas at different pressures, constituted one of the most striking successes for the kinetic theory. It should be mentioned, however, that there are departures from Maxwell's law of the independence of viscosity with pressure at high pressures. This is to be expected as at these pressures the intermolecular forces can no longer be ignored. There are also departures

---

* A more rigid treatment of the problem modifies equation 72 (and thus the subsequent equation 73) by a numerical factor—see Jeans (*op. cit.*).

at extremely low pressures when the mean free path becomes comparable with the dimensions of the apparatus, but otherwise, for wide ranges of pressure, the viscosity of a gas is independent of its pressure.

Since the average velocity of a gas is proportional to the square root of the absolute temperature (section 8.08), equation 73 also indicates that the viscosity of a gas should increase with the temperature. This further remarkable result (it should be remembered that the viscosity of liquids decreases with temperature) has also received experimental verification, although the law of variation observed is actually more rapid than that given by the above theory. By making certain assumptions regarding the inter-molecular forces, Sutherland (1893) has improved the theory and deduced a law of variation with temperature of the form:

$$\eta = \eta_0 \frac{T^{\frac{3}{2}}}{\left[1 + \dfrac{C}{T}\right]}$$

where η_0 and C (*Sutherland's constant*) are constants for the gas. This law has received experimental confirmation over wide ranges of temperature.

By re-writing equation 73 in the form:

$$\lambda = \frac{3\eta}{\rho \bar{c}}$$

and remembering that

$$\bar{c} = \sqrt{\frac{8}{3\pi}} \qquad C = \sqrt{\frac{8p}{\pi\rho}} \qquad \text{(from equation 66)}$$

we have

$$\lambda = \eta \sqrt{\frac{9\pi}{8p\rho}}$$

Thus a determination of the coefficient of viscosity η permits of the evaluation of the mean free path λ. This in turn makes possible an estimate of the molecular diameter using equation 70. The order of magnitude of the quantities λ and σ are given in the previous section.

8.16   The Thermal Conductivity of a Gas

An expression for the thermal conductivity of a gas can be obtained on the basis of the kinetic theory in a manner similar to that in which the viscosity was dealt with in the preceding section. Thus if there is a temperature gradient of $\dfrac{d\theta}{dz}$ in the gas measured in a downward direction perpendicular to the plane AB (Fig. 79) there will be a continual transfer of more energetic molecules downwards across AB, and of less energetic molecules upwards across AB. The net result is thus a transfer of energy downwards through AB along the temperature slope.

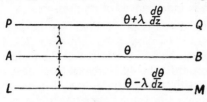

Fig. 79

Consider as before the molecules crossing unit area of the plane AB from two planes PQ and LM situated a distance from AB equal to the mean free path of the molecules—in which planes the molecules made their last collisions. Let the absolute temperature of the plane AB be θ, then the absolute temperatures of the planes PQ and LM will be $\theta + \lambda \dfrac{d\theta}{dz}$ and $\theta - \lambda \dfrac{d\theta}{dz}$ respectively. Now if (as is the case) these two temperatures are nearly equal, the rates at which the two groups of molecules cross the plane AB will scarcely be affected by the slight difference of translational velocity corresponding to these temperatures. Accordingly in one second there will be crossing unit area of AB a mass of $\dfrac{nm\bar{c}}{6}$ from above at a temperature of $\theta + \lambda \dfrac{d\theta}{dz}$, and from below a similar mass at a temperature of $\theta - \lambda \dfrac{d\theta}{dz}$. The net transfer of energy (Q) downward through unit area of AB is thus the same as if a mass $\dfrac{n\,m\bar{c}}{6}$ were cooled through a temperature of $2\lambda \dfrac{d\theta}{dz}$. If c_v is the specific heat of the gas at constant volume, this amount of heat is given by

$$Q = \frac{nm\bar{c}}{6}\, c_v \cdot 2\lambda \frac{d\theta}{dz} \quad . \quad . \quad . \quad . \quad . \quad . \quad [74]$$

If k is the thermal conductivity of the gas, the amount of heat conducted across unit area of AB in one second is

$$Q = k\frac{d\theta}{dz} \quad . \quad . \quad . \quad . \quad . \quad . \quad . \quad . \quad . \quad . \quad [75]$$

Hence identifying equations 74 and 75 we have

$$k = \tfrac{1}{3} nm\bar{c}\, c_v\, \lambda$$
$$= \tfrac{1}{3}\rho\bar{c}c_v\lambda \quad . \quad . \quad . \quad . \quad . \quad . \quad . \quad . \quad [76]$$

since the product nm is the density ρ of the gas. Using this last equation in connection with equation 73 we may write

$$k = \eta c_v$$

and thus a remarkably simple relation is predicted by the kinetic theory between the thermal conductivity k, the coefficient of viscosity η, and the specific heat at constant volume c_v. It also follows that, at a given temperature, the thermal conductivity of a gas is independent of its pressure—a prediction which is confirmed over wide ranges of pressure by experimental observation.

A more complete argument leads to a formula

$$k = B\eta c_v$$

where the coefficient B depends on the atomicity of the gas. Thus Chapman has shown that $B = 2.5$ for a monatomic gas, 1.9 for a diatomic gas, and 1.75 for a triatomic gas.

8.17 Diffusion of Gases

If any two gases (X and Y) at the same temperature and pressure are brought together in a vessel in such a way that the lighter (Y) of the two is uppermost, it is found that ultimately the gases become thoroughly mixed throughout the vessel without the temperature or pressure being in any way affected. The phenomenon, which cannot be due to gravity since the heavier gas rises whilst the lighter one descends, affords a striking demonstration of molecular agitation, and is called diffusion.

The process can be explained on the kinetic theory by considering the transport of mass across a given plane in a manner similar to that in which the viscosity of gases was accounted for by the transport of momentum, and the thermal conductivity by the transport of energy. Consider the transport of the molecules of the heavier gas (X) upwards across unit area of the plane AB (Fig. 80). Let the concentration (num-

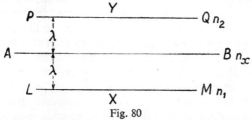

Fig. 80

ber of molecules per unit volume) at the plane AB of gas X be n_x, and if the concentration gradient is $\dfrac{dn}{dz}$ in a direction perpendicular to AB, then the concentrations at the planes PQ and LM distant λ (mean free path) above and below AB will be respectively $n_x - \lambda\dfrac{dn}{dz}$ and $n_x + \lambda\dfrac{dn}{dz}$

Hence the number of molecules crossing unit area of AB in an upward direction in one second is

$$\tfrac{1}{6}\left(n_x + \lambda\frac{dn}{dz}\right)\bar{c} - \tfrac{1}{6}\left(n_x - \lambda\frac{dn}{dz}\right)\bar{c}$$

$$= \tfrac{1}{3}\lambda\bar{c}\frac{dn}{dz}$$

and the mass M transported will be

$$\tfrac{1}{3}m\lambda\bar{c}\,\frac{dn}{dz} = \tfrac{1}{3}\lambda\bar{c}\,\frac{d\rho}{dz} \quad \cdots \cdots \cdots \quad [77]$$

where $\dfrac{d\rho}{dz}$ is the mass concentration gradient. Thus the mass transported per unit area per second is proportional to the mass concentration gradient (this relationship in the case of the diffusion of a solute through a solution is known as *Fick's law*). Thus we may write

$$M = D\,\frac{d\rho}{dz} \quad \cdots \cdots \cdots \cdots \quad [78]$$

where D is the mass transported per unit area per second per unit concentration gradient and is known as the *coefficient of diffusion*. Identifying equations 77 and 78 we have

$$D = \tfrac{1}{3}\lambda\bar{c} \quad \cdots \cdots \cdots \cdots \cdots \quad [79]$$

Now λ varies inversely as the density (section 8.14), and thus at a given pressure, directly as the absolute temperature T. Also $\bar{c}$ varies as the square root of the absolute temperature (section 8.08), and thus D, by equation 79, varies as $T^{\frac{3}{2}}$. Experiment however shows that D varies as a power of T somewhat greater than $\frac{3}{2}$

8.18   Thermal and Electrical Conduction in Metals

On the basis of the electronic theory of matter the difference between electrical conductors and insulators is accounted for by assuming that the electrons remain bound within the atoms of the latter group of substances, whereas in the former there are present a number of free electrons which are the agents responsible for the transfer of electricity through the substance. The theory of conduction in metals has been developed by Drude by treating this "electron gas" as being in thermal-kinetic equilibrium with the atoms of the substance, and on this basis has been able to deduce many of the well-known laws of conduction in metals.

Thus, applying the methods of the kinetic theory, the electrons in the "gas" will possess at a given temperature T (absolute scale) an average random velocity of $\bar{c}$ such that, considering one electron of mass m

$$\tfrac{1}{2}m\bar{c}^2 = \tfrac{3}{2}kT \text{ (see section 8·08)}.$$

If now an electric field of intensity X is applied, there will be a general drift of the electrons along the field direction—or rather, in the opposite direction since the electron carries a negative charge (e). The force on an electron due to the field will be Xe, and this will produce an accelera-

tion of $\dfrac{Xe}{m}$ resulting in the electron acquiring a drift velocity of u after having traversed the mean free path distance λ. Assuming that the velocity of drift is small compared with the velocity c, and further that the drift velocity is not retained after collision with the (relatively) massive atoms, the time elapsing between collisions will be $\dfrac{\lambda}{c}$ and therefore

$$u = \frac{Xe}{m} \cdot \frac{\lambda}{\bar{c}}$$

giving an average velocity of drift along the field of

$$\tfrac{1}{2}\,\frac{Xe}{m} \cdot \frac{\lambda}{\bar{c}}$$

If n is the number of free electrons in unit volume, the number of electrons crossing unit area perpendicular to the field direction in one second is $\tfrac{1}{2}n\,\dfrac{Xe\lambda}{m\bar{c}}$, and hence the current i will be

$$i = \tfrac{1}{2}n\,\frac{Xe^2\lambda}{m\bar{c}}$$

Now the electrical conductivity σ is given by the relation

$$\sigma = \frac{i}{X}$$

and thus we have

$$\sigma = \frac{ne^2\lambda}{2m\bar{c}}$$

$$= \frac{ne^2\lambda\bar{c}}{6kT} \quad \cdot \cdot \cdot \cdot \cdot \cdot \quad [80]$$

a result which is in conformity with Ohm's law since the conductivity is independent of the current.

Drude's theory further supposes that the electrons are also the agents responsible for the conduction of heat through the substance. The analytical treatment of this problem on the basis of the kinetic theory is identical with that given in section 8.16 for the thermal conductivity of gases. The quantity $c_v\rho$ appearing in the expression for the thermal conductivity of a gas (equation 76) is the amount of heat required to raise the temperature of unit mass of the gas by one degree, and in the case of the "electron gas" this quantity will be $\dfrac{1}{J} \cdot \dfrac{dE}{dT}$, E being the energy in ergs of the n electrons contained in unit volume, and J the mechanical equivalent of heat. From equation 63, $E = \dfrac{3}{2}\dfrac{nkT}{J}$ and thus $\dfrac{dE}{dT} = \dfrac{3}{2}\dfrac{nk}{J}$. Hence the thermal conductivity k of the "electron gas" is

$$k = \tfrac{1}{2}\,\frac{n\bar{c}\lambda k}{J}$$

Combining this equation with equation 80 we obtain

$$\frac{K}{\sigma} = \frac{3}{J}\left(\frac{k}{e}\right)^2 \cdot T$$

$$\text{or } \frac{K}{\sigma T} = \frac{3}{J} \cdot \frac{R^2}{N^2 e^2} \quad \cdots \cdots \cdots [81]$$

R being the gas constant and N Avogadro's number.

From equation 81 it thus follows that the ratio of the thermal to the electrical conductivity should be the same for all metals at the same temperature, and further that this ratio should be proportional to the absolute temperature. This is in agreement with the empirical law of Wiedemann and Franz, later extended for temperature variation by Lorentz. The extent to which this law agrees with the experimental results, and certain difficulties associated with Drude's "electron gas" theory of conduction are discussed in section 11.16.

8.19   Radiometric Forces—Crooke's Radiometer

If temperature inequalities are created in a gas at very low pressure —say by introducing two unequally heated surfaces—the thermal molecular forces over the surfaces will be different. The effect of these unbalanced forces is to give rise to a "radiometric force" which is beautifully illustrated by Crooke's radiometer shown in Fig. 81. Four

light mica vanes, each lamp-blacked on one face, are attached to the four arms of a cross which is mounted in almost frictionless pivots inside a highly exhausted glass vessel. The blackened faces of the vanes are so placed that all move round in a forward or backward direction. It is then found that even feeble illumination is sufficient to make the vanes revolve—the blackened surfaces moving as if repelled by the radiation. The simple theory of the action of the radiometer is as follows. The blackened faces of the vanes are more strongly heated by the radiation than the reverse faces. Consequently molecules rebounding from the blackened faces will carry away a greater momentum than from the faces on the reverse sides of the vanes. If then the mean free path of the molecules is comparable with the dimensions of the glass envelope, the reactions to the forces on the two sides of the vane are exerted on the walls, and thus the vane experiences a differential force which produces the motion observed. If however the pressure of the gas is too high, the increased energy of the molecules rebounding from the blackened surfaces is dispersed by inter-molecular collisions throughout the bulk of the gas and the exchanges of momentum

Fig. 81

between the gas and the two faces of the vane tend to become equalised and the motion of the vane ceases. The effect also ceases at extremely low pressures as in this case the number of collisions made by the remaining molecules with the vanes is too few to set them in motion.

8.20 Knudsen's Absolute Manometer

If two unequally heated plane plates are separated by a distance which is less than the mean free path of the molecules of the surrounding gas, a radiometric force of repulsion is set up between them. The law of repulsion has been quantitatively studied by Knudsen who has applied it to the measurement of the pressure of the gas in absolute terms. Consider two plates A_1 and A_2 (Fig. 82) placed parallel to each other and maintained at temperatures of $T_1°$ and $T_2°$ absolute respectively, where T_2 is also the temperature of the surrounding gas. In the equilibrium state let there be n_1 molecules per c.c. moving from A_1 to A_2 with a root mean square velocity of C_1, and let there be n_2 molecules per c.c. moving from A_2 to A_1 with a root mean square velocity of C_2. Then

Fig. 82

since there is no accumulation of molecules at either plate A_1 or A_2, the numbers of molecules striking and leaving either plate per second are the same, and thus

$$n_1 C_1 = n_2 C_2 \qquad \ldots \ldots \ldots \ldots \quad [82]$$

Further, since there is no accumulation of molecules between the plates, the numbers of molecules crossing unit area per second in a direction perpendicular to the surfaces from the space outside the plates must equal the number per second leaving the space between the plates. That is

$$n_1 C_1 + n_2 C_2 = n C \ldots \ldots \ldots \ldots \quad [83]$$

where n is the number of molecules per c.c. of the gas *not* between the plates and C is the root mean square velocity. Hence from equations 82 and 83

$$n_1 C_1 = n_2 C_2 = \tfrac{1}{2} nC = \tfrac{1}{2} nC_2$$

C being equal to C_2 since the surrounding gas is at the same temperature as A_2.

Now if m is the mass of a molecule, the total change of momentum per unit area per second, i.e. the pressure, on the inside surfaces of A_1 and A_2 is by equation 68

$$\tfrac{1}{3} m (n_1 C_1^2 + n_2 C_2^2)$$

and the pressure on their outside surfaces is

$$\tfrac{1}{3} nm C_2^2$$

Thus the excess pressure F between the plates over that of the gas surrounding them is

$$F = \tfrac{1}{3} m (n_1 C_1^2 + n_2 C_2^2 - n C_2^2)$$

$$= \tfrac{1}{6} nm C_2^2 \left[\frac{C_1}{C_2} - 1\right] \quad . \quad . \quad . \quad . \quad . \quad . \quad [84]$$

Then, assuming that the molecules striking the respective plates take up the temperatures of the plates, we have

$$\frac{C_1^2}{C_2^2} = \frac{T_1}{T_2}$$

and equation 84 becomes

$$F = \tfrac{1}{2} p \left[\sqrt{\frac{T_1}{T_2}} - 1\right]$$

where p = pressure of the surrounding gas = $\tfrac{1}{3} nm C_2^2$
Thus

$$p = \frac{2F}{\left[\sqrt{\dfrac{T_1}{T_2}} - 1\right]}$$

Hence the pressure p of the gas is expressed in terms of the pressure F between the plates and their absolute temperatures, and thus p can be determined in absolute terms.

Various absolute manometers of the Knudsen type have been constructed, but we shall not concern ourselves here with the technical details of any one gauge, but will consider the basic principle of construction. Two fixed plates A_1 and A_1' (Fig. 83) are maintained at the

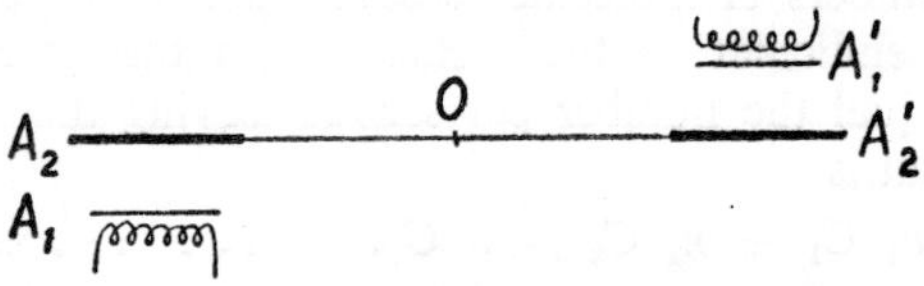

Fig. 83

same fixed temperature T_1 by electrical heating coils. Opposite them are suspended two light vanes A_2 and A_2' of copper or aluminium so that the forces of repulsion between A_1 and A_2 and A_1' and A_2' produce a deflecting couple about O which can be balanced by the torsion in the phosphor-bronze suspension fibre. To ensure a well-defined area A of the plates A_1 and A_1' they are surrounded with guard rings maintained at the temperatures of the plates thus obviating edge effects. If the distance between the centres of the vanes A_2 and A_2' is l, the deflecting couple K will be

$$K = FAl$$

$$= \tfrac{1}{2} pAl \left[\sqrt{\frac{T_1}{T_2}} - 1\right] \quad . \quad . \quad [85]$$

K is obtained from the angle of torsion necessary to annul the deflection of the plates and the couple necessary for unit angular twist of the suspension. This latter quantity can be calculated from the moment of inertia of the suspended system and its period of oscillation. Having thus evaluated K, the pressure p of the gas can be calculated from equation 85.

The special advantages of the Knudsen manometer are that it is an absolute instrument and thus no calibration is required. Its sensitivity is very great, and by suitably increasing the ratio $\dfrac{T_1}{T_2}$, it can give readings of pressures down to 10^{-8} mm. of mercury. The pressure measured is that of all the gases and vapours present, thus the vapour pressure of mercury, which is not measured by the McLeod gauge (section 5.12), is included in the readings. By using a cold trap the partial pressures of condensable vapours can be measured. The instruments are however susceptible to mechanical shock, and due to the low damping (because of the very low gas pressures) there is prolonged oscillation of the suspended system, and the instrument requires a high degree of experimental skill on the part of the manipulator. In view of this the gauge is employed mainly for standardisation purposes.

8.21 Experimental Evidence of Molecular Motion—the Brownian Movement

In 1827 the botanist Robert Brown observed that pollen grains suspended in water exhibited a peculiar erratic motion which appeared to be spontaneous and unceasing. The cause of this motion, which was observed through high power microscopes, was varyingly attributed to vital forces, inequalities of temperature, surface tension, chemical action, vibrations, &c., and although the effect attracted much attention, it was nearly 50 years later before its real significance was realised. Wiener (1863) made the suggestion that the motions observed were caused by molecular impacts on the suspended particles, and this explanation was strengthened by the work of Gouy (1888). Gouy found that the motion was independent of the nature of the suspended particles, being determined only by their size, and from the results of his work there seemed little doubt that the unceasing random motions observed were a result of the molecular agitations.

More recent work, particularly on fine colloidal solutions, has confirmed this point of view, whilst the quantitative work of Perrin (section 8.22) and others leaves little room for doubt that the Brownian movement provides visual demonstration of the thermal movement of the molecules. A particle may be imagined to be under continual bombardment by the molecules of the medium in which it is suspended and, although over a period of time the number of impacts received

H.T.–7+

from all directions will be the same, in any small interval of time the particle will receive more impacts on one side than another. Thus, if the particle is sufficiently small, it will undergo a slight displacement as a result of the instantaneous unbalance in the molecular impacts, and hence the subsequent erratic movements of the particle may be considered as direct evidence of the erratic motions of the molecules themselves. On the basis of Maxwell's equipartition law (section 8.08) the particles will be in thermal kinetic equilibrium with the molecules of the surrounding medium and will accordingly have the same kinetic energy as these molecules. This explains why the more massive particles show a more sluggish movement.

The Brownian movement can be observed in gases; thus smoke particles in air show pronounced movement which may be seen under low magnification. However, the possibility of perturbations—due to convection effects on illuminating the gas for the purpose of taking observations—makes quantitative work more difficult with gases than with colloidal solutions.

8.22 Perrin's Experiments—Determination of Avogadro's Number

The Brownian movement has been quantitatively studied by Perrin (1908), and the success of his experimental work may be considered as providing direct evidence in support of the kinetic theory explanation of the Brownian movement, and also of the kinetic theory itself. From observations on the Brownian movement in colloidal suspensions, Perrin obtained in various ways a value for Avogadro's number N. One of these methods, based on the vertical distribution of the particles in a colloidal solution will be described here.

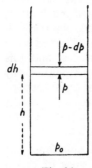

Fig. 84

Consider a vertical cylinder of gas of unit cross section (Fig. 84). The molecules in this cylindrical column will be in gravitational-kinetic equilibrium, the downward gravitational pull on the gas in an elementary section of thickness dh at a height h from the bottom of the cylinder being balanced by the net upward thrust resulting from the differences of pressure on its lower and upper surfaces. If these pressures are p and $p - dp$ respectively, and the density of the gas at the height h is ρ, we have

$$p - (p - dp) = g\rho dh \quad \text{or} \quad dp = g\rho dh$$

To indicate that the pressure falls off with height it will be necessary to include a negative sign in this equation, and hence

$$dp = -g\rho dh$$

If now V is the volume occupied by the gram-molecular weight M of the gas, $\rho = \dfrac{M}{V}$ and therefore

$$dp = -g \cdot \frac{M}{V} \cdot dh$$

and since for a perfect gas $pV = RT$, where R is the gas constant, we have $V = \dfrac{RT}{p}$ and hence

$$\frac{dp}{p} = -\frac{Mg}{RT} \cdot dh$$

On integrating this equation and remembering that $p = p_o$ when $h = o$ it follows that

$$\log_e \frac{p}{p_o} = -\frac{Mg}{RT} \cdot h$$

or $\qquad p = p_o e^{-\frac{Mg}{RT} \cdot h} = p_o e^{-\frac{Nmg}{RT} \cdot h}$ . . . . [86]

where m is the mass of a molecule and N is Avogadro's number.

Since now the pressure of a gas is proportional to n, the number of molecules per c.c., equation 86 can be written

$$n = n_o e^{-\frac{Nmg}{RT} \cdot h} \qquad \text{. . . . . [87]}$$

where n is the number of molecules per c.c. at the height h and n_o the number per c.c. at the bottom of the column.

Equation 87 thus represents the vertical distribution of the molecules of a gas under conditions of gravitational-kinetic equilibrium, and Perrin assumed that this formula would also apply to the vertical distribution of the particles in a colloidal solution. His experiment consisted in taking counts of the numbers of the particles in two planes separated by a known vertical distance h and determining the mass m of the particles from a subsidiary experiment. From these observations he was then able to obtain a value for N using equation 87 above.

In his experiment Perrin used a colloidal suspension of gamboge particles of uniform size obtained after a long continued process of fractional centrifuging. A drop of the emulsion E (Fig. 85) was contained in a hollow microscope slide S over which was placed a cover glass C. The depth of the emulsion was about one tenth of a millimetre, and after equilibrium was established, an instantaneous photograph of the particles was taken with the objective of a high power microscope M in a given position. The microscope was then moved up through a

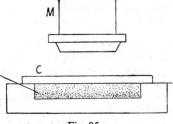

Fig. 85

distance d, measured by an accurate vernier, and another photograph was taken. The number of particles in the two planes were then counted on the photographs, average numbers being taken in each case from a large number of observations. The ratio of these numbers gave the relative concentrations of the particles in two planes separated by a distance h equal to μd where μ is the refractive index of the liquid.

It should be noticed that the effective mass of the particles when in suspension is less than their true mass m on account of the buoyancy of the surrounding liquid. Thus if ρ and ρ' are respectively the densities of the particles of the liquid, the effective mass of a particle is

$$m\left(1 - \frac{\rho'}{\rho}\right)$$

and if n_1 and n_2 are the numbers of particles in the lower and upper planes as counted above, we may express N, using equation 87 as

$$N = \frac{RT\rho}{mg\mu d\,(\rho - \rho')} \cdot \log_e\left(\frac{n_2}{n_1}\right)$$

The density of the particles was obtained by adding potassium bromide to the suspension until the particles neither rose nor sank on being centrifuged, and then determining the density of the solution. The result so obtained was checked by evaporating a known mass of the suspension to dryness and finding the mass of the residue, when its density could be calculated by the usual method for insoluble powders. Knowing the density, the mass of a particle could be obtained by determining its volume. This was done by shaking up the suspension and observing the rate of fall of the particles. By thus obtaining the terminal velocity and applying Stokes' law,* the radius, and hence the volume were found.

All the data was thus available for the calculation of N for which Perrin obtained the value of $6 \cdot 8 \times 10^{23}$. Considering the difficulties of the experiment this is in moderately good agreement with Millikan's value of $6 \cdot 06 \times 10^{23}$ derived from measurements of the electronic charge.

QUESTIONS. CHAPTER 8

1. Outline the kinetic theory of matter and explain the essential differences between the solid, liquid, and gaseous states. How are evaporation and boiling explained on the kinetic theory?

Explain on the basis of the kinetic theory (*a*) why the temperature of a gas rises when it is compressed, (*b*) why the temperature of an evaporating liquid may be less than that of the atmosphere. (C.)

2. Without deriving any formulae, use the kinetic theory of gases to explain (*a*) how a gas exerts pressure, (*b*) why the temperature of a gas rises when the gas is compressed, (*c*) what happens when a quantity of liquid is introduced into a closed vessel.

How are the differences in the behaviour of real and ideal gases explained by the kinetic theory?

If there are $2 \cdot 7 \times 10^{19}$ molecules in a cubic centimetre at 0°C. and 76cm. of mercury pressure, what is the number per cubic centimetre (i) at 0°C. and 10^{-6}mm. pressure, (ii) at 39°C. and 10^{-6}mm. pressure? (N.)

3. Derive an expression for the pressure of an ideal gas in terms of the density and root mean square velocity of the molecules. State clearly the assumptions made.

Give a short account of experimental work on the deviation of gases from Boyle's law, illustrating your answer by means of suitable graphs. (O. and C.)

4. State *Graham's law of diffusion*. Use the simple kinetic theory of gases to derive an expression for the pressure exerted by an ideal gas. Assuming the rate of diffusion to be proportional to the mean molecular velocity, show how Graham's law may be derived theoretically for an ideal gas.

The molecular weights of the fluorides of the isotopes of uranium are 352 and 349 respectively. Assuming they behave as ideal gases, compare their rates of diffusion through a porous barrier under identical conditions. [L.]

5. State the assumptions which form the basis of the simple kinetic theory of gases and show how they lead to the equation $p = \frac{1}{3}\varrho c^2$, where p is the pressure of an ideal gas, ϱ its density, and c the root-mean-square speed of its molecules. How is this equation reconciled with the equation $pV = RT$?

Describe briefly the general nature of the deviations of real gases from ideal-gas behaviour, and discuss to what extent these deviations can be correlated with the inadequacy of the fundamental assumptions of simple kinetic theory. [O.]

6. How does the molecular theory of matter account for the fact that heat is needed to evaporate a liquid at constant temperature?

Explain why (a) a piece of wet cloth is usually at a lower temperature than its dry surroundings; (b) its temperature is lower in a draught than in still air; and (c) it reaches a lower temperature if it is moistened with ether than if it is moistened with water.

A test-tube containing 3gm. of ether is immersed in a beaker of water which is surrounded by melting ice. When the whole is at 0°C. a stream of air at 0°C. is blown through the ether until it has all evaporated. The cap of ice formed round the test-tube has a mass of $3 \cdot 14$gm. Find the latent heat of evaporation of ether.

(Latent heat of fusion of ice = $80 \cdot 0$ cal. gm.$^{-1}$) (O. and C.)

7. Explain the meaning of the terms *ideal gas* and *molecule*.

What properties of a gas such as carbon dioxide distinguish it from an ideal gas and how may these differences from " ideal " be demonstrated experimentally?

According to simple kinetic theory the pressure exerted by a gas of density ϱ is $\frac{1}{3}\varrho \bar{c}^2$ where $\bar{c}^2$ is the mean square molecular velocity. Show how this relation may be correlated with the equation of state for an ideal gas, $PV = RT$, explaining clearly what further assumptions you have to make. (O. and C.)

8. Show how the kinetic theory explains qualitatively (a) the pressure of a gas, (b) the viscosity of a gas, (c) the distinction between solids, liquids and gases (including an explanation of latent heat of vaporization).

Assuming that the root mean square velocity of hydrogen molecules is $1 \cdot 84 \times 10^5$ cm. sec.$^{-1}$ at 0°C. calculate the root mean square velocity of oxygen molecules at 16°C. given that the molecular weights of hydrogen and oxygen are 2 and 32 respectively and that both gases may be considered to be ideal. (N.)

9. State the postulates on which the simple kinetic theory of gases is based and use the theory to derive an expression for the pressure of an ideal gas. What direct evidence is there for the kinetic theory of matter?

Calculate the temperature at which oxygen molecules have the same root-mean-square velocity as that of hydrogen molecules at −100°C., given that the molecular weights of hydrogen and oxygen are 2 and 32 respectively and that both gases may be considered to be ideal. (N.)

10. For hydrogen the density is 0·089 g. litre^{-1} at 0°C. and 760mm. of mercury, the molecular weight is 2·016, the mass of one molecule is 3.34×10^{-24}g. and the ratio of the specific heats is 1·41. For the given pressure and temperature calculate: (*a*) the root-mean-square velocity of hydrogen molecules. (*b*) the velocity of sound in hydrogen. Evaluate also: (*c*) the universal gas constant, (*d*) Boltzmann's constant.
[A.]

11. State the assumptions made in the simple kinetic theory of gases.

Derive from first principles an expression for the pressure of an ideal gas in terms of the gas density and the mean square velocity of its molecules.

Show how this expression accounts for Boyle's Law and Charles' Law.

Indicate how the kinetic theory of gases is modified to take into account the imperfections of a real gas. [A.]

12. Give a qualitative argument in terms of the kinetic theory to show that the number of molecules striking unit area of the containing vessel is proportional to the concentration of molecules and to their mean velocity.

Two vessels at different temperatures, containing the same gas at low pressure, are connected by a small aperture. By assuming that any molecule that falls upon the aperture passes from one vessel to the other, show that the gas is in equilibrium if the pressure in each vessel is proportional to the square root of the absolute temperature of the gas in that vessel.

What would be the ratio of the pressures in the two vessels if the aperture was large? [N.]

13. Discuss the evidence for the kinetic theory of gases and derive an expression for the pressure of a gas in terms of the mean square velocity of the molecules.

A small hole, 0·1mm. in diameter, is opened in a vessel of volume 500 litres which contains hydrogen at a pressure of 10 atmospheres and a temperature of 0°C. If the pressure outside the vessel is kept near zero with a vacuum pump, estimate how long it will take for the pressure in the vessel to fall to 5 atmospheres.   (Camb. Schol.)

14. To what extent may the physical properties of a gas be related to the behaviour of individual molecules? Illustrate your answer with reference to (*a*) pressure, and (*b*) specific heat.

A rocket motor burns 100gm. of fuel per second and the combustion product is of molecular weight 18. If the combustion chamber is maintained at 1,000°K, estimate the thrust exerted by the motor.

($R = 8 \times 10^7$ ergs °K^{-1} mole^{-1}.)   (Camb. Schol.)

15. Describe any experiment which in your opinion shows in rather a direct way that matter is composed of discrete molecules.

Show how the root-mean-square velocity of the molecules of a gas may be calculated, and evaluate this quantity for helium at 0°C. Use your result to estimate the temperature beyond which helium could not be retained in the atmosphere of the earth.

(Atomic weight of helium = 4·00; gram-molecular volume at N.T.P. = 22·4 litres; pressure of one atmosphere = 1.014×10^6 dynes cm.$^{-2}$; radius of earth = 6.38×10^8cm.; g = 981 cm. sec.$^{-2}$)   (Camb. Schol.)

16. Outline the kinetic theory of perfect gases and explain how the theory accounts for Boyle's Law. What assumptions are necessary in order to account for Charles' Law and Avogadro's Law?

A gas is held by a partition in one half of a rigid vessel whose walls are non-conducting, the other half of the vessel being empty. When the partition is suddenly removed the temperature of the gas decreases slightly. What may be deduced from this? (Camb. Schol.)

CHAPTER IX

CHANGE OF STATE

9.01 Change of State

A substance may exist in three states—solid, liquid, and gas—the particular state at any given time being dependent on the temperature of the substance at that time. To bring about a change in the state of a substance, all that is required is the application or withdrawal of heat. According to the kinetic theory (section 8.03) the physical differences between the three states of matter are due to the measure of "freedom" possessed by the molecules of the substance. Thus in the solid state the molecules are constrained to vibrate about fixed positions of equilibrium, in liquids these constraining forces are broken down and the molecules are free to move promiscuously throughout the substance, whilst in gases this freedom of movement is even more marked. It has been estimated that the average distance between the molecules, or mean free path, expressed in terms of the diameter d of the molecule, is from $1 \cdot 5\ d$ to $3 \cdot 0\ d$ in the case of solids, $2\ d$ to $4\ d$ for liquids, and from $30\ d$ upwards for gases.

The heat absorbed when a substance changes its state from solid to liquid (fusion), and again from liquid to gas or vapour (evaporation), is used up to effect this rearrangement of the molecules against their mutually attractive forces. As the state of the substance changes in the reverse direction, this heat is given out by the substance on the re-grouping of the molecules at the gas-liquid (condensation), and liquid-solid (solidification) stages.

9.02 The Lower Change of State—Melting Point

For all pure crystalline substances there is a definite temperature at which the change of state from solid to liquid takes place under a given pressure. At this temperature the solid and liquid forms are in equilibrium under the given pressure, and the temperature is defined as the *melting point* of the solid at that pressure. As stated above fusion is accompanied by the absorption of heat, and in addition there are marked changes in the physical conditions such as for example, rigidity, as the body melts. For amorphous solids such as glass and paraffin wax, the changes accompanying fusion are not so simple as those for a crystalline substance. Thus there is no definite temperature at which the change takes place (see Fig. 87), and the properties of the substance do not undergo an abrupt change as say with ice to water, but there is a relatively protracted "plastic" stage during which the properties of a solid gradually give way to those of a mobile liquid.

With both crystalline and amorphous substances there is always an accompanying volume change with fusion. In the majority of cases this change is an expansion as the solid changes to the liquid state. This type of change is typified by paraffin wax, and substances of this class are often referred to as being *wax-type*. Water, iron, bismuth, and antimony (*water-type* substances) exhibit the reverse effect and contract on melting, or conversely, expand on solidifying. These volume changes are attended by considerable forces. Thus the expansion of water on freezing if opposed can bring about the bursting of the containing pipes. The expansion of molten iron as it solidifies makes it possible to obtain sharp and clear castings of iron, whilst an alloy of lead, antimony and copper, known as "type metal," is employed to produce hard sharp type for printing purposes.

Although a pure crystalline substance possesses a definite melting point under specified conditions of pressure, it is possible to cool the corresponding liquid to temperatures below this temperature before solidification sets in. Thus, if some distilled water contained in a clean test tube and protected from dust, is slowly cooled without stirring, it is possible to reduce the temperature several degrees below 0°C. before ice formation sets in. The water is said to be *super-cooled* and is in an unstable state when in this condition. Vigorous stirring, or the introduction of a small ice crystal, will immediately initiate freezing with the instant release of latent heat which quickly raises the temperature to the normal freezing point.

9.03   Determination of Melting Points

(*a*) *Capillary Tube Method.*—If the substance is available only in small quantities, some of the substance is placed in a short length of glass capillary tubing closed at one end. The capillary tubing is then attached to the bulb of a $\frac{1}{10}$°C. thermometer by means of rubber bands, and the thermometer placed in a bath of suitable liquid so that the open end of the capillary tube is above the liquid surface. The boiling point of the liquid selected should be above the melting point of the solid. The liquid is now slowly heated and the temperature observed at which the solid melts. The source of heat is now removed, and the temperature at which solidification occurs is taken, the desired melting point being the mean of the two temperatures taken.

(*b*) *Cooling Curve Method.*—When larger quantities of the substance are available, the melting point can be conveniently obtained from a cooling curve as follows. Some of the substance is placed in a test tube and completely melted. The tube is now placed inside a beaker to exclude draughts and the readings of a thermometer whose bulb is fully immersed in the liquid are taken every half minute (say) as the liquid cools. The readings are continued for some time after solidification has set in. On plotting temperature values against time of cooling,

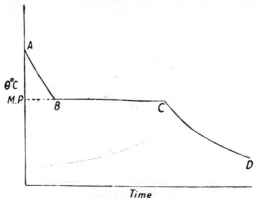

Fig. 86.   Cooling curve for a crystalline substance

a graph similar to that shown in Fig. 86 is obtained. Three distinct stages are revealed by the curve; in the stage *AB* the substance is liquid, the horizontal section *BC* is the solidifying stage, whilst *CD* represents the cooling of the solid substance.

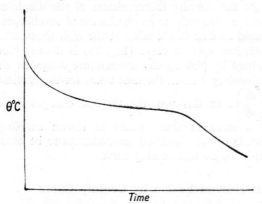

Fig. 87. Cooling curve for an amorphous substance

As the cooling takes place in an enclosure at room temperature, the sections *AB* and *CD* will conform very closely to the form of curve determined by Newton's law of cooling. During the process of solidification however, the temperature is maintained constant as the latent heat released during this process offsets the loss of heat by convection and radiation from the surface of the test tube. The temperature at which this occurs is the freezing point of the liquid substance, or the melting point of the solid substance.

7*

Metallurgists find the cooling curve method of great value for investigating the composition of metallic alloys. The cooling curve for such an alloy will show a series of horizontal stages corresponding to

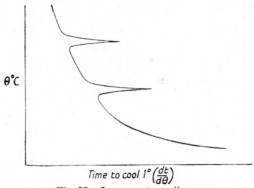

$\theta °C$

Time to cool $1° \left(\frac{dt}{d\theta}\right)$

Fig. 88.   Inverse-rate cooling curve

the melting points of the various constituents, and hence the study of such a curve yields information concerning the composition of the alloy. Since the observations are made at high temperatures, thermo-couples replace the mercury thermometers of the simple experiment described above, and in order to emphasise the phase changes occurring during the rapid cooling of the alloy at the high temperatures of the test, an *inverse-rate* cooling curve (Fig. 88) is usually plotted. This curve is obtained by plotting the temperature θ against the inverse of the rate of cooling, that is, the time taken for the specimen to cool through $1° \left(\dfrac{dt}{d\theta}\right)$. In this way the thermal changes occurring are revealed by a series of sharp peaks as shown enabling a more accurate estimate of the transition temperatures to be obtained than is possible with the normal cooling curve.

9.04   Latent Heat of Fusion by Cooling Curve

An approximate value of the latent heat of fusion of a substance can be obtained by finding the rate of cooling at the start and finish of the solidifying stage. Theoretically these values are the same since the temperature is the same at both points, but it will be found difficult to obtain a gradient exactly at B or C (Fig. 86). If however the rate of cooling (obtained by taking gradients) at a point near B is $\left[\dfrac{d\theta}{dt}\right]_1$ and that at a point near C is $\left[\dfrac{d\theta}{dt}\right]_2$ and s_1 and s_2 are the specific heats in the liquid and solid states respectively, the average rate of loss of heat of M gm. of the substance at the freezing point is

$$\tfrac{1}{2} \left\{ Ms_1 \left[\frac{d\theta}{dt}\right]_1 + Ms_2 \left[\frac{d\theta}{dt}\right]_2 \right\} \text{ cal. per second}$$

(ignoring the thermal capacity of the containing tube).

Hence of the stage BC is complete in t seconds, the total loss of heat during solidification is

$$\frac{Mt}{2} \left\{ s_1 \left[\frac{d\theta}{dt}\right]_1 + s_2 \left[\frac{d\theta}{dt}\right]_2 \right\} \text{ cal.} \quad \ldots \quad [88]$$

and this must equal the total heat evolved, which is

$$ML \text{ calories} \quad \ldots \ldots \ldots \quad [89]$$

Hence from equations 88 and 89

$$L = \frac{t}{2} \left\{ s_1 \left[\frac{d\theta}{dt}\right]_1 + s_2 \left[\frac{d\theta}{dt}\right]_2 \right\} \text{ cal. per gm.}$$

TABLE 21

Melting points of certain elements and common substances in °C.

Substance	Melting Point	Substance	Melting Point
Tungsten	3,387	Carbon	3,500
Platinum	1,773	Silicon	1,415
Iron	1,527	Glass	c.1,100
Nickel	1,455	Common salt	801
Copper	1,083	Cane sugar	189
Gold	1,063	Camphor	176
Silver	961	Resin	135
Aluminium	660	Sulphur	115
Magnesium	659	Naphthalene	80
Zinc	419	Beeswax	64
Lead	327	Paraffin wax	53–55
Tin	232	Phosphorus	44
Sodium	98	Glycerine	17
Mercury	−39	Benzene	5·5

9.05 Effect of Pressure on the Melting Point

Generally, external conditions have little or no effect on the melting point of a substance; change of pressure however produces a slight change in the melting point, a fact which was first deduced theoretically by James Thomson and later confirmed experimentally by his brother Lord Kelvin. The nature of this influence may be stated as follows,

(a) substances which expand on melting experience a rise in their melting points with increased pressure,

(b) substances which contract on melting have their melting points lowered when the pressure increases.

Thus in the case of ice, which belongs to the second class of substances, there is a lowering of about 0·0075°C. per atmosphere increase of pressure. The magnitude of the change for any given substance can be calculated from the Clausius-Clapeyron equation

$$\frac{dp}{dT} = \frac{LJ}{T\,(v_2 - v_1)} \quad \text{(see section 13.09)}$$

Where dT is the change in the absolute temperature T of the melting point for an increase dp in the applied pressure, v_2, v_1 being respectively the specific volumes of the substance in the liquid and solid states, and L the latent heat of fusion in calories per gm.

A simple consideration of the problem reveals the sort of change to be expected. For example, in the case of a substance which expands on fusion an increase of pressure (tending to reduce the volume) will be unfavourable to such a change which can now only be brought about at a somewhat lower temperature. The effect of pressure on the melting point is a good illustration of a general principle of mobile equilibrium known as *Le Chatelier's principle*. This may be stated as follows

If a change occurs in one of the factors, e.g. temperature or pressure, under which a system is in equilibrium, the system will tend to adjust itself so as to nullify the effect of that change.

Other examples of this principle are to be found in Lenz's law in electro-magnetic induction where the induced effects are always of such a nature as to oppose the motion creating them, in the stretching of a liquid film, where the subsequent fall in temperature causes an increase in the value of the surface tension of the film thus tending to reduce the area to its initial value, and in the formation of ammonia gas by the combination of nitrogen and hydrogen where, due to the decrease in volume as the gases combine, the production of ammonia is increased by the application of pressure.

The case of ice-water equilibrium will be considered more closely in the light of the above principle. Consider a mixture of ice and water at 0°C. contained in a heat impervious cylinder closed by a piston head at a pressure of one atmosphere. On compressing the mixture by forcing in the piston a change will take place which, by Le Chatelier's principle, will be such as to nullify the increase of pressure. This can only be done by some of the ice melting so as to reduce the volume. The necessary heat to melt the ice is absorbed at the expense of the rest of the mixture, and in consequence the temperature falls, or in other words the temperature of the melting point of ice is lowered by increase of pressure.

Accurate investigations on the variation of the melting point of ice were carried out by Dewar in 1880 using the apparatus shown in

Fig. 89. A mixture of broken ice and distilled water carefully freed from air was contained in a stout iron bottle *B* immersed in melting ice and connected to a high pressure pump and manometer via the

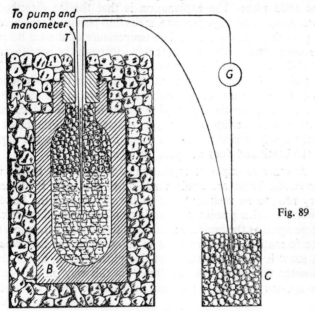

Fig. 89

tube *T*. One junction of a calibrated thermo-couple was immersed in the ice-water mixture in *B*, the other junction being kept at a fixed temperature by surrounding it by melting ice in a beaker *C*. A deflection of the galvanometer *G* indicated a difference in temperature between the junctions which could be accurately obtained in terms of the galvanometer reading. Dewar found that an increase in pressure on the mixture in *B* caused some of the ice to melt thus absorbing heat from the mixture and producing a fall in temperature to the melting point under the new conditions of pressure. Working in steps of 25 atmospheres up to a pressure of 700 atmospheres, it was found that there was a progressive depression of the freezing point with increase of pressure equal to 0·0072°C. per atmosphere—a result in good agreement with the theoretically predicted value.

9.06 Regelation

Faraday first observed that if two pieces of ice are pressed together a solid piece is formed on releasing the pressure. This phenomenon, which is called *regelation*, is accounted for by the fact that the melting point of ice is lowered under the increased pressure thus causing the ice to melt at the point of contact absorbing latent heat from the rest of the ice. On removing the pressure the water so formed is again frozen thereby cementing together the two ice lumps. A well known experiment illustrating the phenomenon is the passage of a weighted copper wire through a block of ice (at 0°C.) which however remains one solid piece. The explanation is that the ice directly below the wire, being under high pressure, melts thus permitting the wire gradually to pass through the block. The temperature of the ice below the wire is reduced below 0°C. and heat flows through the wire from the water (at 0°C.) above, which now freezes again (on giving up its latent heat of fusion). An essential part of the process is the conduction of heat from the water immediately above the wire to the ice below it, and this accounts for the observed fact that iron wire (which has a lower thermal conductivity than copper) passes through the block comparatively slowly, whilst a piece of string merely marks the upper surface of the block and does not pass through at all.

Another example of regelation is to be found in the making of snowballs. These are easily made when the snow is at temperatures very near to its melting point as the pressure of the hand can then lower the melting point sufficiently to melt the small ice crystals which freeze again as the pressure is released thus causing the snow to "bind." The formation of a hard mass of snow under the heel when walking on snow is another example of this phenomenon, although it should be noted that snow will not "bind" in this way on very cold days as the applied pressure is then not sufficient to lower the melting point to the temperature of the snow.

Again in skating, the pressure applied to the ice by the whole weight of the body being supported on the narrow edges of the skates is quite large, and, except on very cold days, will cause the ice in contact with the skates to melt. This permits the blades of the skates to sink a little, and the skates "bite" enabling turns to be made without skidding.

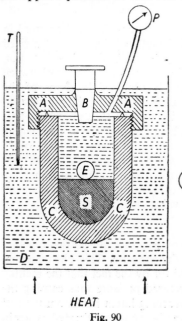

Fig. 90

9.07 Hopkins' Experiments

The apparatus used by Hopkins in his experiments on the investigation of the effect of pressure on the melting points of substances is shown diagrammatically in Fig. 90. The solid substance S was contained in a strong brass cylinder C which was then filled with oil and closed by a screw cap ABA. A pump and pressure gauge P communicated through this screw cap with the interior of the cylinder as shown. The whole was contained in an oil bath D which was heated steadily from below. The chief difficulty was to obtain the exact reading of the temperature at which the substance melted, and this was overcome by Hopkins who placed a small magnetised steel sphere E on the surface of the substance S under test. A small magnetic needle fitted on the back of a mirror F was suspended by means of a silk fibre so as to be level with the ball E. When the substance melted as the temperature of the oil bath was steadily raised, the steel ball sank and caused a sudden movement of the needle F which was recorded by the displacement of a spot of light reflected by the mirror on to a scale. The temperature at which this occurred was then read off on the thermometer T. To repeat the experiment, the steel ball was raised by a strong magnet and the whole apparatus cooled until the substance was again in the solid state. The pressure was then re-adjusted by means of the pump P and a further reading taken. In this way Hopkins obtained values for the melting points of several substances at pressures up to 800 atmospheres.

9.08 Freezing Point of Solutions

The fact that the presence of a dissolved substance lowers the freezing point of water has been known for many years, and as long ago as 1788 Blagden showed that the lowering of the freezing point of a substance was proportional to the concentration of the solute—a statement which is frequently known as *Blagden's law*. As a result of extensive investigations on aqueous solutions of non-electrolytes, and solutions of organic compounds, Raoult (1880) established the fact that the gram-molecular weight of substances dissolved in a given weight of a solvent depressed the freezing point to the same extent (*Raoult's law*). This may be expressed as follows.

$$\varDelta T = KC$$

Where $\varDelta T$ is the depression of the freezing point, C the concentration in gram-molecules per 100gm. of solvent, and K is a constant (*the cryoscopic* constant) which for water is $18 \cdot 5°C$. The depression of the freezing point resulting from known concentrations of a solute is frequently used as a means of estimating molecular weights of substances. For particulars of these experimental methods the reader is referred to books on physical chemistry.

We shall here study in further detail the freezing of an aqueous solution of common salt. If a weak solution is cooled below $0°C$. it is found that pure ice separates out thus increasing the concentration of the remaining solution. If the process is continued, ice continues to separate until a temperature of about $-21°C$. is reached when the whole mixture freezes *en bloc*. The solution at this stage is found to contain 23·6 per cent of the salt. If now a solution with an initial concentration of salt higher than 23·6 per cent is cooled below 0°C. it is found that it is salt and not ice that progressively separates out until, as before, freezing of the mixture occurs at −21°C. with the same concentration of salt, viz., 23·6 per cent. The mixture freezing at this point is called a *eutectic mixture*, the temperature at which it freezes being the *eutectic point*.

9.09 Freezing Mixtures

Any two substances which on mixing produce cooling constitute what is known as a *freezing mixture*. A common mixture of this sort used for laboratory purposes is a mixture of ice and common salt. When salt (at room temperature) is added to ice some of the ice will melt in cooling the salt to 0°C. and a saturated salt solution will be formed. Now we have already seen in the foregoing section that ice cannot remain in equilibrium with a salt solution unless its temperature is below 0°C., and consequently the ice in contact with the salt solution melts drawing the necessary latent heat of fusion from the solution, the temperature of which is lowered. More salt now dissolves (thus keeping the solution saturated), and the process outlined above continues until the temperature of the solution is lowered to − 21°C., at which temperature (eutectic point) ice is in equilibrium with the saturated solution. This also explains why ice on the pavement may be "thawed" by adding salt to it on a frosty day. Except on extremely cold days the resulting salt solution will remain liquid thus preventing the danger of slipping.

The lowest temperature which can be obtained by an ice-salt freezing mixture is − 21°C., and is produced by approximately three proportions of ice shavings or snow to one of salt (eutectic proportions). Another common freezing mixture is obtained by taking four parts by weight of calcium chloride and three of snow. A temperature of − 55°C. is possible with this mixture, the cooling being particularly rapid in this case on account of the large negative heat of solution of calcium chloride.

9.10 The Fusion of Alloys

A study of the fusion of alloys produces results which are very similar to those obtained for solutions of salt in water. When two metals A and B are mixed to form an alloy it is found that the melting point of the alloy so formed is lower than that of the more preponderant metal (A say), the presence of the other metal B acting as a sort of impurity and lowering the melting point (or freezing point) of A. Further there is a certain definite composition for the alloy for which there is a minimum melting point which is below the temperature of

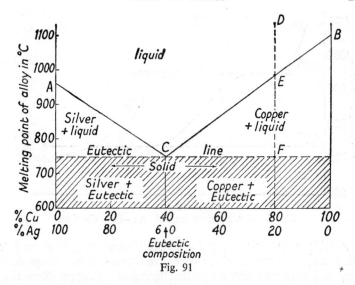

Fig. 91

the lowest melting point of the constituents. We may consider the alloy to be a solid or liquid solution of one constituent in the other, the composition of the homogeneous mixture being continuously variable. As a particular example we will consider the way in which the melting point of a copper-silver alloy varies with its composition (Fig. 91).

The melting point of pure silver is 961°C. and is represented by point A on the diagram, whilst B represents the melting point of pure copper, viz., 1083°C. The curve ACB represents the way in which the melting point of the copper-silver alloy varies with its percentage composition. Consider an alloy of 20 per cent silver and 80 per cent copper at a temperature represented by a point D which is higher than the melting point E (950°C.) for the alloy of this composition. On cooling, the alloy remains liquid until the point E is reached when solidification sets in. The solid that separates out however is pure copper, the "mother liquor" being thus richer in silver. This process continues as the cooling is continued until the point F is reached corresponding to a temperature of 740°C., when the whole mass freezes out as one. The cooling curve for this particular alloy is shown in Fig. 92. The composition of the "mother liquor" at the point F is the same as that of the alloy at C (60 per cent silver, 40 per cent copper) which has the minimum melting point. If a liquid alloy containing more than 60 per cent silver is cooled a similar behaviour is observed, but with silver separating first before the whole mass solidifies at 740°C. as before. This temperature is the eutectic temperature, the alloy at C which melts at this temperature being the eutectic mixture or eutectic alloy.

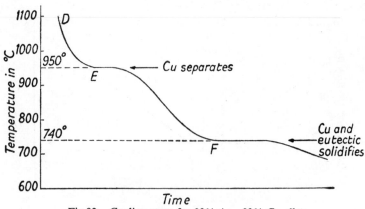

Fig 92. Cooling curve for 20% Ag—80% Cu alloy

A table of eutectic points for simple binary alloys is given below with the melting points of the pure constituents. Other alloys do not show such a simple behaviour as that described for copper-silver alloys. Two or more eutectic points may be shown due to the formation of intermetallic compounds in the alloy.

TABLE 22

Eutectic points for simple binary alloys

Metal A.	M.P. °C.	Metal B.	M.P. °C.	Eutectic Pt. °C.
Antimony	630	Lead	327	246
Bismuth	317	Cadmium	268	146
Copper	1,083	Silver	961	740
Gold	1,063	Thallium	302	131
Lead	327	Tin	232	183

9.11 Change of State from Liquid to Vapour—Evaporation

It is well known that liquids contained in a vessel open to the atmosphere undergo a gradual diminution of mass. This process, which takes place at the free surface of the liquid and at all temperatures, is known as evaporation. This passage of the substance from the liquid to the gaseous state is found to increase with the area of the surface exposed and also with the temperature. It is also controlled by the state of the air to which it is exposed, thus, if the air near the surface is moving, the rate of evaporation is increased, and again, if the air is "damp," that is, if it contains a large proportion of water vapour, evaporation is retarded. It is interesting to note that certain substances evaporate directly from the solid state without passing through the intermediate liquid phase. This process is known as *sublimation*, examples of which are to be found when camphor or iodine are slowly heated. Again on a dry frosty day snow may disappear by direct passage into water vapour (see section 9.19).

It is instructive to discuss the evaporation of liquids from the standpoint of the kinetic theory. As we have already seen in sections 8.02 and 8.03 the molecules of a liquid are supposed to be in continual motion, and on account of the relatively small mean free path they will be continually colliding with one another. The velocity and direction of travel of an individual molecule is thus subject to constant change although the mean velocity of the molecule remains the same at a particular temperature. The molecules, being closely packed together in the liquid state will exert appreciable forces on each other. For a molecule situated within the bulk of the liquid these forces will very nearly balance each other out, but if a molecule is near the surface it will be subject to a force tending to drag it back into the liquid due to the unbalanced attractive forces of the molecules below it. However if a molecule reaches the surface with a sufficiently high upward velocity it may have sufficient kinetic energy to overcome these forces, and it will accordingly escape into the space above the liquid. Some of the molecules thus escaping from the bulk of the liquid will not have sufficient kinetic energy to carry them clear of the "range of molecular attraction" and will accordingly return to the liquid. Others however will not return and will exist in the space above in the form of a vapour. With rise of temperature the average kinetic energy of the molecules will be increased, and hence the number of molecules escaping clear of the liquid in a given time will show a corresponding increase—that is, the rate of evaporation increases with temperature.

If the space above the liquid is limited there will be a gradual accumulation of molecules in it and ultimately there will be a tendency for some of them to return to the liquid. For any given temperature a condition of equilibrium will be attained at which the number of molecules leaving the liquid for the space above is just equal to the number of molecules returning. The space above the liquid is then saturated with vapour, the pressure exerted by the vapour at this stage being a maximum under the given conditions and is known as the *saturated vapour pressure* (S.V.P.), or, frequently, the vapour pressure, at that particular temperature. The condition of dynamic equilibrium described above does not depend on the bulk of the liquid or the space occupied by the vapour, the number of molecules leaving the liquid being proportional to the area of the exposed surface, which area also determines the number of molecules returning. Thus the saturated vapour pressure depends on the temperature only.

When a liquid freely exposed to the air undergoes steady evaporation due to the loss of the more energetic molecules, it is clear that the average kinetic energy of the molecules remaining will be lowered. This means that the temperature of the liquid is reduced, and thus the process of evaporation produces a cooling effect. This effect can be readily demonstrated by blowing air through ether contained in a test tube, the end of which is placed in a little water poured in a watch glass. The cooling produced by the evaporation of the ether quickly freezes the water into a solid block of ice.

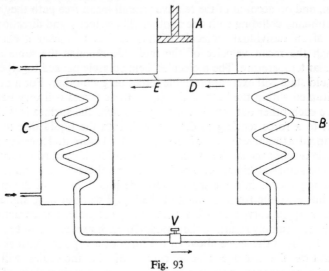

Fig. 93

9.12 Refrigerators

The cooling produced by the rapid evaporation of a liquid is used in refrigerating machines, the usual vapour-compression type being shown diagrammatically in Fig. 93. The working substance or refrigerant as it is called, is a readily liquefiable substance such as ammonia, carbon dioxide, or sulphur dioxide. Ammonia is most commonly used, carbon dioxide requiring much higher pressures. Carbon dioxide however has the advantage that the apparatus can be made more compact which together with its less injurious properties, should there be a leak in the apparatus, results in its adoption as the refrigerant on ships. In modern machines volatile organic compounds such as dichloro-difluoro-methane ($C Cl_2 F_2$), commercially known as "freon," are used.

The action of the refrigerator is as follows. As the piston A descends, valve D closes and the compressed vapour is forced past E into the coil C where it liquefies. The latent heat thus released is removed by a stream of circulating water. The liquefied ammonia (or other substance) now passes through a throttle valve V into the low pressure side of the apparatus where it evaporates in the coil B, abstracting its latent heat of evaporation from the brine bath surrounding *B*. The cooled vapour is now drawn into the compressor as the piston *A* ascends, and the circulating action repeated. The action of the refrigerator is thus to convey heat from one body to another at a higher temperature. The concentrated brine solution which is thus cooled (to a temperature of — 10°C. perhaps) is circulated through pipes to maintain a low temperature in a cold storage room, or can be made to produce ice blocks, as in a freezing machine, by being circulated past tanks containing water.

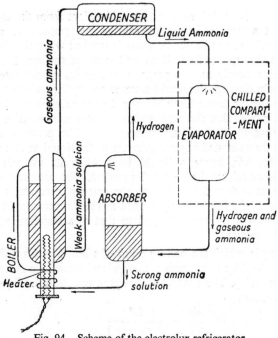

Fig. 94. Scheme of the electrolux refrigerator

The action of the "electrolux" refrigerator, the scheme of which is shown in Fig. 94, is based on the same thermal principle as that described above, the evaporation of the liquid ammonia in this case being effected not by mechanical means, but by the admixture of hydrogen gas which reduces the partial pressure of the ammonia below its saturation value, thereby promoting evaporation. This process is carried out in the evaporator situated in the chilled compartment, and the mixture of hydrogen and gaseous ammonia leaving the evaporator passes to the absorber where it meets a weak ammoniacal solution. The very soluble ammonia gas is at once dissolved, the hydrogen gas rising through the ammonia solution and continuing its circulation to the evaporator again. The strong ammonia solution emerging from the absorber is forced up into the boiler by the action of the heater as shown The same heater also drives off some of the ammonia from the solution, the weakened ammonia solution being returned to the absorber, where it takes up more ammonia issuing from the evaporator. The gaseous ammonia generated in the boiler is passed into a condenser from which the liquid ammonia is delivered again to the evaporator for the process to be continuously maintained.

9.13   Ebullition or Boiling

If a liquid is heated evaporation takes place from the surface and proceeds at a steadily increasing rate as the temperature rises. This process continues until a temperature is reached when the liquid freely evaporates throughout its bulk as well as at the surface. The temperature remains steady at this point, and bubbles of saturated vapour stream freely through the liquid mass to the surface where the vapour escapes to the air. Since these bubbles do not collapse on rising to the surface, it is clear that the pressure inside them must equal the external pressure (usually that of the atmosphere). The pressure inside the bubbles is the saturated vapour pressure of the liquid at that particular temperature, and hence a liquid boils when its saturated vapour pressure is equal to the external pressure. The temperature of the boiling point is thus dependent on the external pressure, the normal boiling point of a liquid being defined as the temperature at which its saturated vapour pressure is equal to that of a standard atmosphere (76cm. of mercury).

The way in which the boiling point of a liquid varies with pressure can be obtained by experimental methods which will be described later (see section 9.18). The Clausius-Clapeyron equation for liquid-vapour equilibrium, *viz.*,

$$\frac{dp}{dT} = \frac{LJ}{T(v_2 - v_1)}$$

where L is the latent heat of vaporisation, T the absolute temperature of the boiling point, v_2, v_1 the specific volumes of the vapour and the liquid respectively at the boiling point, and dp is the change in pressure which results in a change dT in the boiling point, cannot be used directly to obtain a theoretical relationship between the pressure and the temperature of the boiling point as the quantities L, v_2, v_1 are themselves a function of the temperature. The equation is however useful as it enables a value of the latent heat of vaporisation to be calculated at a given temperature if experimental values of v_1, v_2 and $\frac{dp}{dT}$ are available.

Thus for water, using the established data at 99·5 and 100·5°C., $\frac{dp}{dT}$ is 2·72cm. of mercury per °C., *i.e.*, 2·72 × 13·6 × 981 dynes per cm.² per °C. The specific volumes of water and water vapour at 100°C. are 1c.c. and 1,674c.c. respectively, and hence

$$L = \frac{2 \cdot 72 \times 13 \cdot 6 \times 981 \times 373 \times 1,673}{4 \cdot 185 \times 10^7}$$

$$= 541 \text{ cal. per gm. at } 100°C.$$

A value in good agreement with the observed value at this temperature.

9.14 Bumping

When a liquid is heated from below it is noticed that at a certain stage in the heating the bubbles rising from below condense in the somewhat cooler upper layers of the liquid. The "singing" which often precedes boiling is attributed to the collapse of the walls of these bubbles in this way. Another interesting phenomenon which can be observed when an air-free and dust-free liquid is slowly heated is the delaying of the boiling process to a temperature often considerably above that of the normal boiling point. When boiling does commence it is initiated with explosive violence and is followed by alternating periods of quiescence and vigorous "boiling with bumping." The explanation of this is to be found in a consideration of the pressure required to establish a bubble of vapour in a liquid. It can be shown that for a liquid of surface tension T, the pressure inside a spherical bubble of radius r must exceed that in the liquid by an amount equal to $\dfrac{2T}{r}$ if the bubble is to be in equilibrium. It is thus seen that in order to establish such a bubble of vapour the vapour pressure of the liquid must be considerably in excess of the external pressure. In the absence of air bubbles or dust particles by means of which the initial stages of bubble formation may be overcome (see also section 9.27), superheating of the liquid results, and when the bubbles of vapour are eventually formed, the pressure inside them is considerably larger than that in the liquid. Accordingly the bubbles expand at a rapid rate producing violent agitation of the liquid. To prevent "bumping" it is sufficient to put a few pieces of broken porcelain or some sand in the liquid to provide surfaces over which bubble formation can be facilitated and thus bring about tranquil boiling.

9.15 Latent Heat of Vaporisation and Boiling Point

A consideration of evaporation from the standpoint of the kinetic theory (section 9.11) reveals that evaporation entails a fall in temperature of the liquid. Thus if the temperature of the liquid is to be maintained during evaporation, as at the boiling point, it is clear that energy must be supplied to the liquid. The heat so supplied at the boiling point forms the larger part of the latent heat of vaporisation, and represents the energy required to overcome the forces of molecular attraction in the liquid. This quantity is often referred to as the internal latent heat which can be represented by L_i. Measurements of latent heats are generally carried out at constant pressure (the vapour pressure of the system), and accordingly an additional amount of heat equivalent to the external work done in establishing the vapour, with specific volume v_2, from the specific volume v_1 in the liquid state, against the external pressure p, is required. The total latent heat L is thus given by

$$L = L_i + \frac{p(v_2 - v_1)}{J} \quad \ldots \quad \ldots \quad [90]$$

By neglecting v_1 compared with v_2, the external work becomes pv_2, and if we assume the vapour obeys the ideal gas laws, we may re-write equation 90 in the approximate form,

$$L = L_i + \frac{R'T}{J}$$

R' being the gas constant per gram.

It is clear from the foregoing discussion that L_i and hence L will vary with temperature. Regnault was the first to investigate experimentally the variation of latent heat with temperature. He measured the quantity of heat Q necessary to raise the temperature of one gram of water from 0°C. to the temperature t°C. of the boiling point (varied by adjusting the external pressure) and to convert it into steam at this temperature. His results may be stated thus,

$$Q = 606 \cdot 5 + 0 \cdot 305t$$

Now $Q = t + L_t$, where L_t is the latent heat at t°C.
Hence $L_t = 606 \cdot 5 - 0 \cdot 695t$,
showing that the latent heat decreases with temperature.

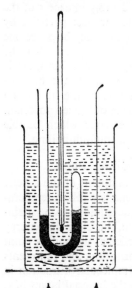

Fig. 95

Regnault's equation takes no account of the variation of the specific heat of water with temperature, and gives values for L which are too large at low temperatures. The equation has been improved by Henning who gave the empirical relation:

$$L_t = 599 \cdot 1 - 0 \cdot 60t$$

9.16 Determination of Boiling Points

We have already seen (section 9.13) that a liquid boils when the pressure of the saturated vapour is equal to that of its surroundings. This forms the basis of a simple laboratory method by which the boiling points of liquids may be readily obtained. The apparatus (Fig. 95) consists of a J-tube containing mercury, the shorter limb being closed whilst the other limb is open to the atmosphere. A few drops of the liquid whose boiling point is required are placed above the surface of the mercury in the short limb, care being taken to shake out all the air from the space occupied by the liquid. The J-tube is then placed with a stirrer and thermometer in a beaker of water (for temperatures up to 100°C.) or oil (for temperatures above 100°C.) which is steadily heated. The temperature is read when the pressure of the vapour arising from the liquid is sufficient to depress the mercury in the shorter limb until its surface is level with that in the longer limb. This temperature, at which the saturated vapour pressure of the liquid is equal to the atmosphere, gives the boiling point of the liquid corresponding to the atmospheric pressure at the time of the experiment. By continuing the heating process so that the level of the mercury in the closed limb sinks below that in the open limb, and taking the temperature reading when the levels are again the same as the beaker is subsequently allowed to cool, the boiling point may be taken as the mean of the two observed temperatures.

The temperatures of the boiling points of a liquid at pressures other than that of the atmosphere may be obtained from the experiments on the variation of saturated vapour pressure with temperature described in section 9.18. The saturated vapour pressure is equal to the pressure of the surroundings when the liquid boils, and hence the values of saturated vapour pressure against temperature also give the values of the boiling point at corresponding values of the external pressure.

TABLE 23

Boiling points in °C. of certain liquids at 76cm. of mercury

Liquid	B.P. °C.	Liquid	B.P. °C.
Ether	34·5	Benzene	80·2
Carbon disulphide............	46·2	Turpentine	159
Chloroform.....................	61·2	Aniline.........................	184
Methyl alcohol	64·7	Paraffin oil...................	280
Ethyl alcohol..................	78·3	Glycerine......................	290

The boiling point of a solution is not the same as that of the pure solvent. The effect of dissolving a substance in a liquid is to cause an elevation of the boiling point which is approximately proportional to the concentration of the solute. Thus the boiling point of an aqueous solution of common salt shows a steady rise up to about 108°C. at saturation. Raoult's law (see section 9.08) $\Delta T = KC$ which was found to apply to the depression of the freezing point, also holds good for the elevation of the boiling point for aqueous solutions of non-electrolytes and solutions of organic compounds. Here ΔT is the elevation of the boiling point, C being the concentration in gram-molecules per 100gm. of solvent. For aqueous solutions K has the value of 5·2°C.

9.17   The Measurement of Saturated Vapour Pressure

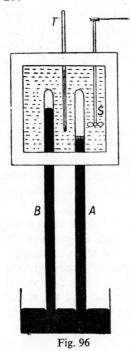

Fig. 96

Statical Method.—The saturated vapour pressure of a liquid may be determined at temperatures which are well below its boiling point by measuring the depression of the mercury column when some of the liquid is introduced in the Torricellian space of a mercurial barometer. The liquid is introduced at the lower end of the barometer tube using a small pipette bent up at the end. As the first drops of liquid rise into the space above the mercury column, they vaporise and the mercury is forced down. This continues as more drops are introduced until a stage is reached when no further evaporation takes place and the mercury is not further depressed. The surplus liquid remains on the surface of the mercury, and is in equilibrium with the saturated vapour in the space above the mercury. The depression of the column at this stage is the saturated vapour pressure of the liquid at the temperature of the experiment.

The apparatus used by Regnault in determining the saturated vapour pressure of water between 0 and 50°C. by this means is shown in Fig. 96. Sufficient water was introduced into a barometer A to saturate the space above the mercury, and the depression of the mercury column relative to that of another barometer B with an unvitiated vacuum was read off by means of a cathetometer. The upper parts of both barometers were surrounded by a water bath with a plate-glass front. The bath was heated from below and was provided with a stirrer S to ensure a uniform temperature which was recorded by the thermometer T. A correction was made for the pressure of the liquid layer above the mercury surface in B by subtracting from the observed depression the amount $h\dfrac{\rho_1}{\rho_2}$ where h is the height of the liquid layer and ρ_1, ρ_2 are the densities of the liquid and of the mercury respectively at the given temperature. Further corrections were made for the differential capillary depression arising from the different inter-facial surface tensions of the mercury in the two tubes (mercury-water in A, and mercury-vacuum in B), and for the fact that the mercury columns were not at 0°C. (see section 4.11).

To determine the values of the saturated vapour pressure at temperatures below 0°C., Regnault used the modified form of apparatus shown in Fig. 97. The barometer A had a bulb C containing the water attached as shown. This bulb was immersed in a freezing mixture of snow and calcium chloride, the temperature of which was recorded by the thermometer T. The saturated vapour pressure was obtained as before by measuring the depression of the mercury in A relative to that in B.

9.18 Measurement of Saturated Vapour Pressure

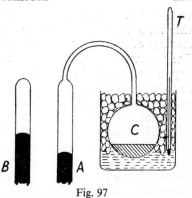

Fig. 97

Dynamical Method.—This method depends on the fact already discussed (section 9.13) that a liquid boils when its saturated vapour pressure is equal to the pressure of the surrounding atmosphere. By varying the pressure of the vapour space above a liquid and noting the temperature at which boiling takes place, a set of values of saturated vapour pressure against temperature can be obtained over a very wide range. This method was first applied by Regnault for values of the saturated vapour pressure of water between 50 and 100°C. He found that for these temperatures the statical method was not reliable as the water bath required became too long to ensure a uniformity of temperature. Regnault's apparatus is shown in Fig. 98. The water was heated in

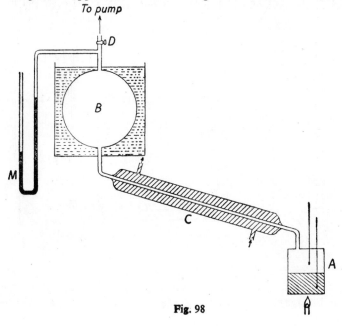

Fig. 98

an air-tight copper vessel A furnished with four thermometers (two only shown), two being placed in the water and two in the vapour. Identical readings of the thermometers indicated the absence of delayed boiling. The thermometers were protected from the high pressures inside the apparatus by containing them in iron tubules closed at the ends and containing mercury to ensure good thermal contact. The copper vessel was connected via a Liebig's condenser C to a large globe B which contained air and was surrounded by a constant temperature water-bath. The function of the large air reservoir B was to smooth out fluctuations of pressure occasioned by the boiling liquid in A. The quantity of air in B was adjusted by a pump connected at D, the pressure being recorded by the manometer M. This pressure gave the saturated vapour pressure of the liquid at the temperature indicated by the thermometers in A when the liquid was boiling steadily. In this way, by adjusting the pressure of the air in B, a range of values of the saturated vapour pressure against temperature was obtained. For temperatures below 100°C., the pressure in B was reduced below one atmosphere, whilst for values of the saturated vapour pressure at temperatures higher than 100°C. air was forced into B to raise the pressure above that of the atmosphere.

Regnault's method has been modified by Ramsay and Young who used the apparatus shown in Fig. 99 to measure the saturated vapour pressures of liquids at temperatures well below their normal boiling points. The liquid under investigation is contained in a dropping funnel F, the end of which is drawn out so that drops of the liquid

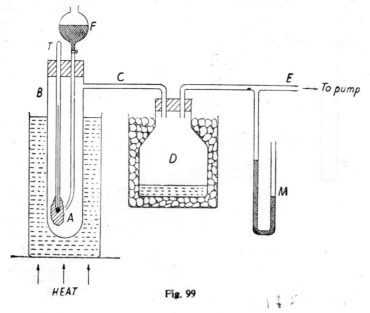

Fig. 99

can fall on to a wrapping of cotton wool *A* surrounding the bulb of a thermometer *T* contained in a boiling tube *B*. A side tube *C* connects *B* to a bottle *D* surrounded by ice, and *D* is in turn connected to a pump at *E*, the pressure inside the apparatus being obtained by subtracting the readings of the manometer *M* from the pressure of the atmosphere. An oil bath surrounds the tube *B* which is always kept at a temperature about 20°C. above that recorded by the thermometer *T*. In performing an experiment the pressure inside the apparatus is reduced by withdrawing air at *E* and some of the liquid in *F* is allowed to flow on to the cotton wool at *A* until it is thoroughly moistened. The liquid on the cotton wool rapidly evaporates, the vapour displacing the air in the lower part of the tube *B* until a stage is reached when the liquid on *A* is in equilibrium with its own vapour. This is the true boiling point of the liquid, and when the conditions are steady, the pressure and the temperature are read. The evaporation takes place perfectly freely, and thus the vapour cannot be superheated, and of course free boiling is impossible. The vapour diffusing into *D* is condensed there and the liquid thus recovered. By successively admitting small quantities of air to steadily raise the pressure, the corresponding temperature of the boiling point can be quickly read off as the liquid attains equilibrium with its vapour under the new conditions. The method gives very satisfactory values for pressures up to about 50cm. of mercury, but for higher pressures the free boiling method already described is more suitable.

The saturated vapour pressures of water at different temperatures are given in table 24. It will be seen that there is a steady rise with temperature up to about 50°C., after which the saturated vapour pressure shows a rapid increase with temperature. This is shown graphically in Fig. 100.

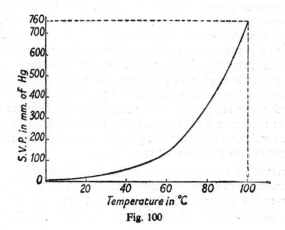

Fig. 100

TABLE 24

Saturated vapour pressure of water in mm. of mercury at
different temperatures

Temp. °C.	S.V.P.	Temp. °C.	S.V.P.	Temp. °C.	S.V.P.
−20	0·78	12	10·51	42	61·30
−10	1·96	14	11·98	44	68·05
− 5	3·02	16	13·62	46	75·43
− 2	3·89	18	15·46	48	83·50
0	4·58	20	17·51	50	92·30
1	4·92	22	18·62	55	117·6
2	5·29	24	22·32	60	149·2
3	5·68	26	25·13	65	187·2
4	6·10	28	28·25	70	233·5
5	6·54	30	31·71	75	288·8
6	7·01	32	35·53	80	355·1
7	7·51	34	39·75	85	433·3
8	8·04	36	44·40	90	525·8
9	8·61	38	49·51	95	633·8
10	9·21	40	55·13	100	760·0

9.19 The Triple Point

The curve showing the saturated vapour pressure of a liquid against temperature, such as that given in Fig. 100 for water, is the locus of the values of pressure and temperature at which the liquid and its vapour may exist in equilibrium, and is known as the vaporisation line of the substance. The line represents the demarcation of the liquid and vapour phases—the substance being liquid at all pressures above the line for each temperature, whilst below it there exists only vapour. Similarly pressure-temperature equilibrium curves can be drawn for the solid-vapour phases (sublimation line), and for the solid-liquid phases (melting line). This latter line represents the variation of the melting point with pressure. It has already been seen (section 9.05) that this variation is very small and consequently the melting line is very nearly vertical, sloping slightly to the left for water-type substances (melting point lowered with pressure), and to the right for wax-type substances (melting point increased with pressure). If the three curves mentioned above are plotted on the same diagram, they are found to intersect in a common point known as the *triple point*. The conditions of pressure and temperature at this point are such that the solid, liquid, and vapour phases may co-exist in equilibrium.

The curves for water are shown in Fig. 101, O being the triple point with co-ordinates of 4·58mm. pressure and 0·0075°C. temperature. The three curves OA (steam line), OB (hoar frost line), and OC (ice line) represent the boundaries of the three phases of the substance; in the space *AOB* it is entirely vapour, in *AOC* entirely liquid, and in *BOC* entirely solid.

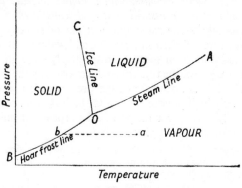

Fig. 101

From Fig. 101 it will be seen that if the temperature of a vapour at a pressure below that of the triple point is sufficiently reduced, direct condensation to the solid form will result. This is represented by the line *ab* on the diagram. Thus a sudden drop in temperature when the partial pressure of the water vapour in the atmosphere is less than 4·58 mm. results in the formation of hoar frost at ground level or snow in higher levels of the atmosphere. Conversely, a solid can be converted directly into vapour (sublimation) with rising temperature if the vapour pressure does not exceed the pressure at the triple point. In the case of substances with high triple points such a change can be effected very readily with steady heating. This accounts for the ease with which iodine (with triple point at 114°C. and 90mm. pressure) sublimes, and also for the condensation of the iodine vapour to solid iodine on a cool surface. Benzoic acid behaves similarly and may be purified by sublimation and subsequent condensation. On heating iodine in a closed space however the partial pressure of the iodine vapour increases and the substance will melt at 114°C. when the vapour pressure is 90mm. Further heating causes the liquid to boil at a temperature 184°C. when the vapour pressure is 760mm. of mercury. On the other hand if the triple point pressure of a substance is greater than 760mm. it cannot have a normal boiling point, *e.g.*, carbon dioxide with triple point at − 56·4°C. and 5·11 atmospheres.

9.20 Dalton's Law of Partial Pressures—Mass of a Volume of Moist Air

If, in the experiment described in section 9.17 to determine the saturated vapour pressure of a liquid by the depression of the mercury in a barometer tube, the space above the mercury had contained a small quantity of air, little, if any, difference to the final result for the saturated vapour pressure would have been observed. This is in accordance with Dalton's law of partial pressures which states that *the total pressure of a mixture of gases and/or vapours, which do not interact chemically, is equal to the sum of the pressures that each constituent would exert if contained separately in the same space at the same temperature.* Thus if three separate litres of nitrogen, oxygen, and water vapour at pressures of 600, 150, and 15mm. of mercury respectively were all contained in the same one litre space, the total pressure exerted by the mixture would be 765mm. The law is not true at high pressures (it is not possible to obtain indefinitely high pressures by mixing a large number of vapours), but is sufficiently correct for all practical purposes at pressures in the neighbourhood of one atmosphere.

As an illustration of the use of the law the following example may be considered, in which the mass of 1 litre of moist air at 20°C. and 770mm. pressure is calculated, using the following data:—

(i) density of dry air at 0°C. and 760mm. pressure is 1·293gm. per litre.

(ii) density of water vapour is five-eighths that of dry air under the same conditions.

At 20°C. the S.V.P. of water vapour is 17·5mm. (Table 24).

Thus by Dalton's law the partial pressure of the water vapour and the dry air in the mixture are respectively 17·5mm. and 752·5mm.

The density of dry air at 20°C. and 752·5mm. pressure is

$$1·293 \times \frac{752·5}{760} \times \frac{273}{293} = 1·193\text{gm. per litre} \quad \ldots \ldots \quad (a)$$

and at 20°C. and 17·5mm. the density of dry air is

$$1·293 \times \frac{17·5}{760} \times \frac{273}{293} = 0·0277\text{gm. per litre.}$$

Hence, since the density of water vapour is five-eighths that of dry air under the same conditions, the density of the water vapour at 20°C. and 17·5mm. pressure is

$$\frac{5}{8} \times 0·0277 = 0·0173\text{gm. per litre} \quad \ldots \ldots \ldots \ldots \quad (b)$$

Thus from (a) and (b) the total mass of 1 litre of moist air under the conditions stated is

$$1·193 + 0·0173 = 1·210\text{gm.}$$

9.21   Water Vapour in the Atmosphere—Hygrometry

Evaporation is constantly taking place from exposed water surfaces —from the sea, rivers, lakes, &c., and in consequence the air must always contain some water vapour as one of its constituents. The exudation of water vapour by trees and plants also serves to add to the moisture content of the atmosphere, and this process of steady accumulation tends towards a state of saturation. Complete saturation may result in consequence of some cooling process when part of the water vapour is condensed out in the form of mist, fog, or dew, or may be precipitated as rain. Further details of these meteorological phenomena will be discussed in chapter XIV.

A common effect due to the presence of moisture in the atmosphere is the feeling of "mugginess"* in a hot overcrowded room or on a sultry day. This is due to the large amount of water vapour present in the atmosphere which retards the normal evaporation rate of moisture from the human body by which means the body temperature is regulated. It is customary to say that the air on such occasions is "humid," although it should be noticed that this "humidity" of the air does not solely depend on the actual amount of water vapour present in it. This will be clear when it is remembered that at midday on a fine summer's day the atmosphere contains far more moisture than in the early morning—the dew and ground mists having been evaporated into the atmosphere as the temperature rises in the course of the day, yet the air at midday feels considerably less "damp" than in the early morning. In the latter case the air is saturated with water vapour (as proved by the presence of dew), whereas later in the day, in spite of the increase in the total quantity of moisture in the air, it is not even nearly sufficient to saturate it at the higher temperatures prevailing. It is this nearness to saturation at a given air temperature which determines the feeling of "dampness" of the air, and is referred to as the *relative humidity*. This is defined as

$$\frac{\text{The quantity of water vapour present in a given volume of air.}}{\text{The amount of water vapour required to saturate the air at the given temperature.}}$$

Assuming Boyle's law to apply to the water vapour, this may be written as

$$\frac{\text{Pressure of water vapour present in the air.}}{\text{Pressure of water vapour required to saturate the air at the given temperature.}}$$

A knowledge of the relative humidity is of importance not only to meteorologists, but also for many other purposes. Thus the air conditioning of buildings requires not only a regulation of the temperature of the air, but also of its moisture content. This latter is particularly important in cotton mills where the humidity must be maintained above a certain level to facilitate the spinning process, but must not rise too high as otherwise conditions become injurious to health. Examples of other cases where a knowledge of the relative humidity is important are in the artificial seasoning of timber and in the cold storage rooms of ships, &c.

The practical measurement of the relative humidity of the atmosphere is carried out by instruments called *hygrometers,* the main types being (a) those depending on the measurement of the dew point,

---

* Mugginess—the body's reaction to humidity—is affected by air movement. Thus, a high relative humidity with moving air does not produce discomfort whereas a lower humidity in still air gives the feeling of mugginess.

(b) wet and dry bulb hygrometers, (c) the chemical or gravimetric hygrometer, and (d) hair hygrometers. A description of these instruments will be given in the following sections.

9.22   Dew Point Hygrometers

If the air is slowly cooled at constant pressure, the partial pressure of the water vapour present will remain constant, assuming it to obey the gas laws. Thus as the temperature is lowered sufficiently for the water vapour present in it to saturate the air, condensation will take place, the saturated vapour pressure at this temperature being equal to the pressure of the water vapour at the original air temperature. The temperature at which the water vapour present in the air will just saturate it is called the *dew point,* and we may restate our definition of relative humidity as

$$\frac{\text{Saturated vapour pressure at dew point.}}{\text{Saturated vapour pressure at the air temperature.}}$$

Thus a determination of the air temperature and the temperature of the dew point enables a value of the relative humidity to be obtained with the aid of vapour pressure tables.

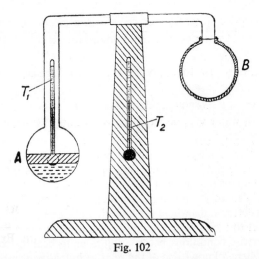

Fig. 102

(a) *Daniell's Hygrometer.*—This is one of the earliest forms of hygrometer and is shown in Fig. 102. It consists of two glass bulbs A and B connected as shown and supported on a stand. A is surrounded by a gold band and contains some ether which is in equilibrium with its vapour throughout the apparatus. A muslin bag is placed round B and saturated with ether the evaporation of which produces local cooling and causes the ether vapour inside B to condense. This disturbs the pressure equilibrium inside the apparatus and causes the ether in A to evaporate and thus cool the glass bulb and the air in contact with it. This continues until the dew point is reached when a film of moisture is deposited on the gold band. The temperature at which this occurs is taken by the thermometer T_1. The instrument is now left to itself and the temperature recorded by T_1 again taken when all traces of the dew have disappeared. The mean of these two readings is then taken as the dew point. The air temperature is obtained from a second thermometer T_2 mounted on the stand of the instrument, the relative humidity of the air being obtained from these two temperatures as indicated in the previous section.

Daniell's hygrometer is not however a very accurate instrument. Amongst the objections to it may be mentioned the errors introduced by the thickness of the glass which being a poor conductor of heat will have an appreciable temperature difference between its inside and outside surfaces. Further, the liquid ether in A is not stirred and thus there may be considerable temperature variation within it, and in addition the presence of ether vapour round B disturbs the hygrometric state of the air. On account of these objections this hygrometer has now been almost completely superseded by Regnault's hygrometer described below.

(b) *Regnault's Hygrometer.*—Most of the objections to Daniell's hygrometer are avoided in the hygrometer designed by Regnault (Fig. 103). Two glass tubes A and B mounted as shown have silver thimbles attached to their lower ends. The tube A contains some ether through which a stream of air is drawn by an aspirator placed some distance from the apparatus to avoid disturbing the hygrometric state of the air near it. The evaporation of the ether produces a steady cooling and the temperature at which a deposit of dew forms on the surface of the silver thimble is read off on the thermometer T_1. The current of air is now reduced so that the ether in A is just kept disturbed. The cooling now produced is insufficient to maintain the temperature much below the air temperature, and as the apparatus slowly warms up the temperature at which the dew disappears is observed. The mean of the two readings of T_1 gives the temperature of the dew point. The second tube B is provided for comparison purposes, the thermometer T_2 giving the air temperature. To prevent the breath of the observer vitiating the readings, observations are taken through a glass sheet placed in front of the apparatus or by means of a telescope situated some distance away.

The special advantages of Regnault's method are that the cooling and heating rates are under the complete control of the experimenter, and thus the temperatures of appearance and disappearance of the dew can almost be made to coincide. Uniformity of temperature is secured by the use of the thin silver caps and by keeping the ether continually stirred as described. Griffiths has further improved this

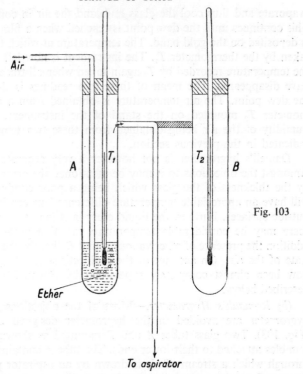

Fig. 103

form of hygrometer to ensure that observations are taken under still air conditions. This was done by enclosing the lower half of the tube A in a cubical box, the lower half of which was hinged so that after being opened and allowed to oscillate a few times, a fair sample of air could be trapped in the box, the observations of the silver thimble being taken through a plate glass window.

9.23 Wet and Dry Bulb Hygrometers

The wet and dry bulb hygrometer, sometimes referred to as Mason's hygrometer, is simple and convenient, and has accordingly found widespread use in industry and for meteorological observations. Two thermometers are placed side by side on a stand (Fig. 104) the bulb of one being exposed to the air whilst the bulb of the other is kept moist by surrounding it with a piece of muslin the lower end of which dips into a small vessel containing water. The evaporation which takes place from the wet bulb will cause its temperature to fall below that of the air temperature, the extent of the fall depending on the rate of evaporation which in turn will depend on the relative humidity of the air. Thus if the air is near to saturation, the temperature difference between the two thermometers will be small whereas for relatively dry air conditions the difference will be large. In order to obtain the relative humidity from the observed readings empirical tables have been compiled showing the relative humidity corresponding to a series of values of the temperature difference recorded for various values of the air temperature. An alternative method is to use an empirical formula of the form

$$p_w - p = AP(t - t_w) \quad . \quad . \quad [91]$$

Where p_w is the saturated vapour pressure at the wet bulb reading t_w, p the actual vapour pressure at the air temperature t, P the atmospheric pressure, and A a numerical factor. The value of p is calculated from the formula, the relative humidity then being obtained by dividing p by the saturated vapour pressure at the air temperature t.

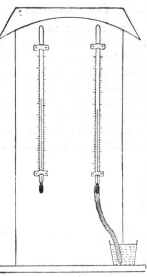

Fig. 104

The value of the numerical factor A depends on the speed of the air flowing past the bulb, and because of this complication the wet and dry hygrometer is regarded as an unreliable instrument. It has been found however that consistent readings can be obtained if the speed of the air exceeds a value of about 3 metres per second—the factor A then assuming a constant value. This led to the introduction of the **ventilated wet and dry bulb hygrometer** or **psychrometer**. One form of such instrument is the sling psychrometer in which the two thermometers and water reservoir are mounted on a frame pivoted about a handle by means of which it can be rapidly whirled through the air—the readings of the thermometers being quickly taken after whirling the instrument for periods of up to one minute. A more widely used form is the Assmann or tubular psychrometer in which the two thermometers are placed inside a metal tube through which a stream of air of the requisite velocity is drawn by means of a clockwork fan. Such an instrument has the advantage of functioning at a distance from the observer, and is capable of giving consistent results of considerable accuracy.

These instruments are in common use in America and some other countries, but have not as yet found wide scale adoption here where the standard wet and dry bulb instrument is of the form shown in Fig. 104 with the bulbs exposed in a Stevenson screen. The ventilating louvres of the screen produce a draught of approximately 1 metre per second and the appropriate value for A is inserted in equation 91 from which the relative humidity values are calculated.

9.24   The Chemical or Gravimetric Hygrometer

In the chemical or gravimetric hygrometer shown in Fig. 105 a measured volume of air is drawn through drying tubes the increase in weight of which gives the mass of water vapour present in the volume of air, thereby enabling the relative humidity to be determined. Although the method leaves nothing to be desired from the point of view of accuracy, it is extremely tedious and is never used in routine work being mainly employed for standardisation purposes. Further, it should be noted that this hygrometer gives a mean value of the relative humidity over the period of the experiment.

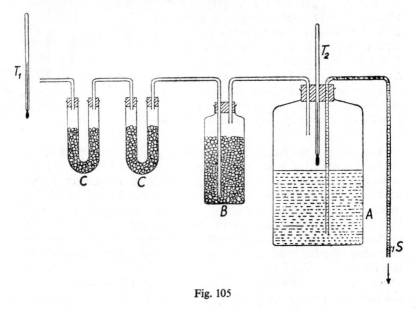

Fig. 105

The apparatus consists of two drying tubes CC containing pieces of pumice stone soaked in concentrated sulphuric acid, connected through a vapour trap B, containing the same drying agent, to an aspirator A. All the corks are thoroughly sealed by means of wax to prevent leaks, and air is drawn through the apparatus by syphoning the water from A, the rate of flow being adjusted by the stop tap S to obtain an airflow rate somewhat less than 1 litre per minute. As already mentioned, the function of the bottle B is to prevent the water vapour in A from passing to the drying tubes CC, the increase in weight of which is determined after a measured volume of water has been drawn from A. The temperature of the air as it enters the apparatus is taken by the thermometer T_1 and a second thermometer T_2 is used to record the temperature of the saturated air in the aspirator.

The method of determining the relative humidity is as follows:—

Let the readings of T_1 and T_2 be $t_1°$C. and $t_2°$C. respectively and let w be the increase in weight of the drying tubes after V litres of air are drawn into A. If then the atmospheric pressure is H (in mm. of mercury), and the S.V.P. at $t_2°$C. is p mm., it is first necessary to find what the volume V of air at a partial pressure of $(H-p)$ mm. and at a temperature of $t_2°$C. corresponds to before entering the apparatus where the temperature is $t_1°$C. and its partial pressure is $(H-f)$ mm., f being the pressure of the aqueous vapour in the air surrounding the apparatus. If this volume is V' we have,

$$V' = V \times \frac{H-p}{H-f} \times \frac{273 + t_1}{273 + t_2} \text{ litres} \quad . \quad . \quad . \quad . \quad [92]$$

Now following the method given in section 9.20 the weight of the water vapour in this volume of moist air is

$$\frac{5}{8} \times 1{\cdot}293 \times \frac{f}{760} \times \frac{273}{273 + t_1} \times V' \text{ gm.}$$

This is the amount measured (w) and thus, substituting for V' from equation 92 above we have

$$w = \frac{5}{8} \times 1{\cdot}293 \times \frac{f}{760} \times \frac{273}{273 + t_2} \times \frac{H-p}{H-f} \times V \text{ gm.}$$

from which f, the vapour pressure of the water vapour actually present in the air, can be calculated thereby enabling a value for the relative humidity to be obtained.

It should further be noticed that having obtained the value of f, the volume V' can be found from equation 92 and hence the mass of water vapour present per litre of air calculated. This latter quantity is known as the *absolute humidity* of the air.

9.25   The Hair Hygrometer

It has been found that human hair has the property of increasing its length slightly as the humidity of the air rises, and hygrometers based on this property have been designed in which the small extension of the hair is suitably magnified and related to the movement of a pointer over a scale calibrated to give direct readings of the relative humidity. The hair should first be freed from all grease as this would otherwise interfere with the absorption of moisture and render the readings of the instrument unreliable. This is done by thoroughly washing it in caustic soda solution before insertion in the hygrometer after which it should not be touched by hand.

The first such instrument was invented by de Saussure in 1783, but the arrangement was somewhat crude and the readings not very consistent. The hair hygrometer has been subsequently improved and Griffiths has found that reasonably accurate readings can be obtained using such instruments provided that the hair is not subject to too great a tension. The hygrometer should also undergo frequent checks as the hair tends to acquire a permanent set under maintained tension which causes the instrument to over-read.

The special advantages of the hair hygrometer are that it can be calibrated to give direct readings, it can be made small and compact, it gives continuous readings of the relative humidity, and further that it can be used for temperatures below 0°C. where the usual wet and dry bulb hygrometer presents special difficulties. Thus it is particularly useful in cold countries and for the measurement of humidity conditions in cold storage rooms.

9.26 Determination of Vapour Density

(a) *Unsaturated Vapours.*—The determination of vapour density is of considerable importance in physical chemistry as it enables a value for the molecular weight of the substance to be obtained. As used in this connection it is usual to consider the density of the vapour relative to hydrogen, and thus the vapour density will be

$$\frac{\text{Weight of a given volume of the vapour.}}{\text{Weight of an equal volume of hydrogen at the same temperature and pressure.}}$$

Now by Avagadro's hypothesis (section 8.09) equal volume of gases under the same conditions of temperature and pressure contain equal numbers of molecules. Hence if the given volumes of vapour and hydrogen contain n molecules we have,

$$\text{Vapour density} = \frac{\text{Weight of } n \text{ molecules of the substance.}}{\text{Weight of } n \text{ molecules of hydrogen.}}$$

$$= \frac{\text{Weight of 1 molecule of the substance.}}{\text{Weight of 1 molecule of hydrogen.}}$$

$$= \frac{\text{Molecular weight of the substance.}}{2}$$

since the molecular weight of hydrogen is 2.

Hence, Molecular weight $= 2 \times$ Vapour density.

A simple method of determining the vapour density which is applicable over a wide range of temperature is that devised by Victor Meyer. The method consists of rapidly evaporating a small mass of the liquid substance and finding the volume of the vapour so formed by measuring the volume of air displaced using the apparatus shown in Fig. 106. The weighed liquid is contained in a small stoppered bottle B, suspended as shown from a wire passing through the stopper closing a long tube A with an elongated bulb at the lower end. A is enclosed by a wider tube C containing a liquid whose boiling point is considerably

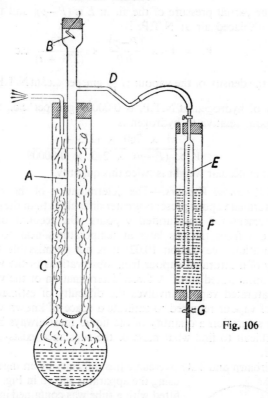

Fig. 106

higher than that of the liquid under investigation. When the liquid in C has been steadily boiling for some time, the bottle B is dropped to the bottom of A (which contains a little asbestos wool to break the fall) by turning the wire support through 180°. The liquid quickly evaporates, and before mixing can take place, the vapour displaces an equal volume of air via the side tube D to the gas burette E which is jacketed by a tube F containing water. The water is accurately adjusted to the same level in E and F by running a little off at G, and the volume of air displaced is read off at atmospheric pressure against the scale on E. The air in E is at the temperature of the water in the jacket F, and it will be noticed in the calculation below that it is not necessary to know the temperature at which vaporisation occurs. The calculation is carried out as follows.

Let m be the mass of the liquid used.

    „ V „ „ volume of air displaced.

    „ t „ „ temperature of the water in F.

    „ p „ „ S.V.P. of water vapour at $t°C$. in mm. of mercury.

    „ P „ „ atmospheric pressure.

8*

Then the partial pressure of the air in E is $(P-p)$, and the volume V of the displaced air at N.T.P. is

$$V_o = V \times \frac{(P-p)}{760} \times \frac{273}{(273+t)} \text{ c.c.}$$

Hence the density of the vapour is $\frac{m}{V_o}$ gm. per c.c. at N.T.P. or, since the density of hydrogen at N.T.P. is 0·00009 gm. per c.c., the density of the vapour relative to hydrogen is

$$\frac{m \times 760 \times (273+t)}{V \times (P-p) \times 273 \times 0·00009}$$

and the molecular weight is twice this quantity.

(*b*) *Saturated Vapours.*—The determination of the vapour density of a saturated vapour presents greater difficulties than the corresponding measurement for unsaturated vapours. On account of the marked deviations from the gas laws at temperatures near to the point of condensation (see section 10.02) it is not permissable to deduce the density of a saturated vapour from observations of the vapour density in the unsaturated state. A direct determination of the vapour density of a saturated vapour involves the difficulty of estimating the exact mass of vapour required to saturate a space of known volume, for to ensure saturation a quantity of the liquid must always be present and it is difficult to find what mass of the substance exists in the form of vapour.

Fairbairn and Tate overcame this difficulty by an ingenious method using the apparatus shown in Fig. 107. A bulb A fitted with a tube was contained in a larger vessel B communication between the two vessels being prevented by mercury as shown. A known amount of the liquid under investigation was introduced into A after it had been exhausted of air, and a larger quantity of the liquid was placed in B which had also been evacuated. Thus both A and B contained only vapour in equilibrium with its own liquid, and the pressures in the two vessels were the same. The mercury level in A stood slightly higher than that in B due to the greater depth of the liquid in B, and these levels were observed by a cathetometer through the glass stem of B as the temperature of the apparatus was steadily raised. The pressures inside both vessels increased equally with rise in temperature so long as sufficient liquid remained in A to keep the space saturated. When all the liquid in A had evaporated, further

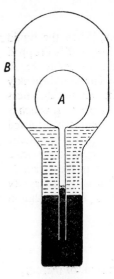

Fig. 107

heating produced a slower rate of pressure increase in the now unsaturated vapour in *A* than for the vapour in *B* which still remained saturated. The exact moment of unbalance was shown by a sudden rise of the mercury in *A*, and the temperature was taken when this movement occurred. At this temperature the amount of liquid introduced into *A* just saturated the known volume of the bulb and stem, and accordingly the density of the saturated vapour at the particular temperature could be obtained. The results of Fairbairn and Tate's experiments showed that saturated and nearly saturated vapours do not obey the gas laws, the density of the saturated vapour being much greater than that deduced from the gas equations.

9.27   Vapour Pressure of a Curved Surface

When a saturated vapour is in equilibrium with its liquid, the pressure of the vapour is found to depend on the curvature of the liquid surface. Lord Kelvin in 1870 was the first to prove that the vapour pressure over a curved surface cannot be the same as that over a plane surface of the liquid, and his calculations showed that there is a reduction in the vapour pressure over a concave surface as compared with that over a plane surface, whilst there was a corresponding increase in the vapour pressure over a convex surface. From this it follows that in a space containing a saturated vapour, condensation will take place most readily on any surface which is concave to the vapour, next on a plane surface, and least, or not at all, on a convex surface. This explains why porous bodies so readily become "damp" in a moist atmosphere, condensation being promoted in the fine capillaries permeating them. Thus the use of charcoal in chemical reactions for absorbing vapours, and the readiness of cotton fabrics to take up moisture from the air. Cotton fibres are very brittle when dry and to facilitate the processing of such materials, the atmosphere of cotton mills is artificially humidified (unless climatic conditions make this unnecessary as in Lancashire where the cotton industry is mainly concentrated)

As a result of the increase in vapour pressure over a convex surface a considerable degree of supersaturation is often necessary before condensation of a vapour is initiated. A small drop cannot remain in equilibrium with its vapour under conditions of normal saturation and will immediately evaporate—if larger drops are present condensation may take place on these, and thus the larger drops grow at the expense of the smaller ones. Further consideration of the process of drop formation will be deferred until after the magnitude of the change of vapour pressure with curvature has been established.

Imagine a capillary tube placed in a liquid contained in a dish inside a vessel from which the air has been exhausted (Fig. 108). When equilibrium is established the whole space will be filled with vapour

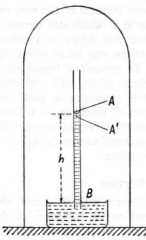

Fig. 108

only whose density is σ (say) at the given temperature (assumed constant throughout the apparatus). Let the level of the liquid in the capillary tube stand at a height h above the plane surface of the liquid in the dish. If p_0 is the vapour pressure at the point B immediately above the plane surface of the liquid, and p the vapour pressure at the point A immediately above the curved surface (of radius r) in the capillary tube, we have

$p_0 - p$ = pressure due to column h of vapour of density σ

$$= g\sigma h \qquad . \quad . \quad . \quad [93]$$

Again, if p' is the pressure at the point A' immediately below the curved surface in the capillary tube, we have

$p_0 - p'$ = pressure due to column h of liquid of density ρ

$$= g\rho h \qquad . \quad . \quad . \quad . \quad . \quad . \quad . \quad . \quad . \quad [94]$$

Now, due to the surface tension T, the pressure at A' is less than that at A by an amount equal to $\dfrac{2T}{r}$, that is

$$p - p' = \frac{2T}{r} \qquad . \quad . \quad . \quad . \quad . \quad . \quad [95]$$

Hence eliminating p' from 94 and 95 we have

$$p_0 - p + \frac{2T}{r} = g\rho h \qquad . \quad . \quad . \quad . \quad . \quad [96]$$

$$= \frac{\rho}{\sigma} (p_0 - p) \text{ substituting for } h \text{ from 93.}$$

Thus $$p_0 - p = \frac{2T}{r} \cdot \frac{\sigma}{\rho - \sigma} \qquad . \quad . \quad . \quad [97]$$

Equation 97 gives the change in the vapour pressure over a curved surface of radius r; for a concave surface the vapour pressure is lower than that over a plane surface by the amount given, whilst for a convex surface equation 97 gives the increase in vapour pressure over the surface.

If r is very small, and hence h large, we cannot assume that the density σ of the vapour is uniform as above. For an increase in height dh at a height h where the density of the vapour is σ the decrease in pressure dp is given by

$$dp = - g\sigma \, dh$$

Thus, assuming the vapour obeys Boyle's law, that is $\dfrac{p}{\sigma} = \dfrac{p_o}{\sigma_o}$ where σ_o is the density of the vapour at the point B, we have

$$\frac{dp}{p} = - g \frac{\sigma_o}{p_o} dh$$

which on integration becomes

$$\log_e p = - g \frac{\sigma_o}{p_o} h + \text{constant}$$

$$\text{or} \quad \log_e \frac{p}{p_o} = g \frac{\sigma_o}{p_o} h \quad . \quad . \quad . \quad . \quad . \quad . \quad [98]$$

the constant of integration being $\log_e p_o$ since $p = p_o$ when $h = 0$.
Eliminating h between equations 98 and 96 we have

$$\rho \frac{p_o}{\sigma_o} \log_e \frac{p_o}{p} = (p_o - p) + \frac{2T}{r}$$

which may be written

$$\log_e \left(\frac{p_o}{p} \right) = \frac{2T\sigma_o}{rp_o\rho}$$

ignoring the quantity $(p_o - p)$ compared with the much larger quantity $\dfrac{2T}{r}$ when r is very small.

The relative change in the vapour pressure is insignificant unless r is exceedingly small. Thus for water the change is less than one per cent for values of r down to 10^{-5} cm., but for values of r less than this the effect shows a very rapid increase, the vapour pressure over a convex surface being many times that over a plane surface at the same temperature. Thus, in the absence of suitable nuclei on which to initiate drop formation, it is possible to have a high degree of supersaturation before condensation of the water vapour is brought about. This was demonstrated by Aitken over 50 years ago, who further showed that if particles of solid matter, e.g. dust, were present, only a slight degree of supersaturation was required before condensation commenced. The dust particles provide surfaces of sufficiently large radius of curvature so that a thin film of water on such a surface has a vapour pressure only slightly above that of the normal saturation pressure. Hence if the vapour pressure is slightly above this, condensation is promoted and the drops will grow.

The presence of ions also facilitates condensation. The quantity $\dfrac{2T}{r}$ appearing in the expressions for the vapour pressure over a convex surface is an inward pressure due to surface tension whose effect, as we have seen, is to give rise to an increase in the vapour pressure over the surface. A drop forming round a charged ion may be considered to be an electrified sphere which will thus be subject to an outward pressure. This acts against the effect of surface tension in opposing drop formation, and it can be shown that when *r* is very small the effect of the electric charge in promoting condensation is greater than that of surface tension in preventing it. This condensation of water droplets on charged ions from supersaturated water vapour freed from all dust particles, has been used by C. T. R. Wilson and later workers in the cloud chamber method of rendering visible the tracks of atomic particles. Photographs of these effects can be seen in most books on atomic physics.

QUESTIONS. CHAPTER 9

1. How is the melting point of ice affected (*a*) by pressure, (*b*) by the presence of a salt? Describe one experiment in support of each of your statements.

Explain how it is possible to obtain temperatures below 0°C. by means of a freezing mixture. What determines the lowest temperature obtainable by mixing ice and salt? (N.)

2. Compare the properties of saturated and unsaturated vapours. Describe an experiment to determine how the saturated vapour pressure of water vapour varies with temperature, and indicate the results you would expect to obtain.

A sample of air at 14°C. and a pressure of 76cm. of mercury contains just sufficient water vapour to saturate it. Calculate its pressure if the volume is (*a*) halved, (*b*) doubled, isothermally. (W.)

3. Explain what is meant by a saturated vapour, and distinguish clearly between a saturated and an unsaturated vapour.

Describe an experiment to show that the saturated vapour pressure of a liquid at a given temperature is not affected by the presence of a gas in the space above the liquid.

A given volume at 100°C. contains a mixture of air and saturated water vapour only. When the temperature of the whole mass is raised to 180°C. without change of volume, the pressure is found to be 1·90 atmospheres. Find, in cm. of mercury, the pressure at 20°C. of the air contained in this volume of the mixture.

(Atmospheric pressure = 76cm. of mercury.) (W.)

4. Distinguish between *saturated* and *unsaturated* vapours. How are (*a*) pressure and temperature, (*b*) pressure and volume, related in each case?

A Boyle's Law experiment is performed with air known to be mixed with water vapour. Initially the air is unsaturated and the pressure is observed to be 52·8cm.; when the volume is halved the pressure becomes 105·2cm. and when the volume is reduced to one-third the pressure is 157·1cm. Determine (*a*) whether the water vapour becomes saturated when the volume is reduced, (*b*) the initial pressure of the water vapour. (N.)

5. Describe an experiment to determine the saturation pressure of ether vapour at temperatures in the neighbourhood of that of the room.

The gas tube of a Boyle's Law apparatus is provided with a tap through which small quantities of liquid can be admitted to the tube. Initially, the tube contains 25c.c. of dry air, the level of the mercury in the reservoir being 33cm. below that in the gas tube. A very small quantity of ether is then admitted so that the volume becomes 35c.c. and the difference in the mercury levels 17cm. The barometric height is 75cm. and the saturation pressure of ether vapour at the temperature of the tube 49cm. Find how the volume of the space containing the air and vapour must be altered so that it may become just saturated. (Assume that the unsaturated vapour obeys Boyle's Law.)   (N.)

6. Explain the meaning of the term *saturation vapour pressure* and describe how the saturation vapour pressure of a liquid at different temperatures may be determined by observations of its boiling point at different pressures.

The space above the mercury meniscus in a barometer tube of uniform bore contains air and water vapour. At 60°C. the space is saturated, the height of the mercury column is 50·0cm. and the length of the tube above the meniscus is 35·0cm. At 20°C. the height is 60·3cm. and the length of the tube above the meniscus 24·7cm. Assuming that the barometric pressure is 75·0cm. of mercury and that the saturation vapour pressure of water vapour at 20°C. is 1·7cm. determine the saturation vapour pressure at 60°C.   (N.)

7. The following data were obtained from a Boyle's Law experiment in which the enclosed space contained a mixture of air and ether vapour:

Volume c.c.	60	50	40	30	20	10
Pressure (mm. of mercury)	284	342	427	540	595	760

Show that the reduction of volume causes the vapour to become saturated and find (*a*) the saturation pressure of the ether vapour at the temperature prevailing during the experiment, (*b*) the pressure exerted by the air alone when the volume is 60c.c.   (N., part ques.)

8. Indicate by means of a graph how the boiling point of water varies with pressure. Give an explanation of the effect, and describe how it may be investigated quantitatively for pressures less than atmospheric.

A flask containing air and a little water is closed when the temperature is 30°C. and heated to 100°C. when there are still traces of water inside. If the barometric pressure is 76cm. of mercury and the saturation pressure of water vapour at 30°C. is 3·2cm., find the pressure inside the flask at 100°C. Would this pressure be greater or less if all traces of water disappeared before the temperature reached 100°C.?
                                                                                       (N.)

9. How is a mercury manometer used to measure the pressure of a gas and what needs to be known to enable its reading to be converted into dynes per square centimetre?

Describe the apparatus you would use and the procedure you would adopt to measure the saturation vapour pressure of water at room temperature (about 1·5 cm. Hg.) in dynes per square centimetre to an accuracy of about one per cent. State the precautions which would be necessary in order to achieve this accuracy. Indicate briefly how your apparatus and procedure would have to be modified in order to determine the s.v.p. of water at a temperature of about 50°C. (about 9 cm. Hg.).   (O. and C.)

10. In what respects do *saturated* vapours differ from *unsaturated* vapours? Show, in a general way, how the kinetic theory accounts for these differences.

Describe in detail how you would carry out an experiment to determine the effect of pressure on the boiling point of water.

The total pressure inside a sealed vessel, which contains air and water, is 755mm. of mercury at a temperature of 20°C. What will the pressure be when the temperature is raised to 100°C., assuming that some of the water remains unevaporated?
(S.V.P. of water vapour at 20°C. = 17mm. of mercury.)   (O. and C.)

11. Write an essay on **either** the measurement of the variation of the vapour pressure of water with temperature, with a discussion of the results obtained, **or** the determination of the specific heat of solids and liquids, pointing out the difficulties that arise.   (O. and C.)

12. What is a *saturated vapour*? How do its properties (*a*) resemble, (*b*) differ from, those of a gas?

50c.c. of air saturated with water vapour at room temperature and 76cm. pressure is compressed slowly to 100cm. pressure. What is its volume and what has happened to the water vapour? If the system is now heated at constant volume what happens, in general terms, to the pressure?

(S.W.V.P. at room temperature = 1·5cm. of mercury.)   (S.)

13. What do you understand by the terms *saturated vapour pressure, unsaturated vapour pressure*?

If you were given a flask containing a mixture of air and water vapour at room temperature, what experiment or experiments would you carry out to find out if the water vapour were saturated or not?

Two similar, wide, uniform glass tubes, closed at one end are inverted over mercury. One contains pure dry air, the other air and saturated water vapour, and in both the mercury levels inside and out are the same when the tops of the tubes are 100cm. above the level of the mercury. Both tubes are now depressed until the tops are 50cm. above the outside level of the mercury. In the first tube the level of the mercury inside the tube is 25·3cm. below the level of the mercury outside and in the second tube 25cm. Calculate the atmospheric pressure and the saturated pressure of the water vapour.   (Camb. Schol.)

14. A mixture of a gas and a vapour in contact with excess of the liquid is contained in a closed vessel. How will the pressure of the mixture vary (*a*) when the temperature is changed at constant volume, (*b*) when the volume is changed at constant temperature?

A mixture of air and the vapour of a liquid is contained in a vessel at constant volume. The pressures in the vessel at 15°C., 45°C., and 60°C. are 80, 101, and 126cm. of mercury respectively. Assuming that excess of the liquid is present and that at 15°C. the vapour pressure of the liquid is 15cm. of mercury, calculate the vapour pressure of the liquid at 45°C. and 60°C.   (Camb. Schol.)

15. A 100c.c. flask contains a small drop of water and is fitted with a tap so that it can be opened or closed to the atmosphere. The flask is heated to 77°C. with the tap open and the tap is then closed. When the flask has cooled to 27°C., it is inverted, immersed in water at 27°C. and the tap opened. It is found that 50c.c. of water enter the flask when external and internal pressures are equalised. Find the saturated vapour pressure of water at 77°C. given that it is 26mm. of mercury at 27°C. and that the atmospheric pressure is 746mm. of mercury.

(Camb. Schol., part ques.)

16. Explain what is meant by the " dew point ", and describe an accurate method of determining its value.

The saturated vapour pressure of water at 11·25°C. is 1cm. of mercury, the density of dry air at 76cm. pressure and 25°C. is 0·001184gm./cm.³, and the density of water vapour relative to that of air at the same pressure and temperature is 0·624. Find the weight of a litre of moist air at 25°C., if the barometer stands at 76cm., and the dew point is 11·25°C.   (W.)

17. Define *relative humidity*, and describe a satisfactory method for finding it.

Find the mass of water vapour per cubic metre of moist air when the temperature is 20°C. and the dew point 10°C.

Use the following data: Density of air at 0°C. and 76cm. pressure = 1·29gm. per litre; density of water vapour = ⅝ that of air at the same temperature and pressure; saturation pressure of water vapour at 10°C. = 9·20mm.   (N.)

18. What is meant by the *hygrometric state* of the atmosphere?

Discuss the merits, or otherwise, of the Daniell, Regnault, and wet and dry bulb hygrometers.

What is the volume of 1gm. of dry atmospheric air on an occasion when the barometric pressure is Bcm. of mercury, the pressure exerted by the water vapour present $p$cm., and the density of dry air at 76cm. pressure and the prevailing temperature ϱgm. per litre? Determine the mass of water vapour present in this volume when $B = 75\cdot6$ and $p = 1\cdot20$ assuming that the density of water vapour is $0\cdot62$ of that of air at the same temperature and pressure. (N.)

19. Describe a method of determining the saturated vapour pressure of water vapour at room temperature.

Calculate the mass of water vapour in a room of volume 800 cu. m. when the temperature is 20°C. and the relative humidity is 70 per cent. The saturated vapour pressure of water at 20°C. is given as 17·5mm. of mercury, the density of air at S.T.P. is 1·293g. per litre and the density of water vapour is 0·62 times the density of air at the same temperature and pressure. (A.)

20. Discuss the processes of evaporation and boiling from the point of view of the kinetic theory, explaining (*a*) the increase of the saturated vapour pressure of a liquid as the temperature rises, (*b*) the variation of boiling-point with pressure.

Describe briefly how you would measure the saturated vapour pressure of water for the temperature range 80°C. to 120°C.

A vessel containing air and a little water is sealed at 25°C. when the atmospheric pressure is 76cm. of mercury. Assuming that the volume remains constant, what will be the pressure inside the vessel when its temperature is raised to 100°C.?

(Take the saturated vapour pressure of water at 25°C. to be 2·4cm. of mercury.) (O.)

21. State Dalton's Law of partial pressures. Draw isothermals showing the relation between pressure and volume for (*a*) a gas, (*b*) a vapour. Give the equations which represent each of these isothermals.

A mercury thread 10cm. long encloses air and a drop of liquid in a uniform tube sealed at one end. The air column is 17·3cm. long when the tube is horizontal and 20cm. long when it is vertical. If the mercury barometer reads 76cm., what is the vapour pressure of the liquid? (W.)

22. Describe an experiment which shows that a liquid which is open to the atmosphere boils when its saturation vapour pressure is equal to atmospheric pressure. Explain why this should be so.

Calculate the relative humidity of the atmosphere in a room where the temperature is 18°C. and the dew point is 10°C.

(The boiling-point of water is 18°C. at a pressure of 1·55cm. of mercury, and 10°C. at a pressure of 0·93cm. of mercury.) (W.)

23. (*a*) Describe how you would determine the saturation pressure of water vapour at various temperatures between 70°C. and 100°C. Draw a diagram of your apparatus and a graph indicating how the saturation pressure of water vapour varies with temperature between 0°C. and 100°C.

(*b*) The relative humidity of the atmosphere on a day when the temperature is 18°C. is found to be 70 per cent. What percentage of the mass of water present in the atmosphere would be deposited if the temperature suddenly fell to 6°C.? The saturation pressure of water vapour at 18°C. is 15·46mm. of mercury and at 6°C. is 7·01mm. of mercury.

State what assumptions you make in your calculation. (N.)

24. 100c.c. of moist air having a relative humidity of 65 per cent is slowly compressed at constant temperature. What will be its volume when dew starts to be deposited? State and discuss the assumption you make in your calculation.
(W.)

25. (a) What is meant by the statement that the dew-point on a certain day is 10°C.? What further facts would you have to know in order to find the relative humidity?

(b) Is the density of a saturated vapour of water at 50°C. greater or less than the density of a saturated vapour of water at 60°C.? Explain your answer.

(c) Calculate the mass of water vapour condensed per cubic metre when saturated air at 15°C. is cooled to saturated air at 10°C. Assume that water vapour obeys the gas laws up to the point of saturation.

SWVP of water at 15°C. = 1·3cm. of mercury.
SWVP of water at 10°C. = 0·9cm. of mercury.
Density of water vapour at 100°C. and 76cm. of mercury = 0·63 gm. litre^{-1}.
(S.)

26. Define *dew point* and describe an experiment to determine its value. What further information would be required to calculate a value for the relative humidity of the atmosphere?

A sample of moist air at 20°C. has a dew point at 12°C. and the barometric height is 76·0cm. of mercury. Assuming that the equilibrium vapour pressure of water at 12°C. is 1·05cm. of mercury and the density of dry air at S.T.P. is 1·293 gm. litre^{-1}, estimate the mass of dry air contained in one cubic metre of the moist air. (L.)

27. Derive an expression showing the change in the saturation vapour pressure which is produced by curvature of a liquid surface. Discuss the effects of this change on meteorological phenomena.

Calculate the change in the saturated vapour pressure due to curvature in the case of a water droplet of diameter 0·002mm. at 15°C., given the following data:
Surface tension of water at 15°C. = 74 dynes per cm.
Saturated vapour pressure of water at 15°C. = 1·3cm. of mercury.
Density of air at 15°C. under standard pressure = 1·23gm. per litre.
Relative density of water vapour to air = 0·625. [N.]

28. Discuss the following:
(a) A small drop of water will evaporate in a space in which water vapour is in equilibrium with a plane water surface.
(b) Boiling by bumping. [N., part ques.]

29. Distinguish between absolute and relative humidity. Which of these is it usually the more important to know? Give reasons.

Describe and explain a method of determining the relative humidity. Air at 20°C. contains 0·012gm. water vapour per litre. What is the relative humidity? The saturated vapour pressure of water at 20°C. is 1·74cm. mercury, and a litre of steam at 100°C. and 76cm. mercury weighs 0·606gm. (Cape Town)

CHAPTER X

CONTINUITY OF STATE

10.01 Introductory—Experiments of Cagniard de la Tour

From the previous Chapter it would appear that there is a distinct line of demarcation between the liquid and gaseous states of matter—a substance being definitely liquid or gaseous depending on the prevailing conditions of temperature and pressure. In general this is true, the transition from the liquid to the gaseous state, or vice versa, occurring abruptly with a sharp distinction between the two states. Experiments performed by Cagniard de la Tour as far back as 1822 however indicated that under certain circumstances the liquid and gaseous states appear to merge into one—the change from one state to the other, by a slight alteration of the conditions, taking place gradually. This continuous transition between the two states is referred to as the *continuity of state*.

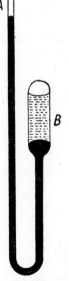

Fig. 109

The apparatus used by Cagniard de la Tour is shown in Fig. 109. It consisted of a bent tube closed at both ends, the end A containing air whilst the other end B contained the liquid under investigation in equilibrium with its own vapour. The space between A and B was filled with mercury. At ordinary temperatures the liquid surface in B was sharp and clear, but on raising the temperature, and with it the pressure of the vapour in B, a stage was reached when the meniscus of the liquid became flatter and less distinct and finally disappeared altogether. The space was then occupied by a homogeneous fluid which, on lowering the temperature once more, exhibited a peculiar "flickering" appearance with the return of the meniscus separating the liquid from the vapour above it. The pressure at which these changes took place was estimated from the volume of air at A.

In addition to demonstrating the continuity between the liquid and gaseous states, Cagniard de la Tour's work was valuable in that it promoted research on the liquefaction of gases. Faraday in 1823 succeeded in liquefying chlorine, sulphuretted hydrogen, carbon dioxide, nitrous oxide, &c., by generating the gases in one end of an enclosed tube, the other end being bent over into a freezing mixture

236 CONTINUITY OF STATE

of ice and salt. Later, in 1845, using a freezing mixture of solid CO_2 and ether he liquefied ethylene, phosphine, &c., but in spite of the low temperature used (—110°C.) such gases as oxygen, nitrogen, and hydrogen could not be liquefied. The efforts of later workers such as Natterer who employed very high pressures in attempts to liquefy these gases, also met with failure, and it was concluded that such gases as these could not be liquefied in any circumstances, and were thus called the "permanent gases." The essential conditions for the liquefaction of gases were first clearly stated by Andrews (1863) whose classical experiments on the compressibility of carbon dioxide will now be described.

10.02 Andrews' Experiments on Carbon Dioxide

Andrews' experiments on carbon dioxide consisted in measuring the volume of a given mass of carbon dioxide contained in a thick walled capillary tube as the gas was subjected to varying pressures when maintained at a series of constant temperatures. The tube containing the carbon dioxide was as shown in Fig. 110 (a) and was

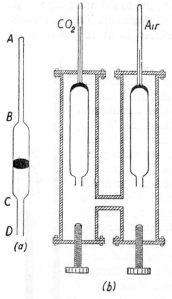

(a)

(b)

Fig. 110

initially open at both ends. AB was the capillary portion which was graduated by measuring the length of a thread of mercury of known volume when at different positions in the tube. The parts BC and CD were of wider bore (2·5 mm. and 1·25 mm. respectively) and carbon dioxide was passed through the tube for 24 hours to get rid of all air traces, both ends then being sealed. The end D was then placed under mercury and opened, and by placing the whole under the receiver of an exhaust pump, the amount of carbon dioxide in the tube was suitably adjusted and cut off from the rest of the tube by a mercury pellet as shown. The tube was then mounted together with a similar tube containing air in two stout copper cylinders which were connected together and filled with water. The whole apparatus was water-tight, the pressure on the gases being applied by two screw plungers fitted at the lower ends of the copper cylinders. The pressure was increased by screwing in either or both of the plungers until the pellets were forced into the capillary portions of the tubes when the volumes of the carbon dioxide and air were read off. The pressure applied was obtained from the readings of the volume of the air assuming Boyle's law to apply in this case, and by maintaining the temperature of the gases constant by surrounding the apparatus with a constant temperature bath, a series of corresponding values of the volume and pressure of the carbon dioxide were obtained. The experiment was repeated at different temperatures and the results represented graphically by a set of isothermal curves, some of which are shown in Fig. 111. Referring

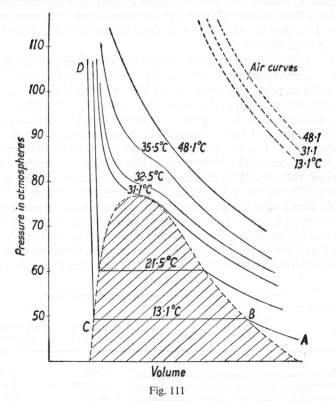

Fig. 111

to the 13·1°C. isothermal starting with the carbon dioxide in the gaseous state represented by *A*—application of pressure causes a diminution of volume along the part of the curve from *A* to *B* at which point, under a pressure of about 49 atmospheres, liquefaction commences and the volume decreases at this pressure until all the gas is liquefied at the point *C*. The remaining part *CD* of the isothermal rises almost vertically and corresponds to the liquid state which shows only slight shrinkage in volume even under very large pressures. For the next higher temperature (21·5°C.) the course of the curve is similar to that at 13·1°C., but a much higher pressure is required before liquefaction commences, and the horizontal portion, over which liquefaction occurs, is shorter. As the temperature is further raised the horizontal portion becomes progressively shorter as indicated by the curve enclosing the area of liquefaction, and at the temperature 31·1°C. there is no horizontal portion to the isothermal. Although the slope changes steeply on this curve for pressures round about 75 atmospheres, thus showing that the carbon dioxide was very compressible in this region, there was no evidence of liquefaction. For still higher temperatures the irregularities in the isothermals become even less pronounced until at a temperature of 48·1°C. they are very similar to those for air shown in dotted line on the same diagram.

It is thus evident that however great the applied pressure, liquefaction of carbon dioxide is not possible at temperatures above 31·1°C. Repeated experiments showed that the highest temperature at which it is just possible to obtain liquid carbon dioxide was 30·9°C. This temperature, above which it is not possible to obtain liquid carbon dioxide, is called the *critical temperature*. The pressure required to produce liquefaction at this temperature is called the *critical pressure*, and the volume occupied by one gram of the substance is known as the *critical volume*. (The critical volume is sometimes defined as the ratio of the volume at the critical temperature and pressure to that it would have had at N.T.P.). A substance existing in the gaseous state at temperatures below its critical temperature is referred to as a "vapour" and can be liquefied by the application of pressure alone. Liquefaction is not possible by compression above the critical temperature, and the substance is then referred to as a "gas." It is thus clear that in order for

TABLE 25

Critical temperatures and pressures of common gases

Gas	Crit. temp. in °C.	Crit. pressure in atmos.
Helium	−268	2·26
Hydrogen	−239·9	12·8
Nitrogen	−146	33·5
Air	−140	39
Argon	−122	48
Oxygen	−118	50
Methane	−82	46
Ethylene	10	51·7
Carbon dioxide	31·1	73
Sulphuretted hydrogen	100	88·7
Ammonia	130	115
Chlorine	146	76
Sulphur dioxide	155·4	78·9
Ether	197	35·8

a "gas" to be liquefied, its temperature must first be reduced to a value below its critical temperature (*i.e.*, it must first be converted into a "vapour") before it is compressed by the application of a suitable pressure. The reason for the failure (section 10.01) to liquefy the so-called "permanent gases" is now clear—it is simply that they had not been cooled sufficiently before compression.

The values of the critical temperatures and pressures for gases can be obtained directly from experiments using the method of Cagniard de la Tour described in the previous section.

10.03 Continuity of State—James Thomson's Hypothesis

The work of Andrews in establishing the essential continuity of the liquid and gaseous states raised the possibility of representing the observed phenomena by some form of general equation of state. The perfect gas equation $pv = RT$ represents the state of a gas at high temperatures and under low compression. Deviations from the gas laws at high pressure have already been discussed in section 5.08, where reference was made to the empirical equation of state introduced by Kamerlingh Onnes to represent the behaviour of real gases under conditions of high compression. From a study of the curves given in Fig. 111 it is quite clear that the perfect gas equation does not even approximately represent the results in the neighbourhood of the critical temperature, and is clearly not applicable to the continuous fluid state at temperatures below this.

An interesting hypothesis put forward in 1871 by James Thomson emphasized further the continuity of the liquid and gaseous states. He suggested that the ideal form of the isothermal might be a continuous curve as shown in Fig. 112, along which there was a gradual change from the gaseous to the liquid states without the necessity of their co-existence. Under ideal conditions it is possible to obtain the section *BC* of the curve. Thus by compressing a vapour in a perfectly clean and smooth vessel the pressure at which liquefaction occurs can be appreciably exceeded. Again, it is possible to reduce the pressure over a pure dust and air-free liquid without evaporation taking place, and thus the portion *FE* of the curve

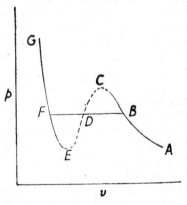

Fig. 112

can be directly observed. These facts are closely associated with the super-cooling of a saturated vapour in the absence of dust particles and the superheating of a pure liquid respectively. The states represented by C and E are essentially unstable—any slight alteration in the conditions producing sudden and violent changes which may be represented by the imaginary dotted section CDE of the curve.

10.04   van der Waals' Equation

The first successful attempt to modify the ideal gas equation so as to obtain an equation of state to represent the behaviour of real gases was made by van der Waals in 1873. The fact that under suitable conditions of temperature and pressure gases can be converted into liquids, the molecules of which possess the property of cohesion, indicated strongly the existence of molecular attractions in a gas. Direct experimental evidence of these molecular attractions had been obtained by Joule and Thomson (Lord Kelvin) in 1853 with their "porous-plug" experiment (see section 10·08), and van der Waals attempted to allow for these intermolecular forces, and also for the finite size of the molecules, in formulating his modified equation of state. The manner in which these two factors are allowed for in van der Waals' equation may be considered as follows:

(a) *Intermolecular forces:* A molecule in the interior of a gas at any instant is surrounded by a large number of other molecules equally distributed in all directions. Consequently for such a molecule the attractive forces of the other molecules balance each other out, and the molecule is thus free to move in the space occupied by the gas with a velocity determined by its temperature (section 8.10). However, when a molecule approaches the sides of the containing vessel, there is no longer such a uniform distribution of molecules round it, and it finds itself subject to an inward pull due to the unbalanced attractions of the molecules behind it. The effect of this inward pull on the molecule at the moment of impact is to reduce its momentum on striking the containing wall, and hence to reduce the effective pressure of the gas. Thus the measured pressure p is less than the ideal pressure postulated by the kinetic theory on the assumption that inter-molecular attractions do not exist (see section 8.06). To obtain this ideal pressure there must be added to the observed pressure p a correcting term p' which is clearly proportional to the inward pull exerted by the bulk of the molecules on the molecules near the wall. This force will depend on the number of molecules in the volume of the gas and consequently upon its density. Further, the force will also depend on the number of molecules striking the surface per second which also depends on the density. Consequently the correcting term p' will be proportional to the square of the density of the gas or inversely proportional to the square of the volume occupied by a given mass of it. Hence the ideal pressure is equal to $p + \dfrac{a}{v^2}$ where a is a constant for the given mass of gas.

(b) *Finite size of molecules:* The effect of the finite size of the molecules is to reduce the effective volume of the gas by an amount b which is proportional to the bulk of the molecules. Thus the volume in which the molecules are free to move, or the effective volume, is $v-b$ where v is the measured volume of the gas. The constant b is called the *co-volume* and van der Waals, and later Jeans, showed that b is four times the actual volume of the molecules, although other slightly different estimates have been given.

It will thus be seen that applying these two corrections to the ideal gas equation, we have, for 1gm. molecule of a gas,

$$\left(p + \frac{a}{v^2}\right)(v-b) = RT$$

R being the gas constant, and a and b being constants for a given gas. This is van der Waals' equation. The agreement between the equation and the observed results is qualitative rather than quantitative for most gases, but in some instances for small ranges of temperature above the critical temperature, where the deviations from the gas laws are small, the agreement is surprisingly good. The analytical treatment of the equation for certain aspects of the behaviour of gases will be given in the sections that immediately follow together with a comparison with other selected equations of state and a relative estimate of their representation of the experimental data.

10.05 The Boyle Temperature

It is interesting to review the experimental results on the deviations from the gas laws discussed in section 5.08 with reference to van der Waals' equation. For low pressures, the volume of a given mass of a gas is relatively large, and we may thus ignore b in comparison to v. We then obtain the more approximate equation:

$$\left(p + \frac{a}{v^2}\right)v = RT$$

or
$$pv = RT - \frac{a}{v}$$

From this last equation it will be seen that for low pressures the value of the product pv is less than that to be expected from the perfect gas equation. This is in accordance with the results (refer to Fig. 53) which show that real gases are in fact more compressible than the ideal gas at moderately low pressures. This was interpreted as indicating the existence of molecular attractions, and we now see on the basis of van der Waals' equation that the low value of pv when p is small is due to the a term which is a measure of the intermolecular forces in the gas.

On the other hand, when p is large and v small, the corresponding approximate formulation of van der Waals' equation is obtained by neglecting $\frac{a}{v^2}$ in comparison to the large value of p, but retaining b which cannot now be neglected in the volume factor. Thus we have

$$p\,(v-b)\ =\ RT$$

or

$$pv\ =\ RT + pb$$

This equation indicates a rise in the value of pv at high pressures (a minimum value must have been reached at some intermediate pressure), and this again is in accordance with the facts. The observed rise in the product pv for high pressures is interpreted on van der Waals' theory as due to the bulk of the molecules which at these pressures results in the gas being less compressible than the ideal gas.

As the temperature rises the minimum of the pv, p curves follows the course of the dotted line indicated in Fig. 53, and at a certain temperature, the Boyle temperature, the minimum occurs at zero pressure. The gas conforms most closely to Boyle's law at this temperature which may be derived from van der Waals' equation as follows. Writing the equation in the form

$$p\ =\ \frac{RT}{(v-b)}\ -\ \frac{a}{v^2}$$

and multiplying by v, we have

$$pv\ =\ RT\left[\frac{v}{v-b}\right]\ -\ \frac{a}{v}$$

Differentiating this equation with respect to p at constant temperature it follows that

$$\left[\frac{\partial\,(pv)}{\partial p}\right]_{T\,\text{const.}}\ =\ \left\{\frac{RT}{v-b}\ -\ \frac{RTv}{(v-b)^2}\ +\ \frac{a}{v^2}\right\}\left[\frac{\partial v}{\partial p}\right]_{T\,\text{const.}}$$

and for a minimum in the pv, p curve

$$\frac{RT}{v-b}\ -\ \frac{RTv}{(v-b)^2}\ +\ \frac{a}{v^2}\ =\ 0$$

i.e.,

$$RT\ =\ \frac{a}{b}\ \cdot\ \left[\frac{v-b}{v}\right]^2$$

or

$$T\ =\ \frac{a}{Rb}\ \cdot\ \left[\frac{v-b}{v}\right]^2$$

Now for the Boyle temperature, the minimum must occur at zero pressure, and as $p\to0$, $\left[\frac{v-b}{v}\right]\to1$ since $v\to\infty$. Thus the Boyle temperature T_B is

$$T_\text{B}\ =\ \frac{a}{Rb}\qquad\cdots\cdots\cdots\cdots\quad[99]$$

10.06 Values of the Critical Constants from van der Waals' Equation

From van der Waals' equation

$$\left[p + \frac{a}{v^2}\right](v-b) = RT$$

we get $(pv^2 + a)(v-b) = RTv^2$
which, on being multiplied out can be expressed as a cubic equation in v, viz.,

$$v^3 - \left[b + \frac{RT}{p}\right]v^2 + \frac{a}{p}v - \frac{ab}{p} = 0 \quad . \quad . \quad . \quad . \quad . \quad [100]$$

and thus gives the general form of isothermal suggested by James Thomson (see section 10.03). For any value of p below the critical temperature equation 100 will have three real roots which correspond to the points B, D, and F in Fig. 112. As the temperature rises, the portion BF of the curve becomes progressively shorter until at the critical temperature the points B, D, and F coincide, or expressed analytically, the three roots of the equation,

$$v^3 - \left[b + \frac{RT_c}{p_c}\right]v^2 + \frac{a}{p_c}v - \frac{ab}{p_c} = 0 \quad . \quad . \quad . \quad . \quad [101]$$

are identical—T_c and p_c being respectively the critical temperature and critical pressure. If v_c is the critical volume, equation 101 may also be expressed in the form

$$(v - v_c)^3 = 0$$

or

$$v^3 - 3v^2v_c + 3vv_c^2 - v_c^3 = 0 \quad . \quad . \quad . \quad . \quad . \quad [102]$$

and hence on equating the coefficients of like powers in equations 101 and 102, we have

$$3v_c = b + \frac{RT_c}{p_c} \qquad 3v_c^2 = \frac{a}{p_c} \quad \text{and} \quad v_c^3 = \frac{ab}{p_c}$$

By elementary algebra, it follows from these three equations that*

$$v_c = 3b \quad . \quad . \quad . \quad . \quad . \quad . \quad . \quad . \quad . \quad . \quad . \quad . \quad [103]$$

$$p_c = \frac{a}{27b^2} \quad . \quad . \quad . \quad . \quad . \quad . \quad . \quad . \quad . \quad . \quad [104]$$

and

$$T_c = \frac{8a}{27Rb} \quad . \quad . \quad . \quad . \quad . \quad . \quad . \quad . \quad . \quad [105]$$

---

* *4liter.* The critical isothermal has a point of inflection at the critical point whose co-ordinates are p_c, v_c. Thus at the critical point both $\frac{dp}{dv}$ and $\frac{d^2p}{dv^2} = 0$

i.e.,

$$-\frac{RT_c}{(v_c - b)^2} + \frac{2a}{v_c^3} = 0$$

and

$$\frac{2RT_c}{(v_c - b)^3} - \frac{ba}{v_c^4} = 0$$

Hence by division of these two equations, we have $v_c = 3b$, and by substitution in either equation and in the original equation we get respectively $T_c = \frac{8a}{27Rb}$ and $p_c = \frac{a}{27b^2}$

Now the values of the constants a, b, and R for a particular gas can be obtained from compressibility measurements, and the values of the critical constants calculated from them using equations 103, 104, and 105. Thus for carbon dioxide van der Waals found the equation

$$\left[p + \frac{0 \cdot 00874}{v^2} \right] (v - 0 \cdot 0023) = 1 \cdot 00646T$$

where p is expressed in atmospheres and v in terms of the volume of 1gm. of the gas at N.T.P. Hence using van der Waals' constants for carbon dioxide in equations 104 and 105, we get, applying the appropriate unit conversion factors,

$$p_c = 61 \text{ atmospheres, and } T_c = 31°C.$$

which are in fair agreement with the observed values of 73 atmospheres and $31 \cdot 1°C$. According to equation 103 the value of $\frac{v_c}{b}$ should be 3 for all gases. Experimental results for a number of gases however give a value nearer to 4 than 3. It would thus appear that the general measure of agreement between van der Waals' equation and the observed behaviour of gases is only fair.

Another way of testing the validity of the equation is to compare the calculated value of $\frac{RT_c}{p_c v_c}$ against the experimental value. The quantity $\frac{RT_c}{p_c v_c}$ is called the *critical coefficient*, and its value as derived from equations 103, 104, and 105 is $\frac{8}{3}$ or $2 \cdot 67$, and is the same for all gases.

TABLE 26

Values of critical coefficients for certain gases

Gas	Critical coefficient
Helium	3·13
Hydrogen	3·03
Nitrogen.............	3·42
Argon	3·43
Oxygen	3·42
Carbon dioxide ...	3·48

The results in table 26 before show that whereas the quantity is sensibly constant for all the cases given, it is appreciably greater than the theoretical value of $2 \cdot 67$*.

10.07   Law of Corresponding States

An interesting result from van der Waals' equation is obtained by expressing the values of p, v, and T in terms of their critical values. These quantities, which are purely numerical, are called the *reduced values* of the pressure, volume, and temperature, and by inserting them in the characteristic equation we obtain a *reduced equation of state.* Thus by writing

$$p = xp_c, \quad v = yv_c \text{ and } T = zT_c,$$

where x, y, z are the appropriate reduced values, and inserting them in the equation of van der Waals, we have

$$\left(xp_c + \frac{a}{y^2 v_c^2} \right)(yv_c - b) = RzT_c$$

Using the values of p_c, v_c, and T_c given in equations 103, 104, and 105, this becomes

$$\left(\frac{xa}{27b^2} + \frac{a}{3y^2b^2} \right)(3yb - b) = \frac{8az}{27b}$$

which reduces to

$$\left(x + \frac{3}{y^2} \right)(3y - 1) = 8z$$

This is van der Waals' reduced equation of state, and being independent of the constants a and b, is true for all substances. It follows from this that the isothermals for all substances when represented in this way should be identical, and that when two of the quantities x, y, z are the same for any two substances, the third quantity will also be the same for those two substances. This is known as the *law of corresponding states.*

It should be noted that any equation of state involving two arbitrary constants only in addition to R can be reduced in this way using the three critical values. Accordingly such equations of state will lead to the law of corresponding states which is not solely a particular consequence of van der Waals' equation.

---

* On van der Waals' theory the Boyle temperature T_B is related to the critical temperature T_c (see equations 99 and 105) by the equation $T_B = \frac{27}{8} T_c = 3 \cdot 375\ T_c$. The ratio between these two temperatures can be used as a further test of van der Waals' equation; the observed ratio $T_B{:}T_c$ being approximately constant for many gases as required by theory, but the value is approximately $2 \cdot 5$ instead of $3 \cdot 375$.

10.08   Other Equations of State

From the review of van der Waals' equation given in the foregoing sections it is evident that although the experimental results receive qualitative representation, the quantitative agreement given by the equation is not very satisfactory. There have been many subsequent attempts to develop other equations of state which shall be more closely in accord with the observed data. Some of these, like the van der Waals' equation, are based on sound theoretical principles, whilst others are purely empirical equations involving a number of arbitrary constants. By suitably adjusting the number and values of these constants this latter group of equations can be made to fit the observed data to almost any degree of accuracy; the extent of the agreement given by the former group of more general equations is revealed by a consideration of some well-known equations given below:

(a) **Clausius.**—One of the weaknesses of van der Waals' equation is that the cohesive force is assumed to be independent of the temperature. Clausius attempted to improve the equation by taking the cohesive force as being inversely proportional to the absolute temperature, and he gave the modified equation

$$\left[p + \frac{a'}{T(v + c)^2} \right] (v - b) = RT \quad . \quad . \quad . \quad [106]$$

The Clausius equation involves four constants a', b, c, and R and thus, as it stands, does not lead to the law of corresponding states. This result can be obtained however if it is assumed that c is proportional to b, and a value of $\frac{RT_c}{p_c v_c}$ results which is somewhat better than that given by van der Waals. On the whole however equation 106 gives little better agreement than does that of van der Waals with the observed data and is much more difficult to deal with mathematically; accordingly it has not found general application.

(b) **Berthelot.**—By eliminating constant c in equation 106 it becomes

$$\left[p + \frac{a'}{Tv^2} \right] (v - b) = RT \quad . \quad . \quad . \quad . \quad [107]$$

Berthelot used this equation and from it developed a second equation in which the constants a', b, and R were expressed in terms of the critical constants. This latter equation fitted the observations at low pressures for a wide range of temperatures, but was not applicable at the critical point. It should be noticed that equation 107, since it involves only three constants, gives a reduced equation of state. Treating it exactly as in the case of van der Waals' equation (section 10.06) it can be shown that

$$v_c = 3b, \quad p_c = \left(\frac{RT_c}{2b} - \frac{a'}{9T_c b^2} \right), \quad \text{and} \quad T_c = \frac{8a'}{27Rb}$$

from which $\dfrac{RT_c}{p_c v_c} = \dfrac{8}{3}$, which gives no better agreement with the observed values than does the equation of van der Waals.

(c) **Dieterici.**—Dieterici proposed a modification of the van der Waals' equation for the difference in density of the gas at the walls due to the higher potential energy the molecules possess there compared with those in the bulk of the gas. By applying the principles of statistical mechanics he obtained the equation

$$p = \frac{RT}{(v-b)} \cdot e^{-\frac{a}{RTv}}$$

which at low pressures leads to van der Waals' equation. The values of the critical constants using Dieterici's equation are

$$v_c = 2b, \quad T_c = \frac{a}{4Rb}, \quad p_c = \frac{a}{4e^2b^2}$$

which give a value for $\dfrac{RT_c}{p_c v_c}$ of $\dfrac{e^2}{2} = 3\cdot69$, which is better than van der Waals'. It should be noted however that the value for v_c is worse than that given by van der Waals' equation.

It will thus be seen that (in the cases of the equations of state considered) there is little better general agreement with the observed facts than was given by van der Waals' equation. Accordingly, in spite of its limitations, van der Waals' equation is still widely used when considering the behaviour of gases and liquids, although the approximate nature of the conclusions to be derived from it should always be borne in mind.

10.09 The Joule-Kelvin Porous Plug Experiment

Before proceeding to a discussion of the methods for the liquefaction of gases, it is necessary to consider an important experiment suggested by Lord Kelvin and carried out by him in collaboration with Joule in 1853 to detect heat changes when a gas is subjected to free expansion. We have already seen (section 7.04) that Joule, in his search for direct evidence of molecular attractions, had carried out experiments designed to measure changes in the energy of a gas by alterations in its pressure and volume at constant temperature. The insensitivity of his apparatus was responsible for his lack of success in these experiments, and the porous plug experiment provided a more delicate means of estimating the magnitude of the intermolecular forces of a gas for which there appeared other strong evidence.

The principle of the experiment was to force a steady stream of gas at constant pressure along a tube from which it escaped at a lower pressure (atmospheric in the actual experiment) via a narrow orifice, and to note the changes of temperature of the gas as the pressure was released. Consider the arrangement shown in Fig. 113. A perfectly heat insulating cylinder C in which is a partition with a small central hole O,

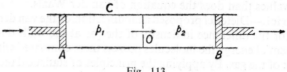

Fig. 113

is closed by two frictionless pistons A and B. A gas contained to the left of the partition is slowly forced through O at a constant pressure p_1 by the piston A, the gas falling in pressure to p_2 to the right of O where it forces back the piston B. The velocity of the gas as it emerges from O must necessarily be large with this arrangement, and consequently there will be an increase in the kinetic energy of the molecules here. This increase in the kinetic energy will take place at the expense of the internal energy of the gas, and so will cause a lowering of the temperature of the gas just beyond O. However, a short distance from the orifice the eddies thus produced will be dissipated by molecular collisions, and in the steady state it may be assumed that the mass kinetic energy of the gas is negligible if observations are taken somewhat to the right of O. If the volumes of one gram-molecule of the gas to the left and right of O are respectively v_1 and v_2, the work done on one gram-molecule of the gas by A is p_1v_1, and the work done by this mass of gas in forcing back the piston B is p_2v_2. Thus the net external work done by the gas is $p_2v_2 - p_1v_1$. In addition, if molecular forces are assumed, an amount of work w is needed to separate the molecules against their mutual attractions in expanding from v_1 to v_2, and this work must be supplied by the gas itself. Hence the total work done by the gas is

$$(p_2v_2 - p_1v_1) + w \qquad . \quad . \quad . \quad . \quad . \quad . \quad . \quad [108]$$

Since there is no heat exchange between the gas and its surroundings, the gas will be heated or cooled according to whether the expression above is negative or positive respectively. With real gases there are three possible cases to be considered depending on the temperature initially. These are:

(a) *Below the Boyle temperature:* Here $p_1v_1 < p_2v_2$ (see Fig. 53), and the bracketted quantity in equation 108 is positive. Thus, since w must be either positive or zero, the gas will be cooled on passing through O.

(b) *At the Boyle temperature:* Provided the pressure p_1 is not very large, $p_1v_1 = p_2v_2$, and any cooling effect observed at this temperature will be solely due to the effect of inter-molecular forces.

(c) *Above the Boyle temperature:* For temperatures above the Boyle temperature the pv, p curve shows a steady rise with p and thus $p_1v_1 > p_2v_2$, and the observed effect will depend on whether the quantity

$(p_1v_1 - p_2v_2)$ is greater or less than w. Thus if the former is the case the gas will be heated as it emerges from O, otherwise it will be cooled. It will thus be observed that the cooling (or heating) of a gas subject to free expansion under the conditions described above, is due to two causes, *viz.*, (*a*) deviations from Boyle's law, and (*b*) inter-molecular forces. In any particular case the magnitude of the cooling effect due to inter-molecular forces alone can be obtained from the observed cooling effect by estimating the effect due to deviations from the gas laws using the experimental results of Amagat or Holborn.

In their actual experiments Joule and Kelvin replaced the narrow orifice by a porous plug P of cotton wool or silk fibres contained between two perforated brass plates (Fig. 114). In this way eddy currents

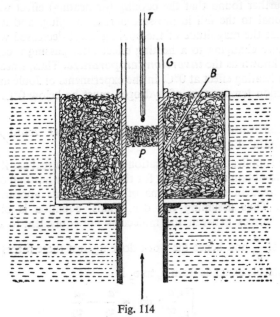

Fig. 114

were overcome and the spurious cooling effect due to the acquired mass kinetic energy of the gas issuing from the orifice was eliminated. The porous plug was placed in a cylindrical boxwood tube B which was surrounded by a vessel containing cotton wool to further reduce heat exchanges with the surroundings. The compressed gas was passed at a slow steady rate through a long spiral copper tube immersed in a constant temperature water bath. The boxwood tube was fitted into the end of the copper tube as shown, and the temperature of the gas as it emerged from the porous plug was taken by a thermometer T placed directly above the plug, the exit tube G being of glass to permit readings of T to be taken. On leaving G the gas escaped to the atmosphere which takes the place of the piston B in Fig. 113. It was found necessary to pass the gas through the apparatus for a considerable time (about one hour) before the initial fluctuation of temperature subsided and steady conditions were obtained.

The experiments of Joule and Kelvin showed that there was a cooling effect for all gases except hydrogen which showed a slight heating effect. This cooling (or heating) effect is known as the *Joule-Kelvin effect*. Thus at 0°C. air showed a cooling effect of 0·275°C. per atmosphere, whilst hydrogen was heated by 0·04°C. per atmosphere. In the case of hydrogen the observed effect was almost equal to that calculated from the Boyle's law deviations for this gas—thus in this case there is very little, if any, work done against the molecular attractions. Joule and Kelvin further found that the cooling (or heating) effect was directly proportional to the fall in pressure across the plug, and it was later shown that the magnitude of the cooling effect decreased with rise of temperature changing to a heating effect after passing a certain temperature known as the *inversion temperature*. Thus, since hydrogen showed a heating effect at 0°C. in the experiments of Joule and Kelvin, the inversion temperature of hydrogen must be below 0°C.

10.10 Estimate of the Joule-Kelvin Effect from van der Waals' Equation

By the application of an equation of state for a real gas it should be possible to obtain an expression for the cooling (or heating) observed in the Joule-Kelvin effect in terms of the constants appearing in the equation. Thus assuming van der Waals' equation is obeyed, the attractive forces between the molecules are equivalent to an internal pressure of $\dfrac{a}{v^2}$ and hence the work done against these forces when one gram-molecule expands from v_1 to v_2 is

$$\int_{v_2}^{v_1} \frac{a}{v^2} \cdot dv = \frac{a}{v_1} - \frac{a}{v_2}$$

This is the term w which appears in equation 108 for the total work (W) done by the gas on passing through the porous plug, and hence on the basis of van der Waals' equation this equation now becomes

$$W = (p_2 v_2 - p_1 v_1) + \left[\frac{a}{v_1} - \frac{a}{v_2} \right] \quad \cdots \cdots \quad [109]$$

Now expanding van der Waals' equation $\left[p + \dfrac{a}{v^2} \right] (v - b) = RT$ we have

$$pv + \frac{a}{v} - bp - \frac{ab}{v^2} = RT \quad \text{or} \quad pv = RT + bp - \frac{a}{v}$$

if the second order term $\dfrac{ab}{v^2}$ is ignored.

Thus $\qquad p_2 v_2 = RT + bp_2 - \dfrac{a}{v_2}$

and $\qquad p_1 v_1 = RT + bp_1 - \dfrac{a}{v_1}$

$\therefore (p_2 v_2 - p_1 v_1) = b(p_2 - p_1) + \dfrac{a}{v_1} - \dfrac{a}{v_2}$

and equation 109 becomes

$$W = b(p_2 - p_1) + 2a\left(\dfrac{1}{v_1} - \dfrac{1}{v_2}\right)$$

Since v_1 and v_2 are not directly measured in the actual experiment, they can be eliminated using the approximate relation $pv = RT$ when we have

$$W = (p_1 - p_2)\left[\dfrac{2a}{RT} - b\right]$$

If δT is the observed cooling effect, then $W = JMC_p\delta T$ where M is the molecular weight of the gas and thus

$$\delta T = \dfrac{(p_1 - p_2)}{JMC_p} \cdot \left[\dfrac{2a}{RT} - b\right]$$

The cooling effect at any temperature is thus shown to be directly proportional to the fall in pressure across the plug as was observed in the actual experiments. Since $p_1 > p_2$ the gas will undergo a cooling effect provided the second bracketted quantity $\left[\dfrac{2a}{RT} - b\right]$ is positive.

The effect will be a heating effect when this quantity becomes negative the inversion temperature T_i being given when

$$\dfrac{2a}{RT_i} - b = 0$$

Thus $\quad T_i = \dfrac{2a}{Rb}$

$$= 2T_B$$

where T_B is the Boyle temperature (equation 99).

Now from equation 105 the critical temperature T_c according to van der Waals' equation is $T_c = \dfrac{8a}{27Rb}$ and hence

$$\dfrac{T_i}{T_c} = \dfrac{27}{4} = 6.75$$

—the experimental value of this ratio for actual gases being a little below 6.

The fact that the inversion temperature is very much higher than the critical temperature (experimental values for hydrogen being 190°K. and 33°K. respectively), has important practical value in the liquefaction of gases—the various methods for which will now be discussed.

10.11   The Liquefaction of Gases

It has already been mentioned (section 10.01) that systematic work on the liquefaction of gases first began in 1823 with the liquefaction of chlorine, sulphuretted hydrogen, and similar gases by Faraday. The liquefaction of these gases was effected under compression with the aid of simple freezing mixtures. Faraday failed to liquefy the so-called "permanent" gases, oxygen, nitrogen, and hydrogen by these methods, and it was not until Andrews in 1863 had discovered the phenomenon of the critical state that further developments were possible. As a consequence of Andrews' discovery that a gas must first be cooled below its critical temperature before being liquefied by compression, attention was concentrated on the problem of temperature reduction. In 1877 Pictet and Cailletet working independently succeeded in liquefying oxygen. In Cailletet's experiments the cooling was produced by adiabatic expansion of the gas, whilst Pictet obtained his low temperature in successive stages employing the ordinary vapour compression refrigeration cycle (see section 9.12) in which the cooling is produced by the rapid evaporation of suitable volatile liquids. Details of the Pictet, or Cascade process as it is sometimes called, are given in the subsequent section. By developing Pictet's method using different cooling agents in the cascade Wroblewski and Olszewski in 1883 succeeded in liquefying nitrogen.

The discovery of the Joule-Kelvin effect in 1853 made possible commercial developments in the liquefaction of gases. Using this effect, together with a heat exchanger (see section 10.13), Linde in Germany in 1895 and Hampson in England in the same year, developed methods for the large scale production of liquid air. Linde's method was modified in 1902 by Claude in France who obtained his temperature reduction by causing the gas under high compression to expand in an expansion cylinder when in addition to the Joule-Kelvin effect there is pronounced cooling due to the expanding gas doing external work. Details of Claude's process are given in section 10.14. Using the Linde-Hampson principle, Dewar, in experiments commenced in 1895, and later Travers, succeeded in liquefying hydrogen, and Kamerlingh Onnes, using the same principle, produced liquid helium in 1908. Thus all the known gases had by this date been obtained in the liquid state. By reducing the vapour pressure of liquid helium using powerful pumps, thus inducing evaporation to produce a further reduction in temperature, and by subsequently applying high pressures, Keesom in 1926 brought to a culmination these researches by obtaining solid helium.

Apart from the commercial developments attendant on the lique-faction of gases, e.g. the preparation of oxygen in large quantities, and the separation of nitrogen and the rare gases from liquid air (see section 10.15), the attainment of the very low temperatures made possible by this means has opened up a very important field of research. Thus measurements made on specific heats, thermal and electrical conductivities, &c., in the low temperature zone have yielded important information on the solid state and the fundamental properties of matter.

10.12   The Pictet or Cascade Process

As mentioned above the Pictet method employs the vapour com-pression refrigeration cycle, a series of refrigerating machines being so connected that the working substance of any one of these when cooled by evaporation is caused to circulate round the condenser of the next machine of the series when a temperature is produced which is lower than the critical temperature of the working substance of the second machine. The vapour in the second machine can now be liquefied by compression, and the cooling produced by the evaporation of this second liquid in a similar way serves to effect the liquefaction of a third vapour, and so on. Thus the temperature reduction is brought about by a step by step process until a sufficiently low temperature is attained to make possible the liquefaction of any particular gas. A double cascade employing methyl chloride and ethylene is shown in Fig. 115. The methyl chloride (critical temperature 143°C.) is liquefied

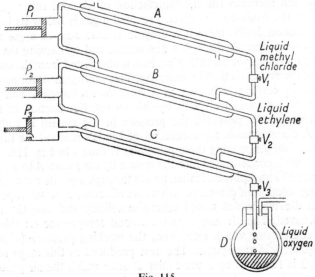

Fig. 115

on the compression stroke of the pump P_1 the heat of compression being absorbed by a stream of cold water circulating round the condenser A. The evaporation of the liquid methyl chloride as it is forced through the valve V_1 cools the vapour which circulates round the condenser B of the second cascade before being drawn into the pump P_1 to be used again and the process repeated. In this way the temperature is reduced to about $-90°C$. in B, enabling liquid ethylene (critical temperature $10°C$.) to be produced under compression by the pump P_2. A similar cycle of operations with the ethylene results in the temperature of the condenser C being reduced to about $-170°C$. at which temperature oxygen (critical temperature $-118°C$.) which passes through the inner tube of C, can be liquefied at modest compression by the pump P_3—the liquid oxygen being collected in a Dewar flask D.

The limit of the Pictet process is reached with nitrogen as there is no liquefied gas which on evaporation will produce a temperature low enough to bring about the liquefaction of hydrogen (critical temperature $- 240°C$.). Apart from this limitation, the chief objection to the Pictet process is that it is cumbersome and is not now-a-days used in the commercial preparation of liquid air (and oxygen). Nevertheless it is more economical than the Linde or Claude processes described in the subsequent sections, a larger return of liquid air per unit of energy used being obtained than with either of the other two methods. It is thus possible that further commercial development of the process may be made in the future.

10.13   The Linde Process

Modern methods for the liquefaction of air and other gases are based on the Joule-Kelvin effect. Provided a gas is below its inversion point (section 10.09) there is a drop in temperature as a gas at high compression expands through a fine orifice, and by causing the cooled gas to pass back over the inlet pipe, heat is abstracted from the incoming gas and the process is thus regenerative. This arrangement for producing cumulative cooling is known as a *heat interchanger,* and as the temperature progressively falls by this means, liquefaction of the gas ultimately results provided the pressure in the system is sufficiently high. A diagrammatic representation of Linde's method for the liquefaction of air embodying these principles is shown in Fig. 116.

The air is concentrated into the system by the pump A to a pressure of about 20 atmospheres. It then passes through a purifier B containing a strong solution of caustic potash where all traces of water vapour and carbon dioxide, which would otherwise solidify and clog the system, are removed. The air now enters a second compressor C: this is the main compressor which establishes the working pressure (about 200 atmospheres) in the system. The heat produced at this stage of compression is removed by a pre-cooler D after which the air passes into the expansion chamber E via a long copper spiral to the nozzle O where it is expanded, the cooled gas passing back over the spiral thereby cooling the oncoming air. This progressive cooling ultimately reduces the temperature of the air issuing at O to such an extent that a portion of the air liquefies on emergence into the chamber— the expanded unliquefied air being returned to enter the high pressure cycle (as indicated) to maintain a continuous circulation. The liquid air thus produced is collected in a Dewar vessel F from which supplies are drawn. In Linde's liquefier the air was expanded down to a pressure of about 20 atmospheres, whereas in a similar apparatus developed by Hampson the air emerged at O at atmospheric pressure. In actual practice the heat interchanger consists of two very long concentric copper tubes wound spirally, the expanded gas returning via the space between the two tubes.

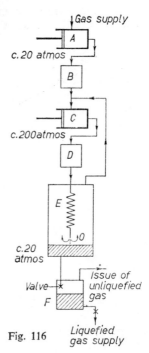

Fig. 116

10.14 The Claude Method

In the Claude method for the liquefaction of gases the compressed gas is allowed to expand adiabatically in an expansion cylinder thereby doing external work. There is thus a much greater temperature reduction than in the Linde process, for in addition to the cooling due to the Joule-Kelvin effect, there is a marked fall in temperature resulting from the loss in energy of the gas as it does external work. This work can be utilised to assist in driving the compressor, and hence the process is thermodynamically more efficient than the Linde process. The apparatus used by Claude is shown diagrammatically in Fig. 117. The compressed gas, purified and pre-cooled to room temperature, is conveyed along the inner pipe of the condenser A and part of it passes into the cylinder B. Here it expands, doing work, and the cooled gas at low pressures is ejected and flows back through the tubular condenser C where it cools the other part of the high pressure gas passing down the inner tubes of C. On leaving C the low pressure gas is conveyed back to the compressor via the outer jacket of A where it also cools the gas flowing from the compressor. In this way the temperature of the compressed gas admitted to *C* is sufficiently reduced to permit of its liquefaction under the pressure of admission, the liquefied gas

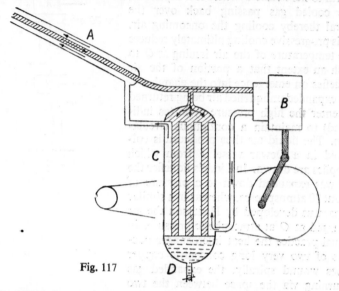

Fig. 117

being collected at *D* from which it can be drawn off through the tap at the bottom of the vessel.

A special problem associated with Claude's method is that of providing suitable lubricants at the low temperatures prevailing. At these temperatures ordinary lubricants solidify, although Claude found petroleum ether to be satisfactory. Once the gas has commenced to liquefy however, it acts as its own lubricant. In practice the Claude method has been found to be only slightly more efficient than the Linde method—due in part to friction in the expansion engine, and the fact that the expansion of the gas is not truly adiabatic. It is also more complicated mechanically. In view of these facts, and the complete absence of lubrication problems in the Linde process, most commercial liquefiers in use to-day are based on the Linde method.

10.15 The Separation of Oxygen and Nitrogen from Liquid Air

One of the most important industrial applications of low temperature technique is the separation of oxygen and nitrogen from liquid air. The preparation of oxygen by the re-evaporation and rectification of liquid air has in fact almost completely superseded the older chemical methods of its production on a commercial scale, mainly because the oxygen obtained is of greater purity. The boiling points of liquid oxygen and nitrogen are —182·9°C. and —195·7°C. respectively; thus starting with liquid air it is possible to remove the more volatile constituents leaving a liquid progressively richer in oxygen. The slow evaporation of liquid air is however a wasteful method of obtaining the oxygen from it as both gases pass off on the free evaporation of liquid air, and although the more volatile nitrogen comes off in greater relative proportion, it is necessary to evaporate two-thirds of the original liquid to obtain a liquid which contains 50 per cent oxygen, whilst for a liquid 90 per cent rich in oxygen, 95 per cent of the original liquid would have to be evaporated. To avoid this loss Linde used a rectifying column in which the escaping gases were "scrubbed" by liquid air passing in the opposite direction, when oxygen is condensed from the up-going stream of gases, and nitrogen evaporated from the down-stream of liquid air. Thus the descending liquid becomes progressively richer in oxygen, whilst the ascending gases become correspondingly richer in nitrogen.

The scheme of Linde's apparatus is shown in Fig. 118. Dry air, purified of carbon dioxide, enters the apparatus at a pressure of 135

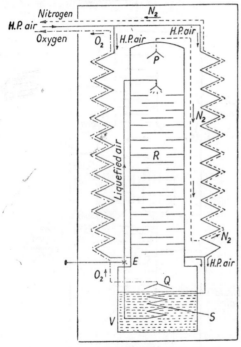

Fig. 118

atmospheres having been pre-cooled to a temperature of about −20°C. by means of an ammonia refrigerator. The air passes through the inner pipes of a heat interchanger (shown symbolically by the zig-zag

9*

lines in the diagram) which is wound spirally round the rectifying column R, and enters the compartment V (the vaporiser) at the base of the column where it is conveyed through a spiral S to the expansion valve E. Cooling takes place here due to the Joule-Kelvin effect, and in conjunction with the heat interchanger, this in due course results in some of the air being liquefied. The liquid air thus formed is sprayed into the top of the rectifying column, and as it descends, the more volatile constituent nitrogen evaporates and leaves the column at P, and a liquid relatively richer in oxygen collects in V at the bottom of the column. Here it is vaporised by abstracting heat from the high pressure air circulating through the spiral S. After the apparatus has been functioning for some time the initial pressure of the air is reduced to less than half its value, and the oxygen gas (which leaves the rectifier at Q) and the nitrogen gas pass from the apparatus through the outer annular tubes (shown in dotted line) of the heat exchanger. To avoid heat exchanges with the atmosphere the entire apparatus is enclosed

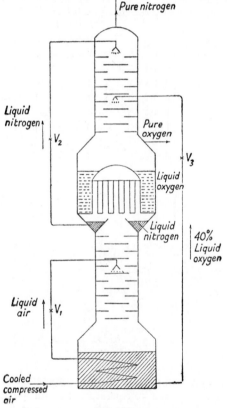

Fig. 119

in a thermally insulating casing. Although oxygen of a high degree of purity can be obtained by this method, it fails to separate the gases completely—the escaping nitrogen containing about 7 per cent of the oxygen. The process is however still in use in small plants, but for a more complete fractionation of liquid air the Linde double column separator shown in Fig. 119 is used.

Compressed air, which has previously been cooled almost to liquefaction point, is liquefied on circulation through a spiral immersed in a liquid bath at the base of the bottom column. From here it is expanded through the valve V_1 and sprayed through the middle of the lower column where it becomes progressively richer in oxygen as it descends, the rising vapours comprising almost pure nitrogen when they arrive at the top of the column. This nitrogen enters a nest of tubes cooled by liquid oxygen boiling under atmospheric pressure, and under the pressure existing in the lower column (5 atmospheres) the nitrogen is liquefied here. Part (c. 50 per cent) of this liquid nitrogen flows back down the lower column where it condenses oxygen from the rising vapours, whilst the remainder is contained in a circular trough situated below the tubes. From here it is expanded through a valve V_2 and sprayed into the top of the upper column condensing oxygen from the vapours ascending this column. The liquid collecting at the base of the lower column contains about 40 per cent oxygen, and for the final separation, this liquid is delivered via a valve V_3 to a point about half-way up the upper column. Pure liquid oxygen collects at the base of the upper column, the vapours rising from this boiling liquid being conducted away as shown (20 per cent of these vapours are led back through the heat interchanger).

In addition to oxygen and nitrogen, which comprise 21 per cent and 78 per cent by volume respectively of the air, there is also about 1 per cent by volume of the inert gases argon ($-$ 185·7°), neon ($-$ 255·9°), helium ($-$ 268·9°), krypton ($-$ 151·7°), and xenon ($-$ 109·1°) in the air—diminishing in relative quantity in the order shown. The figures in parenthesis are the boiling points of the liquids in °C. The first three of these gases can be obtained from liquid air by a low temperature fractionation process similar to that described above. Krypton and xenon have not so far been separated in sufficiently large quantities to have commercial usage.

The main use of oxygen prepared in this way is in metal cutting such as the breaking up of old ships and machinery. It is also extensively used in oxy-acetylene welding, whilst amongst its many other uses may be mentioned its application medically for resuscitation, and its use in breathing apparatus for high altitude flying, deep sea diving and mine rescue work. Nitrogen, which was at one time considered a waste product, is now used on a large scale in the synthetic production of ammonia by the Haber process, and also for the manufacture of cyanamide for use as a nitrogenous fertiliser. The main use of argon is as the inert gas in gas-filled electric lamps enabling higher temperatures to be obtained without the risk of the filament volatilising. Non-inflammable helium (which is in the main derived from sources other than liquid air) has been used in airships as a substitute for hydrogen, and together with neon it is extensively used in discharge tubes for illuminated advertising signs and other purposes. Liquid air itself is used (aided by suitable adsorbants) in the chemical industry for the separation and purification of gases. It has also extensive application in high vacuum laboratory technique by its use in liquid air traps (see Fig. 58) for the removal of water and mercury vapours.

Liquefied gases are stored in silvered vacuum vessels or Dewar flasks, free evaporation at atmospheric pressure being permitted. For the transportation of large quantities of liquid air, &c., large spherical vessels of copper with long necks open to the atmosphere are used. These are not of the vacuum type but are surrounded with such heat insulating materials as slag wool and magnesium carbonate, the daily loss with such containers being only about 5 per cent.

10.16 Liquefaction of Hydrogen and Helium

Hydrogen cannot be liquefied by the normal Linde process as described in section 10.13 as at temperatures above − 83°C. it exhibits a positive (i.e. heating) Joule-Kelvin effect. It must therefore be subject to a considerable degree of pre-cooling before the liquefaction process is initiated. By thus pre-cooling first with a mixture of solid carbon dioxide and ether, and then with liquid air, Dewar succeeded in 1898 in obtaining small quantities of liquid hydrogen using the Linde principle. His method was extended in 1900 by Travers who designed an apparatus for producing much larger quantities of liquid hydrogen. Travers' apparatus has been improved by Olszewski, Nernst and others, and although of more complicated and improved design, modern hydrogen liquefiers are based on the original methods of Dewar and Travers.

The arrangement used by Travers is shown in Fig. 120. Hydrogen at a pressure of about 200 atmospheres is first cooled by passing through a coil B immersed in solid carbon dioxide and alcohol after which it is conveyed through a coil G contained in a vessel filled with liquid air and subsequently through another coil H surrounded by liquid air boiling under reduced pressure. By this means the temperature of the hydrogen on leaving H is reduced to about − 200°C. The gas now passes through the regenerator coil C from which it escapes from the nozzle D the opening of which is controlled by the milled head at K attached to a rod passing down the apparatus through the centre of the coils. The expanded gas, further cooled by the Joule-Kelvin effect at D, sweeps back over the regenerator coil and passes round the outside of the liquid air vessels before leaving the apparatus at L to be conveyed back to the compressor. In due course liquefaction of the hydrogen commences as the temperature at D is progressively lowered, the liquid hydrogen being collected in the Dewar vessel at F. The hydrogen must be carefully purified before entering the liquefying apparatus as otherwise impurities would solidify and cause stoppages. It is also necessary to obtain very complete heat insulation and for this purpose the entire apparatus is contained in a large Dewar vessel as shown.

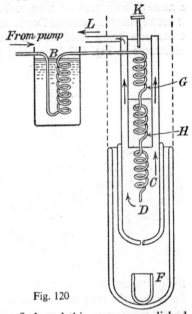

Fig. 120

Helium was the last gas to be liquefied, and this was accomplished by Kamerlingh Onnes in 1908 using the same method as that for hydrogen described above. The inversion temperature of helium is about $- 240°C.$, and Onnes obtained the necessary temperature reduction by pre-cooling with liquid hydrogen boiling under reduced pressure, the temperature of the helium gas being lowered to $- 258°C.$ in this way. Kapitza has liquefied helium pre-cooled by passing through boiling liquid nitrogen using a modification of the Claude expansion engine with a small gap between the piston and the cylinder to overcome the difficulties of lubrication. Simon, using the fact that the thermal capacity of a metal container is extremely small (see section 2.26) at temperatures near the absolute zero, has liquefied helium in small quantities by adiabatically expanding helium (with subsequent cooling) contained in a small metal cylinder immersed in liquid hydrogen. Another interesting method of liquefying helium, also due to Simon, makes use of the fact that heat is evolved when a gas is adsorbed on charcoal, and thus on desorbing (removing) the gas the temperature must fall. The helium is adsorbed in considerable quantities on activated charcoal immersed in liquid hydrogen. The container is situated in an evacuated vessel to minimise heat exchanges with the surroundings, and on pumping off the helium the temperature falls sufficiently to bring about liquefaction.

The storage of liquid hydrogen and liquid helium, on account of the very low temperatures and low latent heats of these liquids, presents a more difficult problem than in the case of liquid air. Liquid hydrogen is usually kept in a double Dewar vessel the space between the evacuated containers being filled with liquid air. Kapitza devised the method illustrated in Fig. 121 for storing liquid hydrogen. Inside the vacuum space of the Dewar vessel containing the liquid hydrogen was a copper shield A (not in contact with the walls). This was placed in thermal contact with some liquid air contained in a second and somewhat smaller Dewar flask by means of a short bar of copper B as shown. As the thermal conductivity of copper is very great at the temperature of liquid air, this method provided a most satisfactory means of reducing the temperature difference between the liquid hydrogen and that of its immediate surroundings with consequent reduction in the evaporation rate of the liquid. To store liquid helium a triple Dewar vessel is generally used, the liquid helium being surrounded by liquid hydrogen which in turn is surrounded by liquid air. To enable the helium to be seen the walls of the vessels are not silvered, and to prevent the condensation of water vapour on the outside of the liquid air flask, the entire vessel is placed in one containing alcohol at room temperature.

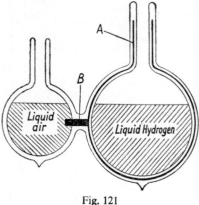

Fig. 121

Liquid helium is a colourless mobile liquid which boils at $4 \cdot 2° K$. It has a density which is about one-eighth that of water and its very flat meniscus indicates a small surface tension. By boiling liquid helium under reduced pressures, the temperature can be further reduced, and Kamerlingh Onnes in this way attained a temperature of $0 \cdot 82° K$ on reducing the pressure as low as $0 \cdot 013$ mm. of mercury. In spite of these low temperatures Onnes never obtained any trace of solidification. Solid helium was first obtained in 1926 by Keesom who previously cooled liquid helium and then applied high pressures. A magnetic stirrer contained in the liquid helium could not be moved when the helium was subjected to a pressure of 86 atmospheres at a temperature of $3 \cdot 2° K$. This indicated that solidification had in fact taken place although the line of division between the solid and liquid forms could not be seen, indicating almost identical refractive indices for the solid and liquid helium.

10.17   The Approach to the Absolute Zero—Adiabatic Demagnetisation

We have seen that by the evaporation of liquid helium under greatly reduced pressures Kamerlingh Onnes attained a temperature of $0.82°K$. The lowest temperature reached in this way is about $0.7°K$ which was obtained by Keesom using extremely powerful pumps. This seems to be about the limit of temperature which can thus be reached on account of the very low vapour pressure (of the order of 10^{-3} mm. of mercury and decreasing rapidly with temperature) of liquid helium at these temperatures. In 1926 Debye, and independently Giauque, suggested the possibility of obtaining still lower temperatures by the method of adiabatic demagnetisation. The magnetic susceptibility χ of a paramagnetic substance (i.e. one that is magnetised in the direction of the applied field) is attributed to molecules possessing a permanent magnetic moment due to unbalanced electron orbits. On the application of an external magnetic field the molecules tend to align themselves in the direction of the field, the tendency being opposed by the thermal motion of the molecules. On the basis of this temperature equilibrium of the molecular magnets, and assuming equipartition of energy for the molecules distributed among the different states under equilibrium conditions, Langevin has developed a theory of paramagnetism (later somewhat modified by the quantum theory), which is in accord with the experimental results found by Curie. From comprehensive observations on the susceptibility of paramagnetics over a wide range of temperature, Curie found that the susceptibility was inversely proportional to the absolute temperature. Thus

$$\chi = \frac{C}{T}$$

where C is a constant. This law of variation is known as *Curie's law*.

The application of the external energy required to produce mass magnetisation in a paramagnetic substance results in a rise of temperature, and when the molecular magnets return to their original random distribution on removing the magnetising field there will be a corresponding fall in temperature. At very low temperatures the susceptibility is, by Curie's law, very large, and since also the specific heat values at this temperature are very much less than at ordinary temperatures, the fall of temperature consequent on the adiabatic demagnetisation at temperatures near the absolute zero should be quite pronounced.

In carrying out the method, the magnetic substance is placed in a chamber and first cooled by surrounding it with liquid helium in a Dewar vessel which is in turn surrounded by another vessel containing liquid hydrogen. A small amount of helium gas is employed in the chamber containing the specimen in order to establish thermal contact with the liquid helium surrounding it. On evaporation of the liquid helium the temperature is reduced to about $1°K$. A powerful electromagnet capable of developing a magnetic field of from 10,000 to 30,000 gauss is then switched on, the heat produced in the process of magnetising the specimen being conducted away by the helium gas. When the specimen has again attained the temperature of the liquid helium it is thermally insulated by evacuating the chamber of helium gas, and on subsequently switching off the magnetic field, the temperature of the specimen falls. In the early experiments rare-earth salts such as gadolium sulphate and compounds of cerium, &c. were used, but more recently good results have been obtained using ferric ammonium alum and potassium chrome alum. In this way de Haas and Wiersma (1935) have obtained the exceedingly low temperature of $0·003°K$.

The extremely low temperature range thus made possible has opened up an entirely new field of research producing results of profound importance in considering the fundamental properties of matter. Thus, to take one example, the electrical resistance of pure metals known to be proportional to the absolute temperature over very wide ranges of temperature has, in many instances, been found to fall suddenly to zero at temperatures somewhat above the absolute zero. This condition, known as *super-conductivity*, was found by Kamerlingh Onnes to occur for solid mercury at a temperature of $4·2°K$., and the temperature at which the effect occurs for many other metals has been found. Thus, lead becomes super-conducting at $7·2°K$., tin at $3·7°K$., aluminium at $1·14°K$., zinc at $0·79°K$., &c. A current established in a substance in the super-conducting state will continue to flow undiminished in intensity for a number of days. The application of a magnetic field affects the behaviour of super-conductors and results in a lowering of the temperature at which the super-conductive state occurs. It should be mentioned that not all metals exhibit the phenomenon, thus it has not as yet been discovered in the case of copper or gold.

10.18   The Measurement of Very Low Temperatures

We shall conclude this chapter with a brief survey of the various methods by which low temperatures are measured in practice. The ideal or thermodynamic scale is most closely realised practically by the use of gas thermometers, and these thermometers are employed as standards over very wide ranges of temperature. For low temperature work hydrogen thermometers (usually of the constant volume type) are used, whilst for very low temperatures helium replaces hydrogen, and by working with the helium at pressures below its vapour pressures for the temperatures concerned, temperatures almost down to $1°K$. can be measured. Vapour pressure thermometers, by which temperatures are obtained from the known variation of the vapour pressure of liquefied gases with temperature, have also been employed for low temperature determinations. A low pressure gauge is used for the measurement of the vapour pressure, and by extrapolating the vapour pressure curve of helium, temperatures can be estimated by this means down to about $0.7°K$. Below this however the vapour pressure of helium is too low to be capable of precise measurement.

Electrical resistance thermometers can also be used for low temperature measurement. Platinum thermometers give very good results down to $-200°C$. and have been used for even lower temperatures, whilst Kamerlingh Onnes has used a lead resistance thermometer for temperatures down to $10°K$. For very low temperatures however these thermometers are somewhat insensitive, and are complicated by the phenomenon of superconductivity at temperatures approaching the absolute zero. Thermo-electric thermometers are also used for low temperature determinations, iron-constantan and copper-constantan thermo-couples being used to record temperatures down to about $18°K$., whilst gold-platinum and silver-platinum couples may be used for still lower temperatures.

Temperatures below $1°K$. are measured from determinations of the magnetic susceptibility of paramagnetic salts for which Curie's law (see previous section) applies. As the susceptibility of these substances increases as the temperature is reduced, the method possesses the advantage of being very sensitive at extremely low temperatures. In determining these "magnetic" temperatures, the apparatus containing the specimen is surrounded by two coils through one of which (the primary) an alternating current is passed. The current induced in the secondary coil depends on the susceptibility of the salt used, which can thus be found by suitably amplifying and measuring this small induced current, and hence the temperature can be determined. Frequently the galvanometer used to detect the secondary current is calibrated to give direct readings of the temperature. Recently work has been done to link magnetic temperature measurements made in this way with gas thermometer readings, and thus to determine the relation between the "magnetic" scale and the thermodynamic scale of temperatures.

QUESTIONS. CHAPTER 10

1. Describe the experiments you would perform to find how nearly sulphur dioxide obeys Boyle's Law for pressures between ½ and 4 atmospheres at the temperature of the laboratory.

Draw diagrams to show how the volume of carbon dioxide varies with its pressure (a) at 0°C., (b) at its critical temperature 31°C., (c) at 100°C.   (O.)

2. Give an account of the work of Andrews on the relation between the pressure, volume, and temperature of a gas. Explain, with the help of diagrams, the terms *critical temperature* and *critical pressure*.

Describe a method of determining critical temperatures.   (Camb. Schol.)

3. Describe the researches which led to the knowledge of the conditions under which a gas can be liquefied.

What modifications have been made in the " perfect " gas equation in order that it may represent more closely the results of these researches? (N.)

4. Describe experiments by means of which the deviations from Boyle's Law shown by gases have been investigated. Draw typical curves of the results showing the relation between the pressure and the product of the pressure and volume. Explain the causes of the deviations. Why does van der Waals' equation give a better representation of the behaviour of gases than that given by Boyle's Law? (N.)

5. What is the law of corresponding states? What is its basis? In what respect is the law useful? The critical temperature and pressure of benzene are 288·5°C. and 47·9 atmospheres respectively, and those of chloroform are 260°C., 54·9 atmospheres; benzene boils at 80°C. What would you expect the vapour pressure of chloroform at 62·1°C. to be? (Oxford Schol.)

6. What experiments have been carried out to investigate the variation of volume of a gas with pressure at various temperatures? How can the kinetic theory of an ideal gas be modified to account for this variation and how far is the modified theory successful? (S.)

7. Describe experiments which have been carried out to determine how the properties of an actual gas, such as carbon dioxide, differ from those of an ideal gas. What results have been obtained, and what explanations have been put forward to account for these results?

What relevance have the results of such experiments on the problem of the liquefaction of gases? (Camb. Schol.)

8. Explain the principles underlying methods of liquefying gases. How may the temperature of liquefaction under a given pressure be measured. (Camb. Schol.)

9. Explain the terms *triple point*, *critical temperature*, illustrating your answers with appropriate graphs.

Describe a process which will yield a continuous supply of liquid air. (N.)

10. Describe the arguments which justify the Van der Waals equation of state for a real gas:

$$\left(p + \frac{a}{V^2}\right)(V - b) = R\theta.$$

If R in the above equation is the gas constant per unit mass, p is the pressure, and θ the absolute temperature of the gas, what are the dimensions of the quantities V, a, and b which appear in the equation?

A vessel of negligible heat capacity is thermally isolated from its surroundings and contains an ideal gas initially at high pressure. When a valve is opened to allow the gas to escape into the atmosphere the temperature of the gas which remains in the cylinder falls. Explain why this is so, and discuss the factors on which the final temperature of this gas depends. [O. and C.]

11. Show how the characteristic equation for unit mass of an ideal gas ($pV = RT$) has been modified by Van der Waals to take into account the finite size of the molecules of real gases and the mutual attraction between neighbouring molecules.

For carbon dioxide Van der Waals' constants are $a = 7·2 \times 10^{-3}$ and $b = 1·9 \times 10^{-3}$. If carbon dioxide of molecular weight 44 has a density of 1·976g. per litre at a temperature of 0°C. and a pressure of 76cm. of mercury, find the universal gas constant R in calories per degC. per gram-molecule, assuming that carbon dioxide is an ideal gas. Show that the error involved in the assumption is very small. Do not attempt to calculate the numerical value of the error. [A.]

12. Describe in detail how you would investigate experimentally the relationship between the pressure, volume and temperature of an imperfect gas and discuss the results that you would expect to obtain.

A gas obeys Van der Waals' equation

$$\left(p + \frac{a}{v^2}\right)(v - b) = RT,$$

where $a = 1\cdot4 \times 10^{12}$ dyne cm.^{4} mol^{-2} and $b = 32$ cm.^{3} mol.$^{-1}$

Calculate the work done when 1 gram molecule is compressed from 20 litres to 0·1 litre at 0°C. How does this amount of work compare with that required to compress an ideal gas in the same way?

($R = 8\cdot3 \times 10^7$ erg mol.$^{-1}$; log$_e$ 10 = 2·3.) (Camb. Schol.)

13. Describe, with the help of a diagram, a method employed for the liquefaction of air. Explain how the cooling of the air is brought about and how, if at all, the method requires to be modified when applied to hydrogen. (N.)

14. Give an account of the methods of liquefying the more permanent gases, and explain briefly how the boiling points may be determined. (Camb. Schol.)

15. Give an account of the liquefaction of gases. Describe the general construction of an apparatus for liquefying hydrogen. Discuss briefly the investigation of one property of matter at low temperatures. (Oxford Schol.)

16. Write an account of methods of liquefying gases, emphasizing particularly the physical principles involved. [A.]

17. Give an account of the principles involved in the liquefaction of gases. Describe some form of air liquefier.
What essential difference is involved in the design of a liquefier of hydrogen? [L.]

18. Distinguish between the cooling of a gas by adiabatic expansion and the Joule-Kelvin effect. How are these two types of cooling explained by the kinetic theory of gases, and how is the latter type used in liquefying gases? [L.]

19. Write a short essay on the liquefaction of gases. (S.)

20. Discuss the terms (i) critical pressure, (ii) critical temperature, (iii) continuity of state, referring to the system of $p - v$ isothermals for a substance.
Discuss the importance of these concepts in the liquefaction of gases.
(Melbourne)

21. (a) Derive van der Waals' equation of state of a real gas and adapt it for use when the unit of pressure is the standard atmosphere and the unit of volume is the normal specific volume.
(b) How far is van der Waals' equation successful in accounting for the behaviour of a gas above and below its critical temperature? (Melbourne)

22. (a) Describe the principles underlying the liquefaction of air, and give a diagram showing how these principles are applied in practice.
(b) Describe the physical changes that take place in materials when they are cooled to the temperature of liquid air. (Melbourne)

CHAPTER XI

CONDUCTION OF HEAT

11.01 Introductory

There are three distinct methods by which a quantity of heat may be transferred from one place to another. These are: conduction, convection, and radiation. In the first two processes the molecules of the body are responsible for the heat transfer, but with the difference that there is no mass movement of groups of molecules in conduction as in the case in the transfer of heat by convection. Thus the heat conducted along a metal bar heated at one end is due to the molecules at the hot end having their vibratory motions increased, which results in the molecules in successive sections of the bar acquiring increased energy by a chain of collision processes—the molecules remaining throughout centred about their equilibrium positions. The transfer of heat in fluids by the process of convection on the other hand is brought about by the actual movement of those masses of the fluid which, having become warmed by contact with the heat source, expand and so rise through the body of the fluid in consequence of the reduction in density. There is an inflow of colder fluid to take the place of the rising heated masses, and the circulating movement so established is known as a convection current. Examples of this process of heat transfer as applied to the atmosphere will be discussed in fuller detail in chapter XIV.

It is clear from the nature of the solid state that the process of convection cannot occur in solids. Conduction, however, is effective in both solids and fluids, although in the latter class of substances convection is far more effective in distributing the heat through the fluid mass than is conduction. Liquids and gases are in fact bad conductors of heat as also are the bulk of non-metallic solid substances. The reader will be familiar with the simple experiments demonstrating these facts and also with the applications of the high thermal conductivity of metals, *e.g.*, the Davy safety lamp, and these will not be discussed further here.

The third method of heat transference, namely, radiation, does not require the presence of a material medium. Thus the heat from the sun travels through vast distances of empty space on its journey to the earth. Heat radiation consists of electro-magnetic waves and has many properties common to other such waves, *e.g.*, speed of propagation, and properties of reflection and refraction, &c. A fuller discussion of radiant heat is given in chapter XII, and it is sufficient for the present merely to recognise the essential difference between this and the two foregoing processes of heat transmission.

11.02 Coefficient of Thermal Conductivity

Consider a parallel-sided slab of material (Fig. 122) of thickness d whose faces P and Q are maintained at steady temperatures of θ_1 and θ_2°C. respectively. Heat will flow through the specimen from the face at higher temperature to that at the lower temperature, and if it is assumed that there is no lateral loss of heat, the lines of flow will be at all points normal to the faces of the slab. In these circumstances, under steady conditions, the quantity of heat H entering face P in a given time will be the same

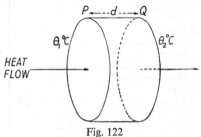

Fig. 122

as that which leaves face Q in the same period, and experiment shows that H is

(i)   directly proportional to the area A of the faces of the slab.

(ii)   directly proportional to the time t of flow,

(iii) directly proportional to the temperature difference $(\theta_1 - \theta_2)$ between the faces,

(iv) inversely proportional to the thickness d of the slab,

(v) dependent on the substance of the slab.

These relationships may be expressed by the following equation:

$$H = k A \left[\frac{\theta_1 - \theta_2}{d} \right] t \quad . \quad . \quad . \quad . \quad . \quad . \quad [110]$$

where k is a constant and is known as the *coefficient of thermal conductivity* of the substance considered. The quantity $\left(\dfrac{\theta_1 - \theta_2}{d} \right)$ is the temperature gradient across the specimen, and hence from equation 110 we may define k as being the quantity of heat per second flowing normally between two opposite faces of a specimen in the form of a unit cube when there is a difference in temperature of 1° between the faces under steady state conditions. The numerical value of k will depend on the units used for measuring the quantities appearing in equation 110. In the C.G.S. system of units, k will be expressed in cals. cm.$^{-1}$ sec.$^{-1}$ deg. $^{-1}$C.

TABLE 27

Thermal conductivities of various substances in cals. cm.$^{-1}$ sec.$^{-1}$ deg.$^{-1}$C.

Substance	k	Substance	k
Pure metals:		*Non-metals:*	
Aluminium	·504	Asbestos	0·0003
Copper	·918	Ebonite.................	0·0004
Gold	·700	Cardboard	0·0005
Iron	·176	Cork	0·0001
Lead	·083	Cotton wool...........	0·00006
Platinum	·166	Felt	0·00009
Silver	·974	Ice	0·005
Zinc	·265	Paraffin wax...........	0·0006
Alloys:		Rubber	0·00045
Brass	·260	Slate	0·0047
Bronze	·099	*Liquids:*	
Constantan	·054	Alcohol.................	0·00043
Manganin	·053	Aniline	0·00041
Gases:		Castor oil...............	0·00043
Air	5·77 × 10^{-5}	Glycerine	0·00068
Ammonia	5·22 „	Mercury	0·0202
Carbon dioxide........	3·43 „	Olive oil	0·00040
Hydrogen	31·8 „	Paraffin oil	0·00030
Nitrogen	5·81 „	Turpentine	0·00030
Oxygen	5·83 „	Water (10°C.)	0·00147

In cases where the temperature gradient is not uniform equation 110 has to be modified to obtain the heat flow across any given plane. Thus if the temperature of two planes, of area A, at distances x and $x + \delta x$ along the direction of heat flow are θ and $\theta + \delta\theta$ respectively, the heat conducted in a time t across the very thin section between these planes is

$$H = kA \cdot \frac{\theta - (\theta + \delta\theta)}{\delta x} \cdot t$$

$$= -kA \cdot \frac{\delta\theta}{\delta x} \cdot t$$

This becomes, as δx tends to zero,

$$H = -kA \cdot \frac{d\theta}{dx} \cdot t \quad . \quad . \quad . \quad . \quad . \quad . \quad [111]$$

this last equation giving the heat conducted across the plane x in time t.

11.03   Temperature Distribution along a Bar in the Steady State

Consider a long cylindrical bar, one end of which is maintained at a steady temperature of θ_0. The temperatures at various points along the bar will show a gradual rise for some time after the initial application of the heat, but ultimately a steady state will be attained at which these temperatures remain stationary. A steady flow of heat will now take place along the bar, and if the bar is exposed to the surrounding air, the quantity of heat passing any given section will be subject to lateral losses before arriving at a subsequent section. The lines of flow along the bar are shown in Fig. 123 where L and

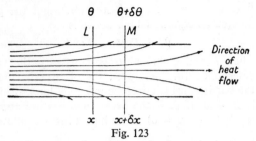

Fig. 123

M are two sections taken at distances of x and $x + \delta x$ from the hot end with temperatures of θ and $\theta + \delta\theta$ respectively. Let the temperature gradient at L be $\dfrac{d\theta}{dx}$. Then if A is the area of cross section of the bar (assumed uniform) and k the thermal conductivity of the material, the quantity of heat flowing per second across the section at L will be

$$- kA \frac{d\theta}{dx}$$

Now, if δx is sufficiently small, the rate at which the temperature varies along the bar between the two planes can be taken to be $\dfrac{d\theta}{dx}$ and hence $\delta\theta = \dfrac{d\theta}{dx} \delta x$. Thus the quantity of heat flowing across section M per second will be

$$- kA \frac{d}{dx} \cdot \left[\theta + \frac{d\theta}{dx} \cdot \delta x \right]$$

Consequently the excess of heat flowing per second across section L over that flowing across section M is

$$kA \frac{d^2\theta}{dx^2} \cdot \delta x = \delta q \text{ (say)} \quad \ldots \ldots \quad [112]$$

In the steady state this amount of heat is lost per second from the sides of the bar between the sections at L and M. Now, if the emissivity of the surface of the bar is E cal. per sq. cm. per unit temperature difference per second, then taking θ as the *excess* temperature of the bar at L and assuming Newton's law of cooling to obtain, we have

$$\delta q = E\theta p \delta x \quad \ldots \ldots \ldots \ldots \quad [113]$$

p being the perimeter of the section. Thus from equations 112 and 113

$$kA \frac{d^2\theta}{dx^2} = E\theta p \quad \ldots \ldots \ldots \ldots \quad [114]$$

or

$$\frac{d^2\theta}{dx^2} = \mu^2\theta \quad \ldots \ldots \ldots \ldots \quad [114a]$$

where

$$\mu^2 = \frac{Ep}{kA}$$

It can be verified by substitution* that the general solution of the differential equation 114a above is

$$\theta = Pe^{\mu x} + Qe^{-\mu x}$$

when P and Q are abitrary constants, the values of which can be determined for any given set of experimental conditions. Thus, if the bar is considered to be infinitely long, then $\theta \to 0$ as $x \to \infty$, that is $P = 0$. If θ_o is the excess temperature above the surroundings at the hot end ($x = 0$) of the bar, it further follows that $Q = \theta_o$

$$\therefore \; \theta = \theta_o e^{-\mu x} \; . \; . \; . \; . \; . \; . \; . \; [115]$$

Hence, for an exposed bar, the temperature falls off exponentially from the heated end in the manner shown in Fig. 124.

In the case of a bar treated so as to prevent lateral losses of heat by lagging it with some suitable heat insulating substance, δq (equation 112) is zero, and hence it follows that

$$kA \frac{d^2\theta}{dx^2} = 0$$

which on integration becomes

$$kA \frac{d\theta}{dt} = \text{constant.}$$

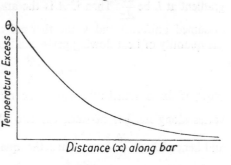

Fig. 124   Temperature distribution along an exposed bar

Thus the temperature gradient $\dfrac{d\theta}{dt}$ is constant for a uniform lagged bar (Fig. 125). The flow lines are parallel to the axis of the bar in this case and to prevent the steady heating of the bar, the quantity of heat, conducted along it in the steady state must be removed (see Searle's method, section 11.06).

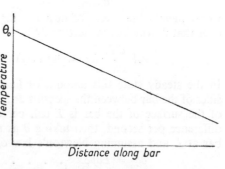

Fig. 125.   Temperature distribution along a lagged bar

---

* Thus

$$\frac{d\theta}{dx} = \mu Pe^{\mu x} - \mu Qe^{-\mu x}$$

and

$$\frac{d^2\theta}{dx^2} = \mu^2 Pe^{\mu x} + \mu^2 Qe^{-\mu x}$$

$$= \mu^2 (Pe^{\mu x} + Qe^{-\mu x}) = \mu^2 \theta$$

11.04 Thermal Diffusivity

Equation 114 applies to the heat conducted along a bar which has attained its steady state. Under these conditions the difference between the heat flow across the sections L and M (Fig. 123) is radiated from the sides of the element of length δx between the two sections. Before steady state conditions are attained, however, the temperature of each section of the bar is gradually rising, and in this initial variable stage the difference between the heat flow across the two sections is accounted for partly by lateral losses in the element, and partly by the amount of heat required to raise the temperature of the mass of the element. If the temperature rises by an amount $d\theta$ in time dt, and the density and specific heat of the material of the bar are respectively ρ and s, the amount of heat thus absorbed per second will be

$$A \; \delta x \rho. \; s. \; \frac{d\theta}{dt}$$

Accordingly equation 114 becomes

$$kA \; \frac{d^2\theta}{dx^2} \;=\; E\theta p \;+\; A\rho s \frac{d\theta}{dt} \quad\cdots\cdots\quad [116]$$

In the case of a specimen heated without lateral losses, e.g., a specimen surrounded with a guard ring (section 11.09), $E = 0$ and equation 116 becomes

$$kA \; \frac{d^2\theta}{dx^2} \;=\; A\rho s \; \frac{d\theta}{dt}$$

from which

$$\frac{d\theta}{dt} = \frac{k}{\rho s} \cdot \frac{d^2\theta}{dx^2}$$

The quantity $\dfrac{k}{\rho s}$ is called the **thermal diffusivity** of the specimen and measures the change of temperature produced in unit volume of the substance by a quantity of heat flowing in unit time through unit area of a layer of unit thickness under unit temperature difference. In all problems concerned with variable temperature conditions (as opposed to the steady state conditions which apply to methods for determining the thermal conductivity discussed in subsequent sections) it is this quantity, and not the thermal conductivity, that must be considered.

11.05 Comparison of Conductivities—Ingenhausz's Experiment

An approximate method of comparing the thermal conductivities of materials in the form of long thin rods was devised by Ingenhausz.

274 CONDUCTION OF HEAT

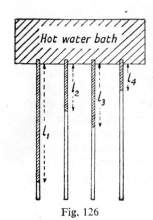

Fig. 126

The rods have identical dimensions and are treated, *e.g.*, by lacquering, so as to have the same surface emissivity. The rods are then thinly coated with wax and placed with one end protruding into a water bath as shown in Fig. 126. The water in the water bath is maintained at boiling point and when steady conditions are attained, the lengths l_1, l_2, &c., of melted wax on the rods are measured. If the rods are considered to be sufficiently long for equation 115 to apply, we have

$$\theta' = \theta_o\, e^{-\mu_1 l_1}$$
$$\theta' = \theta_o\, e^{-\mu_2 l_2}$$

&c.,

where θ_o is the temperature of the hot ends of the bars, and θ^1 the melting point of the wax. Hence,

$$\mu_1\, l_1 = \mu_2\, l_2 = \ldots$$

But

$$\mu_1 = \sqrt{\frac{Ep}{k_1 A}}\,,\; \mu_2 = \sqrt{\frac{Ep}{k_2 A}}\,,\ldots$$

and therefore,

$$\frac{k_1}{l_1^2} = \frac{k_2}{l_2^2} = \ldots$$

or

$$k_1 : k_2 : \ldots = l_1^2 : l_2^2 : \ldots$$

Determination of Thermal Conductivity of Good Conductors

11.06 Searle's Method. A simple method of determining the thermal conductivity of good conductors, such as copper or silver, has been devised by Searle. A typical form of his apparatus is shown in Fig. 127. The specimen under test is in the form of a cylindrical bar about 5cm. in diameter and 20cm. long. One end of this bar fits into a steam chest through which a steady supply of steam is passed, and the heat conducted along the bar is taken off by a stream of cold water from a constant head device circulating through a spiral as shown. The temperature of the water as it enters and leaves the spiral is taken by two $\frac{1}{10}$°C. thermometers T_4 and T_3. Two other thermometers T_1 and T_2 are placed in holes drilled in the bar a measured distance d apart, mercury being placed in the holes to ensure good thermal contact. (Thermo-couples may be used instead of the thermometers T_1 and T_2 with the advantage that smaller holes are required and thus there is less disturbance to the flow of heat along the bar). To prevent loss of heat from the sides of the bar the apparatus is lagged with felt, the whole being contained in a wooden box. The temperature gradient along the bar is then uniform (see Fig. 125).

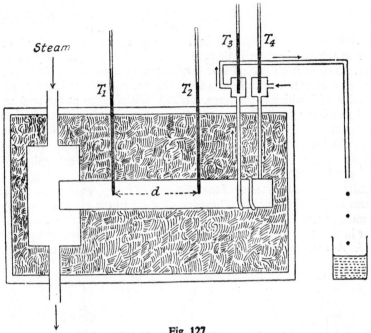

Fig. 127

When the steady state has been attained (usually after 30 minutes or more) the readings of the four thermometers are taken, the constant head device having been adjusted to give a sufficiently slow rate of flow of water through the spiral to ensure a reasonable difference of temperature between T_3 and T_4. Having taken the thermometer readings, T_1 and T_2 and also T_3 and T_4 are interchanged and another set of readings obtained, the appropriate temperatures being taken as the average of the corresponding readings. If θ_1, θ_2, θ_3, θ_4 are the average readings of the thermometers T_1, T_2, T_3, T_4, and if Mgm. of water circulate through the spiral in t seconds, the heat conducted along the bar in this time is

$$= k \cdot \frac{\pi D^2}{4} \cdot \left(\frac{\theta_1 - \theta_2}{d} \right) \cdot t$$

where D is the diameter of the bar. This is equal to the amount of heat absorbed by the water circulating through the spiral, $viz.$,

$$M (\theta_3 - \theta_4)$$

Hence
$$k = \frac{4Md (\theta_3 - \theta_4)}{\pi D^2 (\theta_1 - \theta_2) t}$$

Electrically heated models of Searle's apparatus are available by which it is possible to take observations at the hot end of the bar. Thus, if I amps. are supplied to the heater under a P.D. of E volts, the heat generated in t seconds is $\dfrac{IEt}{4 \cdot 2}$ calories. This amount of heat should equal that extracted in t seconds at the cool end of the bar at the steady state if there is no heat loss from the sides of the bar. As there is no perfect heat insulator such a loss will in fact occur to some extent. Thus readings from the hot end of the bar will give a value of k which is too large, whilst readings taken at the cool end will give k too small. Consequently, a mean of the two results obtained with the electrically heated bar will give a better result for k than that obtained using the steam heated model.

11.07 Forbes' Method. This method, which was originally performed in 1852, aims at determining the thermal conductivity of a substance directly in terms of the definition of k as given in equation 111 from readings taken with a long bar exposed to the air, the bar being steadily heated at one end. The temperature distribution along the bar in the steady state is then of the exponential form give by equation 115. A convenient laboratory arrangement of Forbes' experiment

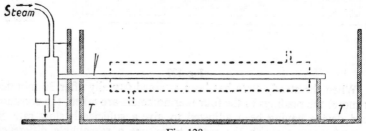

Fig. 128

is shown in Fig. 128 where a cylindrical copper bar about one metre long, steam heated at one end, is drilled at suitable intervals for the reception of a calibrated thermo-couple by means of which the temperatures at the various points along the bar are to be taken in the steady state. The bar is protected from radiation from the heating apparatus by means of wooden screens, and from draughts by means of the trough TT. The experiment is carried out in two parts (a) the *statical experiment*, in which the temperature distribution along the bar is obtained in the steady state, and (b) the *dynamical experiment* to obtain the rate of cooling of the bar at different temperatures from which is determined the lateral heat losses beyond any given section of the bar. The temperature distribution is taken by means of the thermo-couple when the bar has attained the steady state. Care should be taken that the temperatures at various points along the bar are constant, and check readings should be made before the final curve (Fig. 129) is plotted. By taking the gradient $\left[\dfrac{d\theta}{dx}\right]_P$ at a point P' on the curve—corresponding to a position P on the

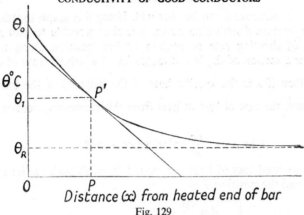

Fig. 129

bar with temperature θ_1—we have, from equation 111, the heat Q conducted across the section of the bar at P per second is

$$Q = -kA \left(\frac{d\theta}{dx}\right)_P \quad \ldots \quad \ldots \quad [117]$$

where A is the area of cross section of the bar and k the thermal conductivity of the material of the bar. In the steady state this quantity of heat is lost per second from the sides of the bar beyond P and may be found from observations of the rate of cooling of the bar at different temperatures obtained in the dynamical experiment as follows.

The bar is heated by passing steam through the jacket provided (shown in dotted line in Fig. 128), the jacket being subsequently removed to permit temperature readings of the bar to be made at measured intervals of time as it cools down almost to room temperature. The data enables a temperature-time graph to be drawn, from which, by takings gradients, a graph of rate of cooling $\left(\frac{d\theta}{dt}\right)$

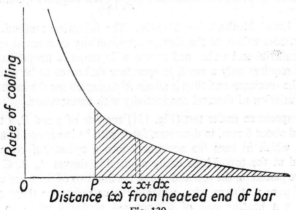

Fig. 130

against temperature θ can be obtained. Using this graph together with the temperature distribution curve, it is then possible to plot a curve (Fig. 130) showing rate of cooling against position along the bar. Consider a section of the bar of length δx at which the rate of cooling is $\dfrac{d\theta}{dt}$, then if s is the specific heat of the material of the bar and ρ its density, the rate of loss of heat from this elementary section of the bar is

$$A \, \delta x \, \rho \, . \, s \, . \, \frac{d\theta}{dt} \text{ calories per second.}$$

Hence the total loss of heat per second from all such sections of the bar beyond the position P is

$$\int_{P}^{End} A\rho s \, \frac{d\theta}{dt} \, . \, dx$$

$$= A\rho s \int_{P}^{End} \frac{d\theta}{dt} \, . \, dx$$

$$= A\rho s S \quad \cdot \quad \cdot \quad \cdot \quad \cdot \quad \cdot \quad \cdot \quad \cdot \quad \cdot \quad [118]$$

where $S = \displaystyle\int_{P}^{End} \dfrac{d\theta}{dt} \, . \, dx$ and is obtained by taking the area (by counting squares) intercepted between the curve and the axis of x by the ordinate drawn through P. Hence from equations 117 and 118

$$- kA \left[\frac{d\theta}{dx} \right]_{P} = A\rho s S$$

from which $\quad k = - \dfrac{\rho s S}{\left[\dfrac{d\theta}{dx} \right]_{P}}$

the minus sign being retained since $\left[\dfrac{d\theta}{dx} \right]_{P}$ is a negative quantity.

11.08 Lees' Method for Metals. The following method, due to Lees, enables values of the thermal conductivity of a small specimen of the material to be obtained over a wide range of temperature. The method requires only a small temperature difference to be established across the specimen and thus is admirably adapted for the investigation of the variation of thermal conductivity with temperature.

The specimen under test (Fig. 131) consists of a rod R, 7 to 8 cm. long and about 6 mm. in diameter, fitted with its lower end in a copper disc D, which in turn fits accurately into a cylindrical copper tube T closed at the top. Three easily adjustable sleeves A, B, C of thin brass are placed on the rod and round the first two of these are wrapped coils of pure platinum wire, which serve as resistance thermometers. C carries a heating coil of silk-covered platinoid wire, an exactly

similar coil PP being wound on the outside of
the copper tube. The electrical arrangements
are such that, when the current is switched
off in C, the current is automatically established
in P and consequently heat is supplied to the
apparatus at a constant rate throughout the
experiment. (A further coil—not shown—is
wound round the outside of T so that the
temperature of the apparatus can be rapidly
raised as required). The whole apparatus is
contained in a Dewar vessel V, into which
liquid air can be placed to reduce the temperature of the apparatus somewhat below the
desired starting point. The liquid air is then
removed and the temperature of the apparatus
adjusted to the exact temperature required by
means of the heating coil previously referred to.

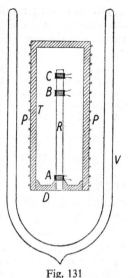

Fig. 131

In carrying out the experiment, the current
is switched on in the heating coil C on the
rod and the temperatures at A and B are taken from resistance
measurements when a steady difference of temperature is established
between these points. The current is now switched to the heating
coil P on the copper tube, and is allowed to flow until the difference
between the temperatures of A and B again becomes steady. Since
electrical energy is continually being supplied to the apparatus,
which is guarded against heat losses, it is clear that the temperatures will show a steady rise, none of them attaining a stationary
value. The conditions are thus not the same as the "steady state"
conditions of the two experiments previously described. It can be
shown, however, that the difference in temperature recorded at A
and B when the current flows round the rod, less the difference in
temperature when the current flows round the tube, is equal to the
difference in temperature which would be established were the same
heating current used on the rod but with the far end maintained at
a constant temperature. Thus, knowing the exact distance d between
A and B, the temperature gradient corresponding to "steady state"
conditions can be obtained which, with the area of cross section
of the specimen, enables the quantity of heat conducted along the
rod per second to be expressed in terms of the thermal conductivity k.

The rate at which heat is supplied at C is $\dfrac{IE}{J}$, I being the current sup-

plied to the heating coil under a P.D. of E volts, and hence k can
be found.

The following results, given by Lees, show the manner in which the thermal conductivity of copper (in cals. cm.$^{-1}$ sec.$^{-1}$ deg.$^{-1}$C.) varies with temperature.

Temp °C.	-170	-160	-150	-125	-100	-75	-50	-25	0	18
Conductivity	1·112	1·079	1·054	·996	·973	·958	·944	·932	·924	·916

Determination of the Thermal Conductivity of Bad Conductors

11.09 Guard Ring Methods. It might appear that a simple method of determining the thermal conductivity of a substance could be devised by maintaining the surfaces of a slab or " wall " of given thickness of the material at two fixed temperatures and determining the rate of flow of heat across the specimen. There are, however, two main objections to this simple method. These are, (a) the difficulty of assessing the exact temperatures of the surfaces—thus a thin film of moisture (across which there can be a large temperature drop) on either of the surfaces will produce marked uncertainties, particularly in the case of good conductors, and (b) the impossibility of obtaining a uniform flow of heat across the specimen—this is the case particularly with bad conductors in which there are divergent lines of flow even if the specimen is carefully lagged.

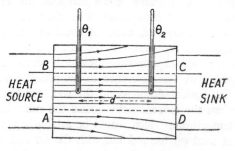

Fig. 132

These objections can be overcome by employing the "guard-ring" principle which is illustrated in Fig 132. A large slab of the specimen is heated (e.g., electrically or by a steam chest) on one face, the heat conducted through the slab in a given time being absorbed and measured at the other face (e.g., by the use of an ice calorimeter or constant flow water calorimeter). To avoid uncertainties regarding the temperatures of these faces, two holes are bored into the specimen (as shown) to receive thermometers or thermo-couples by means of which the temperature drop across a measured length d of the specimen can be directly determined. Uniformity of heat flow is ensured by measuring the heat flow across a given area of an axial section $ABCD$ of the slab, the heat flow across the remaining or "guard-ring" section being disregarded.

The guard-ring method has been applied with considerable success by Berget to the determination of the thermal conductivity of mercury (section 11.12(a)), and it has also been developed at the National Physics Laboratory for the determination of the thermal conductivity of such badly conducting substances as cork, slab-wool, &c. In this latter method an electrically heated hot-plate comprised of suitably insulated heating elements sandwiched between two large square copper plates, is surrounded by a similarly constructed "guard-ring," the whole being placed between two exactly similar walls of the material whose thermal conductivity is required. The outer walls of the material are maintained at a fixed temperature by being placed against metal tanks through which rapid streams of cold water at a thermostatically controlled temperature circulate. Some considerable time elapses before "steady-state" conditions are realised, when the temperatures of the hot and cold plates are taken by copper, copper-constantan thermo-couples inserted in the plates. The rate at which electrical energy is supplied to the hot plate can be found from the current and P.D. measurements for the heating elements, and thus from the observed temperature gradient across the specimen with area equal to that of the central hot plate, the thermal conductivity can be calculated.

11.10 Lees' Disc Method. An accurate method of measuring the thermal conductivity of poor conductors, capable of giving results over a wide range of temperatures, has been devised by Lees. The method is illustrated in Fig. 133. The specimen under test in the form of a flat disc M was placed between two copper discs C_2 and C_3 the faces of which were smeared with glycerine to obtain good thermal contact between them and the faces of M. The discs were about 4cm. in diameter and 3mm. thick. A heating coil H of silk covered platinoid wire was sandwiched between C_2 and a thinner copper disc C_1 (about 1mm. thick), the insulation of the heating elements being improved by mica and shellac. After assembly, the pile of discs was varnished to obtain a constant emissivity over the entire surface, and it was then suspended in an enclosure B maintained at a constant temperature θ_0. Electrical energy was now supplied at a known

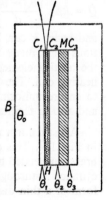

Fig. 133

H.T.–10+

rate to the heating coil, and after a time the temperatures of the copper discs, measured by platinoid-copper thermo-couples inserted in them, became steady. The thermal conductivity k of the specimen was then calculated as follows:—

Let the area of the exposed surfaces of the copper discs C_1, C_2, C_3 be S_1, S_2, S_3 respectively, and the corresponding steady state temperature θ_1, θ_2, and θ_3. The temperature of M may be taken as the mean of the temperatures of the discs C_2 and C_3, and if S_M is the exposed area of M, and E is the emissivity of the varnished surface of the pile, then, assuming Newton's law conditions to apply, the amount of heat lost per second from the whole pile of discs is

$$E\left\{ S_1\left(\theta_1 - \theta_0\right) + S_2\left(\theta_2 - \theta_0\right) + S_M\left[\frac{\theta_2 + \theta_3}{2} - \theta_0\right] + S_3\left(\theta_3 - \theta_0\right)\right\}$$

and at the steady state this is equal to $\dfrac{VI}{J}$ where V is the P.D. across the heating coil and I the current through it. Everything is known here except E, which can thus be calculated.

Now the quantity of heat Q conducted through M per second may be regarded as the mean of that entering and leaving its faces. The amount entering from C_2 is subsequently lost from the surfaces of M and C_3, and the amount leaving is subsequently lost from the surface of C_3. Hence

$$Q = \frac{1}{2}\left[E\left\{ S_M\left[\frac{\theta_2 + \theta_3}{2} - \theta_0\right] + S_3\left(\theta_3 - \theta_0\right)\right\} + ES_3\left(\theta_3 - \theta_0\right)\right]$$

$$= \frac{E}{2}\left\{ S_M\left[\frac{\theta_2 + \theta_3}{2} - \theta_0\right]\right\} + ES_3\left(\theta_3 - \theta_0\right) \quad \ldots \quad [119]$$

If d is the thickness of M and A is the area of cross-section of the disc

$$Q = kA\left[\frac{\theta_2 - \theta_3}{d}\right] \quad \ldots \ldots \ldots \ldots \quad [120]$$

and hence k can be calculated from equations 119 and 120. Corrections were made for the small emission of heat from the edges of the heating element and also for loss of heat along the leads of the heating coil. The function of the copper discs in the experiment was to ensure uniformity of temperature at the surfaces of the specimen and also to facilitate the measurement of these temperatures.

A usual form of the Lees' disc for solid bad conductors is shown in Fig. 134. This consists of a cylindrical slab of metal C (copper or brass) suspended by strings from a heavy stand. On this rests a hollow cylinder A through which steam is passed, and the specimen in the form of a thin disc B of the same diameter is placed between A and C. A hole is bored near the bottom of cylinder A and another in C

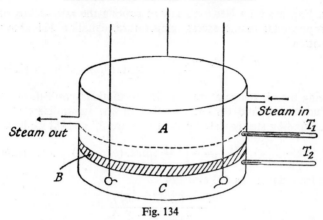

Fig. 134

to take two thermometers T_1 and T_2 which indicate the temperatures of the faces of B in the steady state. The two metal cylinders are nickel plated to give them the same emissive power. When the steady state has been attained, the quantity Q of heat conducted across the specimen per second is equal to the quantity emitted from the exposed surface of the metal slab C per second. Thus

$$Q = k \cdot \frac{\pi D^2}{4} \cdot \left[\frac{\theta_1 - \theta_2}{d} \right] \quad \ldots \ldots \quad [121]$$

where D is the diameter of the specimen, d its thickness, and θ_1 and θ_2 the steady state temperatures given by the thermometers T_1 and T_2.

To find Q the cylinder A is removed and a Bunsen flame is allowed to play on the bottom surface of C until T_2 records a temperature about 10°C. higher than that recorded in the steady state (θ_2). It is now allowed to cool, and readings of the temperature are taken at half-minute intervals until the temperature falls to a temperature about 10° below θ_2. From this data a cooling curve can be drawn from which the rate of cooling $\left[\frac{d\theta}{dt} \right]$ at θ_2 can be obtained. Then, if M is the mass of the slab C and s its specific heat, the rate of loss of heat from the lower face and sides of C when at the temperature θ_2 is

$$Q = Ms \left(\frac{d\theta}{dt} \right) \quad \ldots \ldots \ldots \quad [122]$$

Thus from equations 121 and 122 k can be calculated.

With this apparatus the thermal conductivities of two substances can be compared directly from the readings of T_1 and T_2 in the steady state. For, since by Newton's law of cooling the rate of loss of heat is proportional to the excess temperature, equation 121 above may be written

$$k \cdot \frac{\pi D^2}{4} \cdot \left[\frac{\theta_1 - \theta_2}{d} \right] = \text{const.} (\theta_2 - \theta_o) \quad . \quad . \quad [123]$$

θ_o being the air temperature; and if for another specimen of thermal conductivity k' and thickness d' (but of the same diameter D) the corresponding steady state temperatures are θ_1' and θ_2' we have

$$k' \cdot \frac{\pi D^2}{4} \cdot \left[\frac{\theta_1' - \theta_2'}{d'} \right] = \text{const.} (\theta_2' - \theta_o) \quad . \quad . \quad [124]$$

Thus from equations 123 and 124

$$\frac{k}{k'} = \frac{d \; (\theta_2 - \theta_o)(\theta_1' - \theta_2')}{d' \; (\theta_2' - \theta_o)(\theta_1 - \theta_2)}$$

11.11 Methods of Radial Flow. These methods are interesting in that the problem of lateral losses does not arise as with the wall or slab methods previously discussed. The theory of the methods is as follows:

(i) *Spherical shell method*

The specimen under test, of thermal conductivity k, is contained between two thin spherical shells A and B of radii r_1 and r_2 respectively (Fig. 135). A source of constant heat supply (*e.g.*, an electric heating element) is placed at the centre O of the shells, the heat being conducted through the specimen and subsequently lost by emission from the surface of the outer shell. Let the temperatures of the inner and outer shells be θ_1 and θ_2 respectively when the steady state is attained, and let the temperatures at distances r and $r + dr$ from O be θ and $\theta + d\theta$ as shown. Then the quantity Q of heat conducted per second across the surface of the shell of radius r is

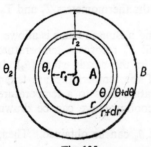

Fig. 135

$$Q = -k \cdot 4\pi r^2 \cdot \frac{d\theta}{dr}$$

It is clear that Q is constant for all values of r (being the rate at which heat is supplied by the heating element at the centre of the shells) and hence re-arranging

$$\frac{dr}{r^2} = -\frac{4\pi k}{Q} \cdot d\theta$$

and integrating

$$-\frac{1}{r} = -\frac{4\pi k}{Q} \cdot \theta + \text{const.}$$

Now when $r = r_1, \theta = \theta_1$

$$\therefore -\frac{1}{r_1} = -\frac{4\pi k}{Q} \cdot \theta_1 + \text{const.} \quad \ldots \ldots \quad [125]$$

and when $r = r_2, \theta = \theta_2$

$$\therefore -\frac{1}{r_2} = -\frac{4\pi k}{Q} \cdot \theta_2 + \text{const.} \quad \ldots \ldots \quad [126]$$

Thus subtracting equation 125 from equation 126 we have

$$\frac{1}{r_1} - \frac{1}{r_2} = \frac{4\pi k}{Q} (\theta_1 - \theta_2)$$

from which

$$k = \frac{Q(r_2 - r_1)}{4\pi \, r_1 \, r_2 \, (\theta_1 - \theta_2)} \quad \ldots \ldots \ldots \quad [127]$$

(ii) *Cylindrical shell*

Let Fig. 135 now represent the section of a cylindrical shell of a material of thermal conductivity k and length l cm., the direction of heat flow and steady state temperatures, &c., being as above. In the steady state the quantity of heat Q conducted per second across the surface of a cylindrical shell of radius r is

$$Q = -k \cdot 2\pi r l \cdot \frac{d\theta}{dr}$$

re-arranging

$$\frac{dr}{r} = -\frac{2\pi k l}{Q} \cdot d\theta$$

and integrating

$$\log_e r = -\frac{2\pi k l}{Q} \cdot \theta + \text{const.}$$

Now when $r = r_1, \theta = \theta_1$

$$\therefore \log_e r_1 = -\frac{2\pi k l}{Q} \cdot \theta_1 + \text{const.} \quad \ldots \ldots \quad [128]$$

and when $r = r_2, \theta = \theta_2$

$$\therefore \log_e r_2 = -\frac{2\pi k l}{Q} \cdot \theta_2 + \text{const.} \quad \ldots \ldots \quad [129]$$

subtracting equation 128 from equation 129

$$\log_e \frac{r_2}{r_1} = \frac{2\pi k l}{Q} (\theta_1 - \theta_2)$$

from which

$$k = \frac{Q \cdot \log_e \dfrac{r_2}{r_1}}{2\pi l \, (\theta_1 - \theta_2)} \quad \ldots \ldots \ldots \quad [130]$$

In the case of thin-walled tubing the value of k may be obtained to a close degree of approximation by considering the heat flow across a section of mean radius $\dfrac{r_1 + r_2}{2}$, thus

$$Q = -k\left\{ 2\pi \left[\frac{r_1 + r_2}{2}\right] l \right\} \cdot \left[\frac{\theta_2 - \theta_1}{r_2 - r_1}\right]$$

i.e.,

$$k = \frac{Q(r_2 - r_1)}{2\pi l \left[\dfrac{r_1 + r_2}{2}\right](\theta_1 - \theta_2)}$$

Experimental details

The spherical shell method has been used by Nusselt in measuring the thermal conductivity of poor conductors. Two thin copper spheres, split hemispherically to permit the introduction of the substance under test between them, were heated by an electric heating element placed inside the inner shell. The temperatures of the spheres in the steady state were taken by thermo-couples, and the quantity Q of heat conducted through the substance per second was obtained from the current and P.D. applied to the heating element. The thermal conductivity is then calculated from equation 127.

As an example of the cylindrical shell method, the following experiment to determine the thermal conductivity of glass tubing may be considered. A length of glass tubing (Fig. 136) is surrounded by a jacket through which steam at a temperature of θ_1 circulates. A slow stream of water from a constant head device passes along the tube which is tilted slightly to eliminate air pockets. By steam heating the glass tube on the *outside* in this way radiation losses from the constant flow water calorimeter are automatically eliminated. The inflow and outflow temperatures of the water are taken by two $\frac{1}{10}°$ thermometers T_1 and T_2 inserted in glass connecting pieces attached to the tube as shown. If the steady state temperatures of these thermo-

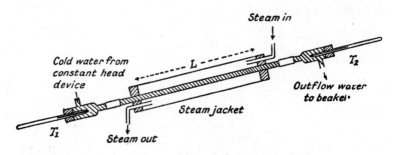

Fig. 136

meters are θ_2 and θ_2' respectively, and the mass of water flowing in a measured time t is M, then the quantity Q of heat conducted through the walls of the glass tube per second is given by

$$Q = \frac{M(\theta_2' - \theta_2)}{t} \quad . \quad . \quad . \quad . \quad . \quad . \quad . \quad [131]$$

Hence substituting for Q from equation 131 in equation 130, and putting $\left[\dfrac{\theta_2 + \theta_2'}{2}\right]$ instead of θ_2, the value of k can be found.

The above theory also applies to the determination of the thermal conductivity of rubber tubing. A length l is coiled in a large copper calorimeter of mass m containing Mgm. of water at a temperature θ_2 (conveniently cooled below room temperature to reduce radiation losses) and is heated by passing steam (at a temperature of θ_1) through the tube until the temperature of the water rises to θ_2' in a time t seconds. Then

$$Q = \frac{(M + 0\cdot1m)(\theta_2' - \theta_2)}{t}$$

This value for Q is inserted in equation 130 instead of that from equation 131 above—otherwise the calculation for k proceeds on identical lines.

11.12 Determination of Thermal Conductivity of Liquids

The transmission of heat in liquids is effected mainly by convection currents and only in part by conduction. This is a complicating factor when measurements of the thermal conductivity of liquids are required as the effect of conduction may be completely masked by that of convection. The effects of convection can, however, be eliminated by arranging for the liquid to be heated from the top, the amount of heat transmitted downwards by the process of conduction being suitably measured to determine the thermal conductivity of the liquid. This method is applied in the experiments described below.

(a) **Berget's Guard Ring Method for Mercury.** The principle of the guard ring method has already been discussed (section 11.09), and in his application of it to the determination of the thermal conductivity of mercury, Berget employed the method of steam heating and used a Bunsen's ice calorimeter to measure the heat conducted along the central specimen.

The details of his experimental arrangements are shown in Fig. 137. The guard ring consisted of a cylindrical column of mercury contained in a tube CD surrounding the test column in the glass tube AB, the lower end of which protruded into a Bunsen's ice calorimeter. The mercury guard ring rested on a sheet iron plate P which formed the cover of the ice box in which the ice calorimeter was situated. The mercury was heated at the top by steam circulating through tubes as shown, and the temperature gradient along *AB* was determined by the thermo-couples 1, 2, 3, 4 inserted through the sides of the tubes *CD* and *AB*. These thermo-couples consisted of short lengths of rubber covered iron wire, so that any pair of them, together with the mercury between them, formed a thermo-couple which gave the difference in temperature between the respective ends. Berget found that the temperature gradient was uniform and, having accurately determined its value, and knowing the area of cross section of *AB* and the heat conducted along it in a given time (obtained from the recession of the mercury thread of the ice calorimeter), he was able to calculate a value for *k*. The mean value obtained by Berget in this way was 0·02015 c.g.s. units.

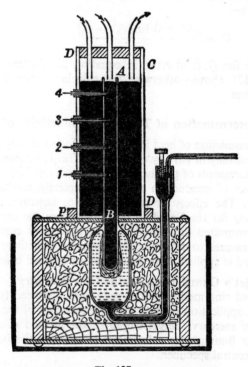

Fig. 137

(b) Lees' Method. This is a modification of the method used by Lees to determine the thermal conductivity of badly conducting solids (section 11.10). The liquid under test is contained in an annulus of ebonite E (Fig. 138) placed between two thin nickel-plated copper discs C_2 and C_3. An electric heating element H is sandwiched between two other copper discs C and C_1 and the sets of discs are arranged as shown with a thin disc of glass of known thermal conductivity between the copper discs C_1 and C_2. The pile of

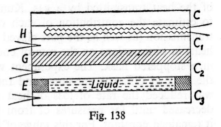

Fig. 138

discs is placed in a constant temperature enclosure and heat is supplied via the heating element until the temperatures of the copper discs C_1, C_2, C_3, measured by thermo-couples inserted in them, become steady. Let these temperatures be θ_1, θ_2, and θ_3 respectively and let the area of cross section of the glass disc G (thermal conductivity k_G) be A_G and its thickness d_G. Then, under steady state conditions, the quantity of heat conducted through G per second is

$$k_G A_G \left(\frac{\theta_1 - \theta_2}{d_G} \right)$$

the temperatures of the copper discs C_1 and C_2 being taken to be the temperatures of the faces of the glass disc.

Assuming no heat losses from the edge of the disc C_2, the above amount of heat will be the same as that conducted across the ebonite ring and the liquid per second. Thus, if the area of the liquid surface is A_L and its depth is d_L, we have

$$k_G A_G \left(\frac{\theta_1 - \theta_2}{d_G} \right) = k_L A_L \left(\frac{\theta_2 - \theta_3}{d_L} \right) + Q \quad . \quad . \quad [132]$$

where Q is the amount of heat conducted through the ebonite ring per second. To determine Q the experiment is repeated with air replacing the liquid in the ring. From the steady state temperature, and the known conductivities of air and glass, the quantity of heat conducted across the ebonite ring per second can be obtained. Thus, knowing Q, the conductivity k_L of the liquid can be calculated from equation 132.

11.13 Determination of the Thermal Conductivity of Gases

The thermal conductivity of gases is very low. The low conductivity of felt, woollen materials, &c., is very largely due to the air enclosed in the open texture of the fabrics. Under high pressure woollen material is found to lose much of its heat insulating properties as a result of squeezing out the tiny air spaces in the material.

The problem of measuring the small thermal conductivities of gases is complicated by the effects of both convection and radiation

10*

which under normal conditions account for by far the larger part of the heat transmitted by a gas. Kundt and Warburg endeavoured to estimate the contribution of each of the three processes of heat transmission (as well as the part played by the apparatus in absorbing heat) in observations on the rate of cooling of a thermometer placed in a stoppered flask. The experiment was performed with the air in the flask at different pressures, and they found that as the pressure was reduced the rate of loss of heat recorded by the thermometer decreased until for pressures of from 150mm. to 1mm. of mercury it remained constant. For this range of pressures, therefore, the effects of convection were absent, and in order to estimate the effect due to convection alone, the contribution of the radiation losses was determined by readings taken when the flask was completely exhausted. These experiments indicate the conditions necessary when making determinations of the thermal conductivity of a gas—the value of which is not affected (see section 8.16) by working at low pressures.

(a) **Hot Wire Method.** This method due to Andrews has been subsequently developed by Schleiermacher, Weber and others in measuring the thermal conductivity of gases. The gas is contained at low pressure in a narrow tube and the rate of dissipation of heat from an electrically heated wire running along the axis of the tube is observed. The effects of convection are eliminated by working at low pressure, using the tube in the horizontal and vertical positions, and also by using tubes of varying widths (the convection effects being much more marked in the wider tubes). The radiation losses are determined by repeating the observations when the tube is highly evacuated. The tube can be suitably maintained at a constant temperature and the temperature of the axial wire can be calculated from resistance measurements. The equation of the radial heat flow from the wire to the walls of the tube is the same as that given in section 11.11 (ii), from which the thermal conductivity of the gas may be calculated. An important correction is necessary for the "end effects"—the wire being cooled at its extremities on account of the heat conducted by the thick connecting leads by which the current enters and leaves. The correction may be made by repeating the experiment with a tube of different length, the difference in heat losses in the two cases being that from a uniformly heated wire of length equal to the difference in lengths of the two wires.

(b) **Method of Laby and Hercus.** A very satisfactory method of measuring the thermal conductivity of gases employing the guard ring principle has been devised by Laby and Hercus. The scheme of their apparatus is shown in Fig. 139. A, B, and C were three circular copper plates each consisting of two thin copper sheets soldered together, the surfaces in contact being grooved to permit of the insertion

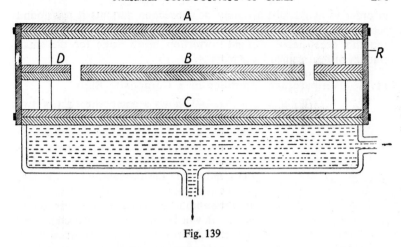

Fig. 139

of heating elements and thermo-couples. B was surrounded with a guard ring D from which it was supported in the same horizontal plane by three I-shaped ivory buttons. The plates were separated from each other by glass distance pieces, the whole being made airtight by clamping a strip of stout rubber R round the edges by two steel bands. The gas under test was that contained between the plates B and C, the heat supplied to B by electrical heating being conducted downwards through the gas to the plate C which was maintained at a constant temperature by a stream of water flowing through a brass chamber attached to its under surface. The guard ring D was maintained at the same temperature as B by an independent heating coil inserted in it, and to prevent any loss of heat from the upper surface of B, the plate A was also at the temperature of the plate B. The guard ring D ensured that the heat flow from B was at all places normal to it, and if a is the cross section of the lower surface of B, and the temperatures of B and C in the steady state are θ_1 and θ_2 respectively, the heat conducted per second across the air film of thickness d is

$$k\, a \left[\frac{\theta_1 - \theta_2}{d} \right]$$

This is equal to the rate at which heat is supplied to B and this can be found from the current and P.D. applied to the heating coil, and thus k can be calculated.

A correction must be applied for the heat received from B by radiation. This effect was reduced by silvering and polishing the surfaces of the plates and its value was estimated from the results of separate experiments on silver walled Dewar flasks. Convection was eliminated by arranging for a *downward* flow of heat, although in order to avoid all possibility of convection currents it was found necessary to maintain the temperature of A slightly above that of B, and this necessitated a slight correction being made for the small amount of heat gained by B from A.

11.14   Temperature Gradient in a Composite Slab

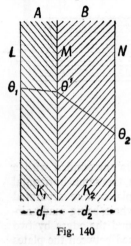

Fig. 140

Fig. 140 shows a composite slab of two different materials A and B with thermal conductivities of k_1, k_2 and of thicknesses d_1, d_2 respectively. The face L of material A is maintained at a fixed temperature θ_1 and in the steady state the temperature of the face N attains a value of θ_2 (say). In this steady state the amount of heat conducted per second across B will be the same as that conducted per second across A, and the interface M will be at some temperature θ' such that, if A is the better conductor, the temperature gradient in A will be less than that in B. The value of θ' and the rate of flow across the composite slab may be found as follows. If Q is the rate of flow across A per unit area of surface, we have

$$Q = k_1 \frac{\theta_1 - \theta'}{d_1} \quad \ldots \ldots \ldots \quad [133]$$

In the steady state this must equal the rate of flow across B per unit area of surface, i.e.,

$$Q = k_2 \left(\frac{\theta' - \theta_2}{d_2} \right) \quad \ldots \ldots \quad [134]$$

Hence from equations 133 and 134

$$\theta' = \frac{\dfrac{k_1}{d_1} \cdot \theta_1 + \dfrac{k_2}{d_2} \cdot \theta_2}{\dfrac{k_1}{d_1} + \dfrac{k_2}{d_2}}$$

Also re-writing equations 133 and 134 as

$$\theta_1 - \theta' = Q \frac{d_1}{k_1} \quad \text{and} \quad \theta' - \theta_2 = Q \frac{d_2}{k_2}$$

we have, eliminating θ' from these two equations,

$$\theta_1 - \theta_2 = Q \left(\frac{d_1}{k_1} + \frac{d_2}{k_2} \right)$$

from which

$$Q \text{ (cals. per sq. cm. per sec.)} = \frac{\theta_1 - \theta_2}{\dfrac{d_1}{k_1} + \dfrac{d_2}{k_2}} \quad \ldots \quad [135]$$

The problem of conduction through composite walls will be illustrated by considering the conduction through the plates of a steel boiler which has become encrusted with "boiler scale."

If the thickness of the metal plates (thermal conductivity 0·10) is 1cm. and that of the scale (thermal conductivity 0·001) is ·5cm. and the amount of water (latent heat of evaporation 540 calories per gm.) evaporated per sq. metre of surface per hour is 50kgm., the difference in temperature between the flue gases and the water in the boiler is, from equation 135 above,

$$\theta_1 - \theta_2 = Q \left[\frac{d_1}{k_1} + \frac{d_2}{k_2} \right]$$

$$= \frac{50 \times 1,000 \times 540}{100 \times 100 \times 3,600} \left(\frac{1}{0·10} + \frac{0·5}{0·001} \right)$$

$$= 382·5°C.$$

Without the encrustation of scale for the same rate of heat flow per sq. metre of surface, the temperature difference required is much less, *viz.*,

$$\theta_1 - \theta_2 = Q \frac{d_1}{k_1} = \frac{50 \times 1,000 \times 540}{100 \times 100 \times 3,600} \times \frac{1}{0·10} = 7·5°C.$$

Thus owing to the bad conductivity of scale a much higher temperature difference must exist between the flue gases and the water if the evaporation rate is to be maintained. This means that those parts of the boiler surface in contact with scale will be raised to much higher temperatures than the rest of the boiler. This unequal heating produces differences in expansion of the various parts of the boiler which may become weakened by the stresses thus set up.

11.15 Accretion of Ice on Ponds

The abstraction of heat from the water of a pond by the cold air above it results in the formation of an ice layer, the growth of which may be obtained as follows. Consider a layer of ice $x$cm. thick already formed on a pond at 0°C. the temperature of the air above it being $-\theta°C$. (Fig. 141). If A

Fig. 141

is the area of the pond, L the latent heat of fusion of ice and ρ its density, the heat given up when the ice layer increases in thickness by an amount δx is

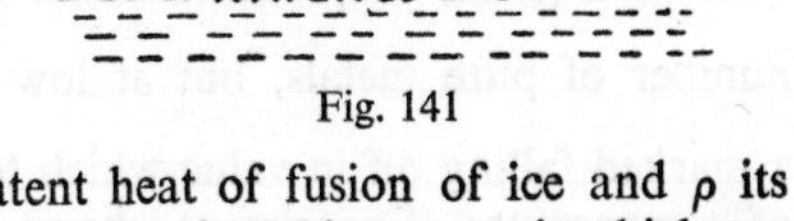

$$A \rho \, \delta x \, L \text{ calories}$$

If this quantity of heat is conducted upwards through the ice layer in δt seconds, we have

$$A \rho \, \delta x \, L = k A \frac{\theta}{x} \delta t$$

k being the thermal conductivity of the ice. Thus the rate of growth of the ice layer is

$$\frac{dx}{dt} = \frac{k\theta}{\rho Lx} \quad \cdot \quad \cdot \quad \cdot \quad \cdot \quad \cdot \quad [136]$$

To find the time elapsing for the layer to grow to a given thickness x, equation 136 can be integrated whence

$$t = \frac{\rho L}{2k\theta} \cdot x^2$$

the constant of integration being zero since $x = 0$ when $t = 0$.

It is of interest to insert numerical values, thus taking $\theta = -10°C.$, $\varrho = 0.92$gm. per cm.³, $L = 80$ cals. per gm., and $k = 0.005$ cals. cm.$^{-1}$ sec.$^{-1}$ deg.$^{-1}C.$,

$$t = 726 \, x^2$$

Hence the first cm. thickness of ice takes 726 secs (12·1 mins.) to form, the second cm. a further 2,178 secs. (36·3 mins.), and so on.

11.16   The Wiedemann-Franz Law

As a result of their experimental investigations on thermal conductivities, Wiedemann and Franz put forward in 1853 an interesting empirical generalisation connecting the thermal and electrical conductivities of metals. This generalisation, known as Wiedemann and Franz's law, states that the ratio of the thermal and electrical conductivities is the same for all metals at the same temperature. The law was extended in 1872 by Lorenz, who suggested that the ratio should be proportional to the absolute zero. Thus if k is the thermal conductivity and σ the electrical conductivity, and T the absolute temperature, then

$$\frac{k}{\sigma T} = \text{const. for all metals.}$$

This statement includes both the law of Wiedemann and Franz and that of Lorenz, and has been found to be approximately correct over a range of temperatures from about $-100°C.$ to $100°C.$ for a number of pure metals, but at low temperatures the ratio $\frac{k}{\sigma T}$ shows a marked falling off in value which tends to zero at the absolute zero of temperature. Experiment shows that the value of the thermal conductivity of pure metals increases as the temperature decreases; there is a corresponding increase in the value of the electrical conductivity, but in this case the change is more marked—the electrical conductivity tending to an infinite value (super conducting state) at absolute zero. Table 28 indicates the extent to which the Wiedemann-Franz law applies in the cases of certain pure metals for the range of temperature given.

TABLE 28

Values of $\dfrac{k}{\sigma T} \times 10^8$ (k measured in watts per cm. degree, and σ in reciprocal ohms per cm.[3])

Substance	−170°C.	−100°C.	−50°C.	0°C.	18°C.
Aluminium	1·50	1·81	1·98	2·09	2·13
Copper	1·85	2·17	2·26	2·30	2·32
Silver	2·04	2·29	2·36	2·33	2·33
Zinc	2·20	2·39	2·40	2·45	2·43
Lead	2·55	2·54	2·52	2·53	2·51
Iron.....................	3·10	2·98	2·93	2·97	2·99

The evidence of the experimental results suggests strongly that the same agents are responsible for both electrical and thermal conductivity. Electrical conductivity has been explained assuming the existence of free electrons in the metal which, on the application of an electric field, drift in the opposite direction (since electrons carry a negative charge) to that of the field. Treating these electrons as an "electron gas" in temperature equilibrium with the molecules of the substance, and using the methods of the kinetic theory of gases, it has been found possible to explain Ohm's law in a satisfactory manner. It can thus be argued that the agents of thermal conductivity in pure metals are also free electrons, a temperature gradient along the metal producing a flow of heat energy along it conveyed by these electrons in the same manner as with a gas. Applying the kinetic theory in this way Drude obtained the relation

$$\frac{k}{\sigma T} = \frac{3R^2}{JN^2 e^2} \qquad \text{(see section 8.18).}$$

R being the gas constant, N the number of molecules in 1 gram molecule, and e the charge on the electron. The value of the ratio $\dfrac{k}{\sigma T}$ is thus related to three universal constants and is in fair agreement with the experimentally obtained value at ordinary temperatures. A more rigorous application of the kinetic theory to the problem by Lorentz gave a result differing only by a numerical factor from that given by Drude.

A weak point in the argument is the assumption of the same equipartition laws to the electrons as to the atoms or molecules of the substance. If this were the case it would follow that the energy of these electrons would have an important part to play in the specific heats of metals whereas the experimental results (see Table 9) clearly indicate that non-conducting substances (i.e., substances with no free electrons present) have the same atomic heat values as conductors. The difficulty of the specific heat of the free electrons has been removed by the application of the newer quantum mechanics treating the electrons as a degenerate gas.

QUESTIONS. CHAPTER 11

1. How would you measure the coefficient of thermal conductivity of a good conductor? Discuss the main errors in this experiment and how they affect the value of the result.

Explain in detail the reasons for (a) lagging both the cold water and hot water pipes in a house, (b) spreading a layer of glass wool on the floor of an attic.   (S.)

2. What is meant by the *thermal conductivity* of a substance? Describe a method for the determination of its value for an asbestos board.

The outside wall of a room is 12ft. by 8ft. and is 10in. thick. Inset is a window 6ft. by 4ft. made of glass $\frac{1}{8}$in. thick. Calculate the ratio of the amounts of heat conducted out of the room by the glass and the bricks.

(Thermal conductivity for glass $= 2 \cdot 5 \times 10^{-3}$ c.g.s. units. Thermal conductivity for brick $= 0 \cdot 3 \times 10^{-3}$ c.g.s. units.)   (S.)

3. A uniform metal rod is steam heated at one end. Explain how the temperature will vary along its length when the steady state is reached, (a) if the rod is exposed to the atmosphere, (b) if the sides of the rod are well lagged but the unheated end is exposed.

Heat flows at the rate of 63,000 cals. per hour along a well-lagged copper rod of cross-section 15cm.^{2} At a certain transverse section of the rod, its temperature is 70°C. At what section will the temperature be 50°C.?

(Thermal conductivity of copper $= 0 \cdot 9$ cal. per cm. per sec. per degC.)   (W.)

4. What factors determine the temperature distribution along a metal rod heated at one end, when the steady state has been reached?

A long brass rod 1cm. in diameter has one end kept at a constant temperature above its surroundings. In the steady state, the temperatures at distances 9, 11, 29 and 31cm. from the hot end are respectively 37·2, 35·8, 29·7, 29·3°C. Calculate the average rate of loss of heat per unit area from the surface between points 10cm. and 30cm. from the hot end. The conductivity of heat for brass is 0·26 cal. cm.$^{-1}$ sec.$^{-1}$ degC.$^{-1}$.   [N.]

5. A cylindrical bar of iron of length 10cm. and radius 1cm. has a small heating coil of resistance 10 ohms wound round one end. The bar is jacketed so as to prevent loss of heat from the sides or from the heated end. The far end is exposed to the air and is found to acquire a temperature of 30°C. when a current of 0·5 amp. is passed through the coil, the temperature of the air being 20°C. What is the temperature of the heated end of the bar? If Newton's law of cooling is obeyed at the free end of the bar, what is the temperature at the heated end when the current through the coil is 1 amp.?

Conductivity of iron $= 0 \cdot 1$ c.g.s. units (gm.-cal.).
Mechanical equivalent of heat $= 4 \cdot 2 \times 10^7$ ergs. per cal.   (Camb. Schol.)

6. Define *thermal conductivity*.

Heat is supplied to a slab of compressed cork, 5cm. thick and of effective area 2 sq. metres, by a heating coil spread over its surface. When the current in this coil is 1·18amp. and the potential difference across its ends 20 volts the steady temperatures of the faces of the slab are 12·5°C. and 0°C. Assuming that the whole of the heat developed in the coil is conducted through the slab, calculate the thermal conductivity of the cork.

Draw a diagram showing how you would propose to carry out the experiment suggested in this example. (N.)

7. Define thermal conductivity and describe critically a method of measuring it for a bad conductor.

A wire of resistivity 2×10^{-4} ohms per cm.^{3} and 1mm. in diameter carries a current of 10 amps. If it is covered uniformly with a cylindrical layer of insulating material having a coefficient of thermal conductivity of 6×10^{-4} cals. cm.$^{-1}$ deg.$^{-1}$ sec.$^{-1}$ and a diameter of 1cm. what is the temperature difference between the inner and outer surfaces of the insulator?

(1 cal. = 4·2 watt-secs.) (Camb. Schol.)

8. Describe and give the theory of a method for determining the thermal conductivity of a *bad* conductor.

The hollow space between two very thin concentric copper shells is filled with sand of thermal conductivity $0·13 \times 10^{-3}$ c.g.s. centigrade units. The radii of the spheres are 6cm. and 4cm., and the temperature of the hollow space inside the inner sphere is maintained at 100°C., whilst the outer sphere is kept at 0°C. Find the quantity of heat passing per minute through the sand when a steady state has been reached. [N.]

9. Describe an experiment for the determination of the coefficient of thermal conductivity of copper.

The operating temperature of the tungsten filament of a 60 watt electric lamp is 2,200°C. The filament is supported by two nickel wire leads each 5cm. long and 2mm. in diameter. Calculate approximately the percentage of the power supplied to the lamp that is lost by conduction along these leads, given that the thermal conductivity of nickel is 0·3 watt cm.$^{-1}$ degC.$^{-1}$ and that the cool ends of the leads are at 150°C. (A.)

10. Describe with the relevant theory a method for determining the thermal conductivity of glass which is available in the form of a tube.

The outside wall of a room is 5m. long, 3m. high and is made of brick 25cm. thick. The temperature of the inside surface is 15°C. higher than the outside. Find the rate at which heat is conducted through this wall and the increase in the rate when a glass window, of area 3 square metres and thickness 4mm., is inserted in it, assuming that the temperature difference is still the same. The thermal conductivities of brick and glass are 2×10^{-3} and $2·5 \times 10^{-3}$ calorie cm.$^{-1}$ sec.$^{-1}$ degC.$^{-1}$ respectively. (A.)

11. Define thermal conductivity.

Give an account of a method for measuring the thermal conductivity of copper. Discuss the distribution of temperature along (a) an *unlagged* metal bar, (b) an ideally *lagged* bar, this second bar consisting of two different metals of the same length and cross-section, joined end to end. In each instance it may be assumed that the steady state has been reached with the hot end at 100°C. and the cold end at 20°C. It may also be assumed that in (b) the thermal conductivity of the metal forming the hotter part of the composite bar is one-third that of the cooler part. (L.)

12. Define *thermal conductivity.*

Describe and explain how the thermal conductivity of a material of a rock, or other badly conducting substance available on the form of a slab, may be determined experimentally.

A greenhouse has walls and roof of glass 0·4cm. thick and total area 50 metre.^{2} Estimate the heat lost through the walls and roof in one hour if the temperatures inside and outside (*not* the temperatures of the inner and outer surfaces of the glass) are 25°C. and 5°C. respectively.

(For glass assume its thermal conductivity to be 0·0025 cal. cm.$^{-1}$ sec.$^{-1}$ degC.$^{-1}$, and its heat transfer coefficient, i.e. the heat transferred per unit area per second per degree temperature difference, is everywhere 0·00020 cal. cm.$^{-2}$ sec.$^{-1}$ degC.$^{-1}$)

[L.]

13. Discuss the analogy between the flow of heat along a metal bar whose two ends are maintained at constant temperature, and the flow of electricity along a metal wire when a steady potential difference is maintained across it. Why is it usually more difficult to measure thermal conductivity than electrical conductivity?

Describe how you would determine the coefficient of thermal conductivity of rubber tubing.

A 750 watt electric iron has a base area 150 sq. cm., which is 8mm. thick. The thermal conductivity of its material is 0·1 cal. sec./degC./cm. Assuming that 60 per cent. of the heat generated flows out normally through the base, and that its outer surface is at 120°C., find the temperature of the inner face.

(Take the value J to be 4·2 joules per calorie.) (O.)

14. Give a detailed and critical account of the standard laboratory experiment for determining the coefficient of thermal conductivity of copper.

The total rate of loss of heat from unit area of a plane vertical copper surface at θ°C., exposed to essentially still air at θ_0°C., may be taken to be $10^{-2}(\theta - \theta_0)^{5/4}$ cals./sec./sq. cm. A horizontal cylindrical copper bar of diameter 4cm. and length 20cm. has one end maintained at 100°C. in a steam chest, the whole of the curved surface perfectly lagged, and the other plane face exposed to the air at 20°C. Taking the thermal conductivity of copper to be 0·9 cal./sec./°C./cm., obtain a formula for the temperature of the exposed face, and show that it is approximately 70°C. Estimate also the rate of flow of heat down the bar.

(Assume that the air temperature does not rise significantly.) [O.]

15. Describe the measurement of the thermal conductivity of a good conductor, such as a metal, in the form of a rod. Give an account of the theory of the method and point out the sources of inaccuracy.

An aluminium saucepan which has a base 16cm. in diameter and 4mm. thick contains boiling water and rests on an electric hotplate. The water boils away at the rate of 13·7 gm. min.$^{-1}$ What is the temperature of the underside of the saucepan, assuming it to be uniformly heated and neglecting heat losses from the sides?

(Latent heat of evaporation of water at 100°C. = 540 cal. gm.$^{-1}$ Thermal conductivity of aluminium = 0·49 cal. deg.$^{-1}$ cm.$^{-1}$ sec.$^{-1}$) (O. and C.)

16. Describe how you would measure the thermal conductivity of a poor conductor, such as glass or rubber, indicating how you would calculate the result from the observations made. What are the principal sources of error in the experiment?

A hot-water tank, which has a total exposed area of 1·5 sq. m. is protected by lagging 2·5cm. thick. An immersion heater in the tank dissipates 126W. continuously and it is found that, in the steady state, the difference of temperature between the inside and the outside of the lagging is 50°C. Calculate an approximate value for the coefficient of thermal conductivity of the material of the lagging. (O. and C.)

17. Much of the heat lost from a closed room, in the absence of draughts, is through the glass of the windows. Discuss the mechanisms involved in the transfer of heat from a reflector-type electric fire inside to the cold air outside.

The area of the windows of a room is 5 metre^2, and their thickness is 3 mm. Given that the heat transferred to a vertical surface by natural convection is

$$1·7 \times 10^{-4} \times (\triangle t)^{5/4} \text{ joule cm.}^{-2} \text{ sec.}^{-1}$$

where $\triangle t$ is the temperature difference between the surface and the main body of air, calculate the rate of loss of heat through the windows when the air inside is at 24°C. and the air outside is at 0°C., and also the temperature of the inner and outer surfaces of the glass. The temperature drop across the glass may be assumed to be very small compared with 24°C.

(Thermal conductivity of glass is 76×10^{-4} joule cm.$^{-1}$ sec.$^{-1}$ °C.$^{-1}$)

[O. and C.]

18. Explain what is meant by *coefficient of thermal conductivity*.

Draw a labelled diagram of the apparatus you would use to measure the coefficient for a good conductor. Why must the form of the apparatus be changed for a material which is a bad conductor?

A window of area $1·0m.^2$ consists of two plane parallel sheets of glass each of 0·40cm. thick, enclosing a layer of air 0·20cm. thick. If the outside and room temperatures are steady at 6·5°C. and 20°C. respectively, calculate (a) the temperatures of the inner surfaces of the glass, (b) the rate of loss of heat through the window.

Assume that the coefficients of thermal conductivity of glass and air are $2·5 \times 10^{-3}$ and $5·0 \times 10^{-5}$ c.g.s. centigrade units respectively.

What assumptions do you make in your calculation? (N.)

19. Describe how you would find the thermal conductivity of cardboard. Justify your choice of the shape of the specimen used.

Steam at 100°C. is passed through a cylindrical metal tube 8·0cm. in diameter which is covered with a layer of plastic material 2·0mm. thick. If the outside of this material is maintained at 0°C., find the mass of the steam condensed each minute per metre length of the tube.

(Thermal conductivity of the plastic material $= 4·0 \times 10^{-4}$ cal. cm.$^{-1}$ sec.$^{-1}$ degC.$^{-1}$ Latent heat of vaporization of water at 100°C. $= 540$ cal. gm.$^{-1}$) (N.)

20. Define *thermal conductivity* and describe how you would find its value for copper.

An iron plate, 2·5cm. thick, has its faces covered with films of water each 0·10mm. thick. If the total fall of temperature through the plate and films is 80 degC., calculate (a) the fall of temperatures through the plate, (b) the quantity of heat passing through 1 sq. metre of surface per hour.

(The thermal conductivity of iron is 0·112 and that of water 0·00154, both in c.g.s. centigrade units.) (N.)

21. Define conductivity for heat and describe briefly how it may be measured experimentally either for a good conductor or for a bad conductor.

If the air temperature remains constant at -10°C. find the increase in thickness of the ice on a pond between 7 p.m. and 7 a.m. if the initial thickness was 10cm.

(Conductivity of ice 0·005cal. per sq. cm. per unit C. temp. gradient per sec. Density of ice $= 0·917$gm. per c.c. Latent heat of ice $= 80$cals. per gm.)

(Camb. Schol.)

22. Define *thermal conductivity* and from your definition derive a unit in terms of which it may be expressed.

Give an account of a method of measuring the thermal conductivity of glass, assuming the glass to be available in a form suitable for the method you describe.

The ice on a pond is 4·9cm. thick and the temperature of its surface in contact with the air is -1°C. It is found that in 200 minutes the thickness of the ice increases by 0·2cm. Obtain a value for the termal conductivity of ice.

(Density of ice 0·9 gm. cm.$^{-3}$, latent heat of fusion of ice $= 80$ cal. gm.$^{-1}$) (L.)

23. A strong thin spherical vessel of diameter 20cm., full of water at 0°C., is placed in a bath maintained at −20°C. Calculate the time taken for the water to become completely frozen, assuming that the ice forms in spherical layers.

(Latent heat of ice = 330 joules/gm., density of ice = 0·92 gm./cm.³. Thermal conductivity of ice = 0·016 joules/sec. cm. degC.) (Oxford Schol.)

24. How may the thermal conductivity of a metal be measured? What is the accuracy of the method you describe?

At low temperatures, the thermal conductivity of a metal is proportional to its absolute temperature T. For copper, the thermal conductivity is equal to $0·2T$ cal. sec.$^{-1}$ cm.$^{-1}$ deg.$^{-1}$. What will be the temperature distribution along a copper rod, 10cm. long, the ends of which are maintained at 4°K. and 1°K. If the radius of the rod is 2mm., what will be the rate at which heat is conducted along it? Heat losses from the surface of the rod may be ignored. (Oxford Schol.)

25. What are the essential differences between heat conduction in a gas and a solid?

The rate of heat flow across a surface of area A is $-KA \dfrac{dT}{dx}$, where $\dfrac{dT}{dx}$ is the temperature gradient parallel to the flow and K is the thermal conductivity of the material. A uniform rod of length l, whose conductivity is independent of temperature, is thermally insulated along its sides, and has its ends maintained at temperatures T_1°K. and T_2°K. respectively. Show that the rate of heat flow from T_1 to T_2 is $KA \dfrac{T_1 - T_2}{l}$ when a steady state is reached.

If the thermal conductivity is inversely proportional to the absolute temperature, find the heat flow in a similar case and the temperature at the mid-point of the rod. (Oxford Schol.)

26. Define the coefficient of thermal conductivity of a substance and describe a method of measuring it for a poor conductor.

The earth consists of a core of molten material, which is a perfect thermal conductor, at a temperature of 5,000°C. and having a radius of 3,500km., surrounded by a spherical shell of rock 3,000km. thick. If the surface of the earth remains at 0°C. calculate the time in years for the temperature of the core to drop by 1°C. The specific heat of the rock may be neglected.

(Density of core = 10; specific heat of core = 0·1; coefficient of thermal conductivity of the rock = 0·005 cal. cm.$^{-1}$ degC.$^{-1}$) (Camb. Schol.)

27. How would you attempt to measure the coefficient of thermal expansion of a gas?

A spherical vessel, radius r, at a temperature T_1, is insulated with a covering of thickness d of material of thermal conductivity K. The outer surface of the covering attains a temperature T_2, and under these conditions the rate of loss of heat from unit area of its surface is $E(T_2 - T_3)$, where E is a constant and T_3 is the temperature of the air. Find an expression for T_2 in terms of T_1, T_3, E, k, r and d.

(Camb. Schol.)

CHAPTER XII
RADIATION

12.01 Introductory

As already indicated in the previous chapter (section 10.01), the transfer of heat energy by radiation differs from the processes of convection and conduction in that it takes place independently of matter. Thus radiant heat can be transmitted through a vacuum, or through an intervening material medium without affecting the temperature of the medium. It is by this means that the heat from the sun reaches the earth, and the fact that both the heat and the light are simultaneously cut off from the earth during a total eclipse of the sun, indicates that light and radiant heat travel through space with the same velocity and suggests a similarity of nature between them. This point will receive further attention in later sections of this chapter, although mention may be made here of the fact that if a thermometer with a blackened bulb is used to explore the spectrum formed by passing the sun's rays through a quartz or rock-salt prism, an appreciable rise in temperature is recorded at and *beyond* the red end of the spectrum.

The similarities between light and heat radiation suggest that the mode of generation and transmission are the same, and it is now considered that both forms of radiation are propagated by transverse waves in the ether known as electromagnetic waves. Other forms of radiation such as Hertzian waves, X-rays, ultra-violet waves, &c. are also members of the family of electromagnetic waves, the essential differences between the various types of radiation being due to their different wavelengths which account for their widely different properties. The section beyond the red end of the spectrum where the thermal radiations are detected is called the *infra-red* region, and its relation to other forms of radiation in the electromagnetic spectrum is shown below in Fig. 142:

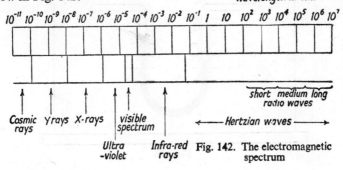

Fig. 142. The electromagnetic spectrum

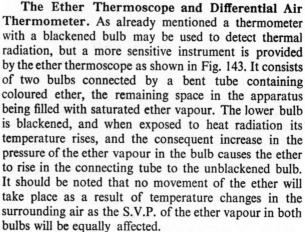

12.02   Detection of Infra-Red Radiation

The Ether Thermoscope and Differential Air Thermometer. As already mentioned a thermometer with a blackened bulb may be used to detect thermal radiation, but a more sensitive instrument is provided by the ether thermoscope as shown in Fig. 143. It consists of two bulbs connected by a bent tube containing coloured ether, the remaining space in the apparatus being filled with saturated ether vapour. The lower bulb is blackened, and when exposed to heat radiation its temperature rises, and the consequent increase in the pressure of the ether vapour in the bulb causes the ether to rise in the connecting tube to the unblackened bulb. It should be noted that no movement of the ether will take place as a result of temperature changes in the surrounding air as the S.V.P. of the ether vapour in both bulbs will be equally affected.

Fig. 143

The differential air thermometer shown in Fig. 144 was extensively used by Leslie in his researches on thermal radiation. Two equal bulbs containing air are connected by a U-tube in which is placed some coloured sulphuric acid to serve as a manometric index. Any difference in temperature between the two bulbs creates a pressure difference in the air in them which is evidenced by a movement of the liquid index round the U-tube. The vertical limbs of the U-tube can be graduated and the instrument calibrated to read off temperature differences directly.

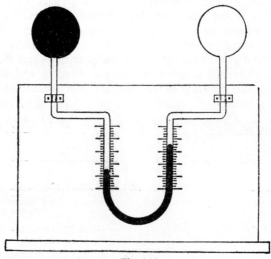

Fig. 144

12.03 The Thermopile

The ether and air thermoscopes described above have now been almost completely superceded by electrical instruments which are capable of greater precision and can be constructed to possess quite remarkable degrees of sensitiveness.

A well-known instrument which finds wide general application in the laboratory is the thermopile. This instrument was invented in 1829 by Nobili and was subsequently improved by himself and Melloni. It makes use of the fact, already discussed in chapter I that an e.m.f. is set up round a circuit of two dissimilar metals when there is a temperature difference between the two junctions. The metals employed in the construction of the thermopile are antimony and bismuth, and in order to increase the sensitiveness of the instrument, a number of these thermo-couples are arranged in series as indicated in Fig. 145.

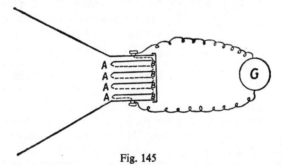

Fig. 145

The couples, adequately insulated from each other, are arranged in cuboid formation and contained in a cylinder of brass supported on a stand. One set of junctions B, B, B, B is covered by a brass cap to ensure constancy of temperature, and the other set A, A, A, A are coated with lamp-black to receive the radiation which is concentrated on the exposed face by a "reflecting cone" of brass. For maximum sensitivity the resistance of the galvanometer used should be approximately equal to that of the thermopile, and the deflection of the galvanometer is very nearly proportional to the temperature rise of the exposed face.

For the investigation of the thermal spectrum Rubens has devised a *linear thermopile* in which alternate junctions are arranged one above the other in a straight line. These junctions are exposed through a linear slit in a metal case which also serves to shield the other junctions. Silver-bismuth, iron-constantan, and copper-constantan couples have been used in the design of these thermopiles, the sensitivity of which can be greatly increased by enclosing them in a vacuum by which means the heat losses due to conduction and convection can be greatly diminished.

12.04   Boys' Radio-Micrometer

A much more sensitive instrument than the thermopile is the Boys' radio-micrometer which is a combination of a thermo-couple and sensitive galvanometer. The details of the instrument are shown in Fig. 146.

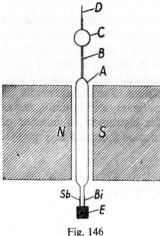

Fig. 146

A loop of fine copper wire A is situated between the pole pieces N, S of a powerful permanent magnet. The loop is supported by a fine glass tube B, carrying a small reflecting mirror C, the glass tube being suspended by a fine quartz fibre D. At the lower end of the loop are two thin bars one of antimony and the other of bismuth, the bases of these being soldered to a small piece of blackened copper foil E. This arrangement completes the electrical circuit and comprises the hot junction of the thermo-couple. Thermal radiation directed towards E is absorbed there producing a rise in temperature which results in a thermo-electric current circulating round the copper loop. This causes the loop to rotate, the tendency to set itself perpendicular to the magnetic field being opposed by the torsion in the suspension fibre. The resulting deflection, which is proportional to the intensity of the radiation falling in E, is read by the movement of a spot of light reflected from the mirror C on to a scale. The instrument is very responsive and exceedingly sensitive, being capable of recording temperature changes of the blackened copper foil of the order of a millionth of a degree centigrade.

12.05   The Bolometer

The bolometer was originally invented by Langley in 1881 and depends on the change of resistance of platinum with temperature to which reference has been made in chapter 1. Two identical strips of very thin platinum foil arranged in the form of gratings as shown in

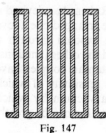

Fig. 147

Fig. 147 are coated with a thin layer of lamp-black and placed in two adjacent arms of a Wheatstone's bridge network. The other two arms of the bridge contain resistance boxes, and when the bridge is balanced the two gratings are at the same temperature. If now one of the gratings is exposed to thermal radiation, its temperature rises increasing its resistance and the bridge becomes unbalanced. By adjusting the bridge the increase in resistance of the grating can be found, and this is a measure of the amount of radiation falling on it per second. The second grating remains shielded whilst taking the observations, and serves to compensate for the changes in resistance brought about by local temperature fluctuations which produce equal effects in both gratings, and thus do not disturb the balance of the bridge.

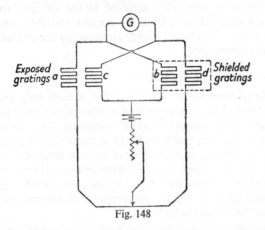

Fig. 148

In an improved form of bolometer due to Lummer and Kurlbaum, four gratings are used and these comprise the four arms of the Wheatstone bridge. One pair (a, c) of gratings connected in opposite arms of the bridge (see Fig. 148) are mounted one behind the other in such a way that the lengths of platinum foil of one grating cover the spaces of the other. These two gratings thus arranged are exposed to the incident radiation, the other pair (b, d) being shielded. The advantages of this arrangement are that twice the amount of thermal radiation falls on the bolometer as with a single grating, and also the resistance changes are doubled with the bridge connections used. A further advantage of this bolometer is the complete compensation that it affords for local fluctuations in temperature. The instrument is capable of giving very accurate relative measurements of radiation incident on the surfaces of the gratings. For the detailed investigation of the energy for particular wavelengths of the thermal spectrum however a linear bolometer, in which the gratings are replaced by a pair of narrow strips (one shielded) of platinum foil arranged in adjacent arms of the Wheatstone bridge is used.

To obtain an absolute measurement of the amount of incident radiation *Ångström's pyrheliometer* can be used. Two identical strips of blackened platinum foil are attached to the junctions of a thermocouple in the circuit of which is connected a sensitive galvanometer. With both pieces of platinum foil shielded the junctions are at the same temperature and the galvanometer shows no deflection. On exposing one of the strips a temperature difference is established between the junctions and a thermo-electric current is recorded by the galvanometer. The temperature of the shielded strip is then raised by electrical heating until the galvanometer is again undeflected. The temperatures of the two strips are then the same, and it follows that they must be receiving energy at the same rate. That supplied to the shielded strip can be calculated (in ergs per sec.) from the current and P.D. applied to it, whence the amount of radiant energy absorbed by the exposed strip can be absolutely determined.

12.06   Laws of Heat Radiation

(*a*) *Velocity of propagation.* Reference has already been made in this chapter to the fact that during a total eclipse of the sun both the heat and light are simultaneously cut off—showing that the velocity of propagation of thermal radiation is the same as that of light (3×10^{10} cm. per sec.). Further, since the heat of the sun reaches the earth after passing through vast distances of empty space, it is evident that no material medium is required for its transmission. There is thus a similarity between light and thermal radiation which will receive additional emphasis in the following sections.

(*b*) *Rectilinear propagation.* This may be demonstrated by the following simple experiment. Three metal screens S_1, S_2, S_3 (see Fig. 149) are placed at suitable distances from each other and arranged so that vertical slits cut in them are accurately in line. A source of heat *H*,

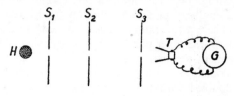

Fig. 149

e.g., a small cylindrical heating coil, is placed behind S_1, and with the screens so arranged the galvanometer attached to a thermopile *T* behind the third screen S_3 shows a large deflection due to the incidence of thermal radiation on the receiving surface of *T*. On displacing S_2 so as to destroy the collinear arrangement of the slits, the galvanometer is no longer deflected.

(*c*) *Reflection.* The laws of reflection of thermal radiation can be illustrated by the apparatus shown in Fig. 150, in which *A* and *B* are two metal tubes mounted horizontally and inclined as indicated towards a polished metal screen *M* which is capable of rotation about a vertical axis. Thermal radiation from a heat source *H* is directed down *A* on to *M*, which is rotated into such a position that it reflects the radiation

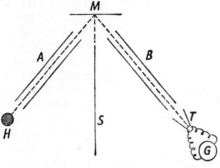

Fig. 150

along the axis of B where it is received by a thermopile T protected
from the direct radiation from H by the screen S. The correct position
of M will be indicated by the galvanometer attached to the thermopile
showing a pronounced deflection, and it will then be found that the
angles of incidence and reflection of the thermal radiation are equal.

Total internal reflection of thermal radiation may be shown by the
following experiment due to Tyndall. A lens L (Fig. 151) converges the
radiation from an arc lamp through a filter F on to a plane mirror M.

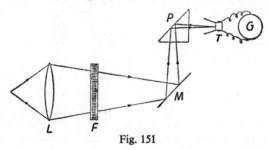

Fig. 151

The filter used by Tyndall contained a solution of iodine in carbon
disulphide which is opaque to light but permits the thermal radiations
to pass through. From M the radiation is reflected to a prism P, and a
thermopile placed at T records a large deflection indicating that the
radiation has undergone total internal reflection at the hypotenuse
face P.

(d) *Inverse square law.* That the intensity of thermal radiation falls off
with distance according to the inverse square law may be demonstrated
as follows. A large tank containing water which is kept steadily boiling
by an immersion heater, has one of its faces lamp-blacked and directed
towards the receiving face of a thermopile. It is then found that on
moving the thermopile along a line normal to the radiating surface of
the tank, the deflection of the galvanometer attached to the thermopile

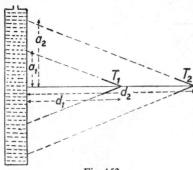

Fig. 152

remains constant provided the prolongation of the cone of the thermopile wholly intersects on the face of the tank. Let T_1, T_2 (Fig. 152) represent two such positions of the thermopile, distant d_1 and d_2 respectively from the radiating surface. In the first position the thermopile will be receiving radiation from a circular area of radius a_1, and the effect at the thermopile will be proportional to this area and to some inverse nth power (say) of the distance d_1. Thus the influx of radiant energy at T_1 may be expressed as $\dfrac{k\pi a_1^2}{d_1^n}$ where k is same constant, and at T_2 as $\dfrac{k\pi a_2^2}{d_2^n}$. But the deflections in the two positions are the same,

$$\therefore k\frac{\pi a_1^2}{d_1^n} = k\frac{\pi a_2^2}{d_2^n} \quad \ldots \ldots \ldots \quad [137]$$

Now from the properties of similar triangles it is clear that

$$\frac{a_1}{d_1} = \frac{a_2}{d_2} \quad \ldots \ldots \ldots \ldots \quad [138]$$

and thus from equations 137 and 138 it follows that $n = 2$, thereby establishing the inverse square law for thermal radiation.

(e) *Refraction and polarisation.* Evidence of the refraction of thermal radiation is afforded by the fact that a simple convex lens may be used as a "burning" glass with the sun's rays. The thermal radiation is converged on passing through the lens and may be focused so as to scorch or burn a piece of paper. The thermal radiations from a hot body may be refracted and dispersed by means of a rock-salt prism, and reference has already been made (section 12.01) to the thermal spectrum so obtained in the case of solar radiation.

To complete the identity of light and thermal radiation, Tyndall has demonstrated that heat rays can be polarised. A consideration of these experiments is beyond the scope of this book, but the results show in a conclusive manner that thermal radiation, like light, is a transverse wave motion.

12.07 The Behaviour of Bodies towards Thermal Radiation

When thermal radiation is incident on a body the subsequent effects will vary considerably with the nature of the body. Thus a highly polished surface will reflect most of the radiation, whereas a lamp-blacked surface will absorb almost the whole of it. On the other hand substances like quartz or rock-salt transmit a considerable proportion of any thermal radiation incident on them. In general, if unit quantity of radiant energy is incident on a body and a, r, t represent respectively the fractions absorbed, reflected, and transmitted, we have

$$a + r + t = 1$$

It follows therefore that if for any given medium any one of these quantities is large, the sum of the remaining two quantities must be small. Thus a good absorber will reflect badly, and contrariwise, a good reflector will be a poor absorber.

A substance which transmits the bulk of the incident thermal radiations is said to be *diathermanous,* but if the radiation is absorbed it is said to be *adiathermanous* (or *athermanous*). These terms are equivalent respectively to *transparent* and *opaque* when referring to luminous radiation. The absorption of thermal radiation has been investigated by Melloni using the apparatus shown in Fig. 153. A screen S with a small aperture was placed between a source of heat H and a thermopile T. The reading of the galvanometer attached to the thermopile was taken before and after interposing a plate P of the substance

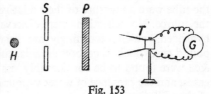

Fig. 153

under test, the ratio of the second to the first of these deflections giving the diathermanency of the substance for the thickness considered. The diathermanency of liquids was also examined by Melloni by enclosing them in a parallel sided glass cell. The results of these experiments show that the diathermanency of a substance depends on the nature of the source, *i.e.*, on the quality of the radiation incident on the substance. Thus a plate of mirror glass of thickness* 2·6mm. transmits 24 per cent of the radiation from incandescent platinum, but only 6 per cent from a blackened copper surface at 390°C., the corresponding figures for a similar thickness of borax being 12 per cent and 8 per cent respectively. A plate of rock-salt on the other hand appeared to be equally transparent to radiation from all sources, transmitting 92 per cent of the incident radiation. Alum and ice were found to be almost completely athermanous for radiation from any source.

Melloni also examined the diathermanency of a body for the radiation already transmitted by other bodies, and found that any substance was particularly transparent to the radiation which had passed through a slab of that particular substance. The results for liquids reveal that

---

* The effect of thickness on the transmission of radiation of a given kind is such that as the thickness of a transmitting medium is increased in arithmetical progression, the amount of radiation transmitted decreases in geometrical progression. Thus if unit thickness transmits a fraction t of the incident q_0 units of radiation, the quantity emerging will be $q_0 t$, which will be reduced to $q_0 t^2$ on passing through a second layer of unit thickness, and so on. Thus, after passing through a layer n units thick, the quantity of radiation emerging will be $q_0 t^n$. The quantity t has been called the *coefficient of transmission.*

pure water is very opaque to thermal radiation, but that the presence of a dissolved salt causes the diathermanency to increase; even a solution of alum is more diathermanous than pure water.

Experiments on the diathermanency of gases are difficult to carry out owing to the effects of conduction and convection, and to the small amount of radiation absorbed by them. Measurements for gases and vapours have been made by Tyndall who enclosed the gas in a long cylindrical tube closed at both ends by plates of rock-salt. A cube containing boiling water was placed near one end of the tube, and a thermopile at the other. The thermopile was provided with a double cone, and another cube containing boiling water was placed on the other side of the thermopile in such a position as to exactly compensate for the effect of the radiation received from the main source when the cylinder was exhausted of air. The pure dry gas was then admitted into the tube, and the resulting deflection of the galvanometer attached to the thermopile noted. The deflections obtained with different gases in the tube were a measure of their relative opacities to the radiation from the source. No deflections were observed with air, oxygen, nitrogen, or hydrogen, which accordingly are very highly diathermanous. Compound gases such as carbon dioxide, sulphuretted hydrogen, ammonia, &c., were found to be considerably less diathermanous than the simple gases, and the presence of water vapour greatly increased the absorption of the radiation.

12.08 Some Practical Considerations

The effects described in the foregoing section have certain important practical applications, a few of which will receive mention here.

(a) *The "Thermos" flask.* The double-walled vacuum flask was originally designed by Dewar for the purpose of storing liquefied gases, the evacuated space between the walls of the vessel precluding the transfer of heat from the surroundings by the processes of conduction and convection which require a material medium. The well-known "Thermos" flask is such a vessel, and is used for maintaining the temperatures of hot (or cold) liquids for long periods. To minimise the loss (or gain) of heat by radiation, the surfaces facing the evacuated space are silvered. Thus radiant heat from the hot liquid is reflected back at the outer wall whilst, when storing cold liquids, heat radiation from the surroundings is reflected back from the silvered face of the inner wall. Any radiant heat entering the vacuum space is accordingly "trapped" there, the leakage across it being very slight.

(b) *The greenhouse.* The selective absorption of glass finds practical application in the greenhouse and "cold" frame. When the sun's rays fall upon the glass the visible portion of the radiation and the infra-red rays of shorter wavelength are transmitted, the rest of the infra-red rays being absorbed by the glass. The objects inside the greenhouse absorb the radiation transmitted through the glass, and as the temperature of these objects rises, they lose heat by radiation. This radiation, having its origin in a low temperature source, is however of long wavelength, and as such cannot be transmitted by the glass. Most of it is absorbed by the glass and radiated back into the interior of the greenhouse which thus acts as a trap for the energy of the sun's rays.

Another example of this selective absorption of glass is provided by glass fire-screens. The visible radiations and a small part of the infra-red rays are transmitted by the glass, but the longer wavelength radiations, which comprise the bulk of the fire's radiant energy, are absorbed by the screen. Thus the screen will cut off the majority of the heat of the fire while still allowing it to be seen.

(c) *The screening action of clouds.* It is well known that it is colder on clear nights than when the sky is thickly covered with cloud. The radiation from the earth escapes freely in the former case, whereas on cloudy nights this radiation will in part be irregularly reflected back to the earth from the base of the clouds, the rest of the radiation being absorbed by them owing to the fact that water vapour is very adiathermanous to thermal radiation. The radiation thus absorbed causes a rise in temperature of the cloud which thereby emits radiation on its own account, some of which will arrive at the earth's surface. Smoke clouds produce the same "blanketting" effect as clouds of water vapour, and use is sometimes made of this fact for the protection of fruit blossoms against late frosts.

12.09 Emission and Absorption of Radiation

Considerations of the problem of radiation of heat by bodies have shown that the quantity of heat energy emitted in a given time depends on the nature and size of the surface of the body, and also upon its temperature. Different sources of the same size under the same temperature conditions have different radiating or emissive powers. It is as well here to distinguish between the *emissivity* and the *emissive power* of a surface.

The *emissivity* of a surface is defined as the amount of heat energy emitted by unit area of the surface per second per unit excess temperature over its surroundings.

The *emissive power* of a surface is the ratio of the amount of heat energy radiated per unit area in unit time to the amount of heat energy radiated from unit area of a perfectly black body in unit time under the same conditions.

A comparison of the emissive powers of surfaces may be carried out using a Leslie cube and thermopile. The Leslie cube is a hollow metal box, the four vertical faces of which are differently treated. Thus one surface may be lamp-blacked, another polished, a third painted dull white, the remaining face being coated with black lacquer. The cube is filled with water which is kept steadily boiling by means of an immersion heater, and the deflections of the thermopile taken when placed with its receiving surface a fixed distance from each face of the cube in turn. The deflections so obtained are proportional to the emissive powers of the surfaces concerned, and the results may be given relative to the lamp-blacked surface taken as unity.

The *absorptive power* of a surface is defined as the ratio of the amount of radiation absorbed by unit area of the surface per second to the amount of radiation falling per second on the same area. As with the emissive power this quantity depends on the nature of the surface and the quality of the radiation incident on it. The absorptive powers of different substances may be compared by the following method due to De la Provostaye and Dessains. A thermometer bulb, having been coated with one of the surfaces under test, is heated to a temperature somewhat higher than that at which it will be subsequently used. It is then allowed to cool when placed in a draught-free enclosure and data for a cooling curve obtained. Radiation from a suitable source is now concentrated on to the bulb of the thermometer by means of a lens when a rise in temperature is recorded due to the absorption of the radiation by the coated bulb. The bulb itself now commences to radiate heat to the surroundings, and after a time the temperature will have risen to such a point that the rate at which radiation is emitted by the thermometer is just equal to the rate at which radiation is absorbed by it from the source. The thermometer will thus record some stationary temperature θ_1 (say), and the rate of cooling at this temperature can be obtained from the preliminary cooling curve. If this rate of cooling is θ_1' and W is the thermal capacity of the thermometer bulb, then the rate of loss of heat at the equilibrium temperature is $W\theta_1'$, and this must clearly also be the rate at which heat is absorbed by the bulb at this temperature. The bulb of the thermometer is now coated with another substance (e.g. lamp-black) and the two experiments, for the cooling curve and the equilibrium temperature when exposed to the radiation stream, described above, are repeated. If the equilibrium temperature for this second surface is θ_2 and the corresponding rate of cooling θ_2', the rate of absorption of heat by this surface at the steady state is $W\theta_2'$. Hence the ratio of the absorptive powers of the two surfaces

$$= \frac{\text{rate of heat absorbed by first surface.}}{\text{rate of heat absorbed by second surface.}}$$

$$= \frac{W\theta_1'}{W\theta_2''} = \frac{\theta_1'}{\theta_2'}$$

The results obtained from the foregoing experiments on the relative emissive and absorption powers are very striking and show that if the surfaces are arranged in decreasing order of emissive power, this also represents the order of decreasing absorptive powers of the surfaces. In other words it appears that good emitters are also good absorbers of radiation, bad emitters being bad absorbers.

This relationship between the emissive and absorptive powers can be verified by the following experiment which is a modification of an experiment originally performed by Ritchie. Two tin plates A and D (Fig. 154) arranged vertically have copper wires soldered to them, the wires being connected through a sensitive galvanometer G. The soldered junctions comprise two thermo-couples, and the electric circuit is completed by connecting the tops of the plates by an iron wire. The inside surface of A is lampblacked, whilst that of D is polished, and symmetrically between the plates is placed a Leslie cube containing boiling

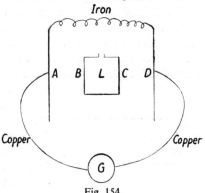

Fig. 154

water, the lamp-blacked surface C of the cube being opposite to the polished surface D, and the polished surface B of the cube being opposite the lamp-blacked surface A. When so arranged the galvanometer shows no deflection indicating that the surfaces A and D are at the same temperature (as otherwise a thermo-electric e.m.f. would be established across the junctions which would result in G being deflected). For this result to be obtained it is clear that the total amounts of radiation absorbed by A and D must be the same, for although D receives a good deal of radiant heat from C, it absorbs very little of it on account of its low absorptive power, whilst on the other hand, face B emits very little radiation, but the bulk of this is absorbed by the black surface of A.

If now E_1, E_2 represent the emissivities of the polished and lampblacked surfaces respectively, and A_1, A_2 their absorptive powers, then the amount of energy absorbed per square cm. per second by $A = A_2 E_1$, and the amount absorbed per square cm. per second by $D = A_1 E_2$. But since the surfaces A and D are receiving energy at the same rate, we have

$$A_2 E_1 = A_1 E_2$$
$$\text{or} \quad \frac{E_1}{E_2} = \frac{A_1}{A_2}$$

That is, the emissive power of a surface is proportional to its absorptive power.

H.T.–11+

12.10 Prévost's Theory of Exchanges

The modern theory of radiation is founded on the **Theory of** Exchanges originally enunciated by Prévost in 1792, and later quantitatively developed by Kirchoff and Balfour Stewart. This theory, which consistently accounts for all the observed facts so far considered, asserts that every body emits radiation in all directions at a rate which is determined only by its temperature and the nature of its surface, and that it absorbs radiation from all surrounding bodies. As an illustration of the theory we shall discuss a simple experiment, the consideration of which is said to have led Prévost to propound his principle of exchanges. If a block of ice is placed at the focus of a pair of conjugate mirrors (Fig. 155), the temperature recorded by a

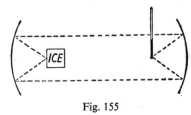

Fig. 155

thermometer at the other focus will fall perceptibly. The effect is similar to the heating effect observed if a source of heat replaced the ice at the focus of the first mirror. At first sight it would appear as if the ice were emitting a "cold" radiation, and indeed this explanation was at one time generally accepted. The complications involved in this supposition are removed however if we consider the thermometer to be in a state of thermal equilibrium with its surroundings in which the rate at which it is emitting radiation is exactly balanced by the rate at which it is absorbing from its surroundings. By placing the ice block in the proximity of the thermometer, this latter rate is reduced as the ice will screen the thermometer from some of the radiation it originally received, and this will not be supplemented by the thermal radiation from the ice itself on account of its lower temperature. Thus the thermometer will be emitting radiation at a greater rate than that at which it receives it from its surroundings, and in consequence its temperature falls until a new state of balance is attained.

According to the theory of exchanges therefore we must consider the temperature of a body to be determined by a state of dynamic equilibrium between the radiation emitted and absorbed by the body. It is not to be assumed that when a body has cooled to the temperature of its surroundings that radiation, unlike the processes of conduction and convection, has ceased. Exchange still takes place, the two streams of radiation from and towards the body exactly balancing in quantity and quality at the equilibrium temperature. If this balance is upset in any way, the net change is evidenced by a rise (or fall) in the temperature of the body.

12.11 Uniform Temperature Enclosures—" Black Body " Radiation

Let us now consider a heat impervious enclosure whose walls are maintained at some constant temperature. If a body at a different temperature is placed in the enclosure, we know from experience that it will ultimately attain the temperature of the enclosure irrespective of its nature or original temperature. On the theory of exchanges the body will radiate to the walls of the enclosure and at the same time receive radiation from the walls. Its temperature will accordingly rise or fall until these two amounts balance one another when its temperature becomes steady, the enclosure being crossed by two equal streams of radiation.

The temperature of the body is independent of its position from which it follows that the quantity of radiation must be the same at all points within the enclosure. Further, the quality of the radiation must be constant throughout the enclosure. For suppose the nature of the body to be such that it only absorbs (and emits) radiation of a definite wavelength. If then it is moved from some position P where it has attained temperature equilibrium with the enclosure, and at which the energy density for the particular wavelength is a maximum for the given temperature, to some other position at which the amount of the particular radiation emitted by the body is less than that at P, the temperature of the body must fall since it will now be radiating more than it receives. This is contrary to experience, and we therefore conclude that each kind of radiation is constant in quantity (which, from the above reasoning, must also be a maximum for the particular temperature concerned) throughout the enclosure.

If other bodies of different natures are placed in the enclosure each will take up the temperature of the enclosure irrespective of the presence of the others. Thus A (Fig. 156) may be a dull black body, B highly polished, C a body transparent to the radiation e.g. a piece of rock-salt, and D a body which selectively absorbs a particular kind of radia-tion. Each will ultimately take up the tempera-ture of the enclosure, and it will then be impossible to distinguish the boundary lines of the different bodies, the result being one

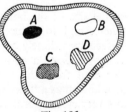

Fig. 156

uniform glare. A will absorb most of the radiation falling on it, and since its temperature is constant it must radiate equally to the enclosure. Similarly B, which reflects the bulk of the radiation incident on it, will only emit a sufficient quantity of radiation to restore back to the energy stream the small amount absorbed from it. C on the other hand transmits the radiation falling on it and thus does not upset the balance of radiation in the enclosure, whilst D will strongly emit the particular radiation it absorbs. Thus each body in the enclosure radiates exactly as much energy to the enclosure as it takes from it—from which it clearly follows that the emissive and absorptive powers of a body are equal.

The quantity and quality of the radiation inside a uniform temperature enclosure are therefore not affected by the nature of the bodies contained in it, nor indeed by the nature of the walls of the enclosure. The radiation is determined solely by the temperature of the enclosure and is known as *temperature radiation* or *full radiation*. Now a perfectly black body is one which will completely absorb all the radiations of whatever wavelength which are incident of it. If such a body is placed inside an isothermal enclosure, it is clear from the above that it must emit the full radiation of the enclosure on attaining temperature equilibrium with it. Hence the radiation in a uniform temperature enclosure is often referred to as *" black body "* *radiation*.

There is however no perfectly absorbing surface available in practice. Even a lamp-blacked surface, although it absorbs practically all the visible part of the spectrum and near infra-red region, reflects the longer infra-red rays quite strongly. Nevertheless it is possible to construct an absorber which will approximate extremely closely to the ideal conditions. Thus by taking a hollow copper sphere (Fig. 157)

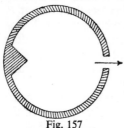
Fig. 157

lamp-blacked on the inside and making a small hole in it, any radiation falling on the hole will suffer complete absorption at the interior surface. The fraction of the radiation of any particular wavelength not absorbed when first striking the lamp-blacked surface will ultimately suffer complete absorption after continued reflection at this surface. Only a very small proportion of the radiation (depending on the size of the hole) will be reflected back from the interior of the sphere, and to reduce the amount of direct reflection the interior surface immediately opposite the hole has an inward conical projection as shown. Such a body when maintained at a given temperature will also emit from the aperture the corresponding full radiation. It is then known as a " cavity radiator" and emits full black body radiation to a very close approximation.

12.12 Kirchoff's Law

We have seen that the intensity of the radiation of any particular wavelength inside a uniform temperature enclosure is determined solely by the temperature of the enclosure, and is a maximum for that particular temperature. Consider now a body with any type of surface placed within the enclosure and let dQ be the amount of energy of the radiation between the wavelengths λ and $\lambda + d\lambda$ incident on unit area of the body in one second. If a_λ is the absorptive power of the body for the wavelength λ, then the amount of energy absorbed per unit area per second by the body will be $a_\lambda\, dQ$, the remainder $(1 - a_\lambda)\, dQ$ of the incident energy being reflected or transmitted. To maintain the stream of radiation inside the enclosure, the body must emit an amount of energy equal to that which it absorbs. Thus if it emits an amount e_λ of energy per unit area per second at the wavelength λ, the amount of energy emitted by the body for the wave range λ to $\lambda + d\lambda$ per unit area per second is $e_\lambda\, d\lambda$. The total energy sent out by unit area of the body per second is therefore $(1 - a_\lambda)\, dQ + e_\lambda\, d\lambda$, and this must equal the amount of energy received, and hence

$$(1 - a_\lambda)\, dQ + e_\lambda\, d\lambda = dQ$$

Now for a perfectly black body the absorptive power is unity, and if E_λ is the amount of energy it emits per unit area per second at the wavelength λ, we have

$$E_\lambda\, d\lambda = dQ$$

Thus from these two equations we get

$$(1 - a_\lambda)\, E_\lambda\, d\lambda + e_\lambda\, d\lambda = E_\lambda\, d\lambda$$

$$\text{or} \qquad \frac{e_\lambda}{a_\lambda} = E_\lambda$$

Since E_λ is constant at a given temperature, it therefore follows that

$$\frac{e_\lambda}{a_\lambda} = \text{const.}$$

for all substances at the same temperature. This is Kirchoff's law.

The ratio $\dfrac{e_\lambda}{E_\lambda}$ is the emissive power of the body as defined on page 311 and therefore we see that the emissive and absorptive powers of a body are equal. This relationship already demonstrated experimentally in section 12.09 is thus seen to be quantitatively exact.

The classic example of Kirchoff's law is provided by the emission and absorption spectra of sodium vapour. Thus when the temperature of sodium vapour is raised sufficiently for it to emit visible radiations, it is found that the emission spectrum consists only of two lines very close together in the yellow region. According to Kirchoff's law, sodium vapour should then be a good absorber of light of these two wavelengths. This was demonstrated by Bunsen and Kirchoff by examining spectroscopically the light from a high temperature source after it had passed through a cooler cloud of sodium vapour. The continuous spectrum of the white light was then found to be crossed by two dark lines corresponding exactly in position to the emission lines of the sodium spectrum. This selective absorption by a gas or vapour of the characteristic radiation it emits when at a higher temperature accounts for the Fraunhöfer lines in the solar spectrum and provides a means of analysing the constitution of the sun and stars.

As other examples of Kirchoff's law may be mentioned the apparent reversal of the pattern of china when heated. The coloured pattern which absorbs more strongly than the white background at ordinary temperatures, shines out brightly against the apparently much darker background when heated in a furnace. The same effect too is seen when a piece of iron wire and a length of glass tubing are heated side by side in a bunsen flame. The glass, which is not as good an absorber as the iron, appears less bright than the iron wire when in the flame. Another example is afforded by copper and gold which at low temperatures are reddish in colour indicating a low absorptive power at the red end of the spectrum. When the temperature is raised to incandescence however, both these metals glow with a greenish light thereby indicating a low emissive power in the red region in agreement with Kirchoff's law.

12.13 Stefan's Law

We shall now examine how the quantity of the radiation emitted by a black body depends on its temperature, and in the following section we shall discuss the question of the quality of the radiation, i.e. how the total energy emitted is distributed amongst the various wavelengths for different temperatures of the radiating source.

The first systematic attempt to establish the law connecting radiation and temperature was made by Dulong and Petit in 1817. They investigated the rate of cooling of a thermometer suspended in a copper sphere surrounded by water at a constant temperature. To eliminate the effects of conduction and convection the air pressure in the sphere was reduced to about 2 mm. of mercury. Their results were embodied in an empirical law of an exponential form which seemed to fit the observations very satisfactorily, and for many years this was accepted as the law governing the dependence of radiation on temperature.

In 1879 Stefan suggested that the total radiation from a body was proportional to the fourth power of its absolute temperature. He was led to this result from a consideration of the observations on the rate of emission of radiation from a heated platinum wire made by Tyndall who had found that the radiating powers of platinum at the temperatures 1200°C. and 525°C. were in the ratio 11·7 to 1. Stefan showed that the ratio $\left(\dfrac{1200 + 273}{525 + 273}\right)^4$ is very nearly 11·7, and thus the two radiations were proportional to the fourth power of their absolute temperatures. Stefan then examined the results of Dulong and Petit and found that they were very well represented by the fourth power law. In 1884 Boltzmann, following up Bartoli's work on the application of thermodynamics to radiation, established the fourth power law on theoretical grounds and pointed out that it was only valid for an "ideal black body." The law, which is frequently referred to as the *Stefan Boltzmann law,* may be expressed as

$$E = \sigma T^4$$

where E is the total energy of the radiation emitted at the absolute temperature T, and σ is a constant known as *Stefan's constant,* the mean value of which, as taken from the mean of recent observations, is

$$\sigma = 5 \cdot 75 \times 10^{-5} \text{ ergs per cm.}^2 \text{ per sec. per deg.}^4 \text{K}$$

In using the law account must be taken of the radiation received by the body from its surroundings. Thus if a body at a temperature $T°K$. is cooling in an enclosure maintained at an absolute temperature of $T_o°K$, the rate of cooling will be determined by the net rate of loss of energy which is given by

$$E \text{ (net)} = \sigma (T^4 - T_o^4).$$

The validity of the fourth power law has been tested by a number of investigators, notably by Lummer and Pringsheim who made observations of the radiation from a "black body" over temperatures from 100°C. to 1300°C. For temperatures between 100°C. and 600°C. the "black body" used was a copper sphere coated inside with platinum black, whilst for the higher range of temperatures an iron cylinder coated within with platinum black and heated inside a double-walled gas muffle furnace was used. The energy radiated by these bodies was measured by means of a Lummer-Kurlbaum bolometer (see section 12.05). The results of these experiments served to verify the fourth power law for the range of temperatures specified within the accuracy limits of the gas scale to which all the temperatures were reduced.

12.14 Distribution of Energy in the Spectrum of a Black Body

The distribution of energy amongst the wavelengths of the thermal spectrum of a full radiator was investigated by Lummer and Pringsheim in 1899. For these experiments they used as their source a carbon plug situated in the middle of an electrically heated carbon tube which was thermally insulated by surrounding it with co-axial tubes of fireclay and asbestos. A current of nitrogen was passed down the carbon tube thus permitting of the attainment of very high temperatures without the destruction of the tube by oxidation; the temperatures were measured by means of a thermocouple. The radiation from the carbon tube fell on a slit and was dispersed by a prism of fluor-spar*

---

* Fluor-spar has two strong absorption bands in the extreme infra-red, but as these were outside the limits of the wave-length range of Lummer and Pringsheim's experiments, it did not affect the results.

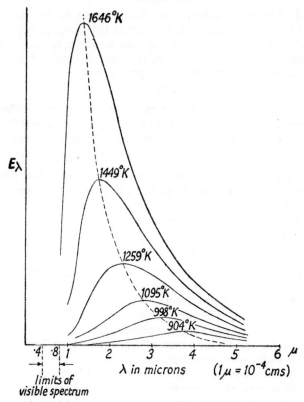

Fig. 158

which is very transparent to infra-red radiation. Corrections were made using the known dispersion curve of fluor-spar to convert the prismatic to the normal energy curves. The thermal spectrum was focused by means of silvered concave mirrors, and the energy at various parts of the spectrum was measured using a linear Lummer-Kurlbaum bolometer of width 0·6 mm. The results obtained by Lummer and Pringsheim are shown in Fig. 158. Each curve relates the intensity of the radiation (E_λ) to the wavelength for a given temperature of the source. The general shape of the curves is the same for all temperatures, each possessing a well defined maximum which increases rapidly in magnitude as the temperature rises and at the same time undergoes a displacement towards the shorter-wavelength end of the spectrum. (The locus of the maxima is shown by the dotted line). The total energy E emitted by the source per second at a given temperature is given by

$$E = \int_0^\infty E_\lambda \, . \, d\lambda$$

and this is represented by the area intercepted by the curve and the axis of λ. These areas are found to be directly proportional to the fourth power of the absolute temperatures of the curves—a result which is in accord with Stefan's radiation law discussed in the previous section.

Prior to the experimental work of Lummer and Pringsheim, Wien (1896) had shown on thermodynamical grounds that for full radiation the wavelength λ_{max} of the maximum radiation for a given value of the absolute temperature is given by

$$\lambda_{max} \times T = \text{const.} = 0 \cdot 2892 \text{ cm. deg. } K$$

This is known as *Wien's displacement law*. He also showed that the energy E_{max} of the maximum radiation was related to the temperature by the law

$$E_{max} = \text{const.} \times T^5$$

These two laws can be verified from the results of the experiments of Lummer and Pringsheim by plotting the values of E_λ as ordinates against the product λT as abscissae. The maxima of the curves then lie immediately above one another showing that for the maxima of the curves the product λT has a constant value—thus verifying the first of Wien's laws. The ordinates representing the values of E max. can also be shown to be proportional to the fifth power of their corresponding absolute temperatures—as demanded by the second of Wien's laws.

Wien was not successful however in obtaining an exact law on thermodynamical grounds alone for the distribution of the energy amongst the wavelengths. Wien's radiation law (from which the two above laws are derived) is of the form

$$E_\lambda = \frac{1}{\lambda^5} \cdot f(\lambda T)$$

where $f(\lambda T)$ is some undetermined function. The form of this function can only be determined by making some specific hypothesis concerning the process of emission. Taking a gas as his black body, and assuming that each molecule emits radiation with a wavelength depending only on its velocity Wien, by applying Maxwell's law for the distribution of velocities among the molecules, and the principle of equipartition of kinetic energy, obtained a form for $f(\lambda T)$ from statistical calculation and gave as his radiation formula

$$E_\lambda = c_1 \lambda^{-5} e^{-c_2/\lambda T}$$

where c_1 and c_2 are constants.

Paschen and Wanner from observations made in the visible region using a photometer instead of a bolometer, showed that Wien's formula represented their results very satisfactorily. The experimental work of Lummer and Pringsheim in the infra-red however revealed that the

11*

law held only in the region of *short* waves, there being systematic discrepancies for the longer wavelengths.

The problem of spectral distribution of energy was next tackled by Rayleigh (1900) who applied the principles of statistical mechanics and electrodynamics to obtain the formula

$$E_\lambda = \frac{8\pi kT}{\lambda^4}$$

where k is Boltzmann's constant. The work of Rubens and Kurlbaum (1900) on radiations of wavelengths between 25 and 30 microns obtained by successive selective reflections from fluor-spar, confirmed the formula for these wavelengths, although it is clearly in disaccord with the general experimental data as it indicates that practically all the energy is concentrated in the very short-wave region.

Thus neither Wien's formula nor Rayleigh's formula (later confirmed by Jeans) adequately represented the facts of experiment, and it thus appeared that the laws of classical dynamics, which also included the equipartition principle, were incapable of providing the correct interpretation of the problem. It was at this juncture that Planck (1901) took up the subject; and he was able to obtain an expression which fitted the experimental results by abandoning the equipartition principle, but only by making entirely new assumptions regarding the emission and absorption of radiation by matter. According to his theory a vibrating system was no longer to be supposed to emit and absorb radiation continuously, but only in certain multiples of the fundamental frequency of the "resonator." Thus a vibrating electron can receive or give out "quanta" of energy equal to $h\nu$, $2h\nu$, $3h\nu$, &c., where ν is the frequency of oscillation of the electron and h a constant known as Planck's constant, but not in intermediate amounts. By an application of statistical methods based on these assumptions, Planck derived the formula:

$$E_\lambda = \frac{8\pi hc}{\lambda^5 (e^{ch/k\lambda T} - 1)}$$

where c is the velocity of electromagnetic waves (3×10^{10}cm. per sec.), k Boltzmann's constant, and h the constant referred to above.

This expression represented very satisfactorily the spectral distribution of energy, and also gave the shift of the maximum to the period of shorter wavelengths with increase of temperature. For short waves $\left(\frac{ch}{k\lambda T}\text{ large}\right)$ the formula passes into Wien's expression, and for long waves $\left(\frac{ch}{k\lambda T}\text{ small}\right)$ Rayleigh's formula is obtained. Thus both Wien's and Rayleigh's formulæ are included in Planck's relation as limiting cases.

The revolutionary concepts of the quantum theory thus enunciated by Planck subsequently received strong support by the remarkable successes obtained by its application to other branches of physics—as for example in dealing with specific heats of solids to which reference has already been made in section 7.17. Its ability to explain these widely divergent topics gives justification for the radical departure from classical electrodynamics demanded by the theory.

12.15 Radiation Pyrometry—Féry's Total Radiation Pyrometer

An important application of the laws of radiation discussed in the preceding section is to the measurement of high temperatures. Thus the Stefan-Boltzmann law:

$$E = \sigma(T^4 - T_o^4)$$

permits of the evaluation of the temperature T of the radiating source if σ is known and the energy E emitted per cm.² per second by the hot source to the receiving body at a known temperature T_o is measured. Temperatures defined in this way are independent of the specific properties of any particular substance, and are therefore *absolute* temperatures in the same sense as those defined on Kelvin's thermo-dynamic scale of temperatures (section 13.07)—the ideal black body serving the same functions as the perfect gas of the latter scale. Departure from the ideal conditions will of course produce errors and discrepancies, although the radiation emerging from a small hole in the wall of a furnace may be considered as approximating very closely to that from an ideal black body (see section 12.11).

A special advantage of radiation pyrometers is that it is not necessary to have any part of these instruments in contact with the hot body whose temperature is to be determined, thus avoiding the difficulties encountered in the use of resistance and thermo-electric thermometers at temperatures of 1300°C. and above. In addition there is no upper limit to temperatures which can be measured by this means.

A convenient form of instrument based on the fourth power law is that devised by Féry and is shown in Fig. 159. A concave mirror C of copper, plated with nickel or gold situated at the end of a short tube focuses the radiation from the hot body on to a small blackened strip S to the back of which is attached one junction of a thermo-couple. The focusing is done visually by the aid of two small semi-circular mirrors MM inclined to one another at an angle of about 5°, and placed immediately in front of S which receives the radiation through an aperture provided by two small semi-circular holes in the mirrors. The mirrors are sighted by an eyepiece E through a hole in the concave mirror, and if incorrectly focused the appearance will be as given in Fig. 159a. The position of C is then adjusted by means of a

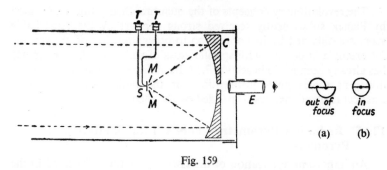

out of focus   in focus

(a)   (b)

Fig. 159

rack and pinion until the appearance through the eyepiece is as indicated in Fig. 159*b*, when the radiation will be accurately focused through the aperture between the mirrors on to *S*. The cold junctions of the thermo-couple are screened from the direct radiation by a tongue and box (not shown) surrounding the couple, and having obtained the "in focus" position, the reading of a millivoltmeter connected to the terminals *TT* is taken.

The readings of the millivoltmeter are independent of the distance of the pyrometer from the hot source provided the heat image completely fills the aperture between the mirrors *MM*. This is because the reading of the instrument depends only on the intensity of the heat image and this remains constant up to the limiting distance. Thus if the distance from the source is doubled, the amount of radiant energy received by *C* is reduced to one fourth, but this reduction is compensated for by a similar reduction in the size of the image, and so the intensity is unaltered. From the geometry of the optical system of a given instrument it is possible to calculate the limiting distance up to which the readings remain constant in terms of the diameter of the source.*

The radiation pyrometer is calibrated by comparison with a standard platinum-rhodium thermo-couple over the range 500—1400°C. The instrument is focused on the inside of an electrically heated muffle furnace the temperature of which is determined by one or more of these standard thermo-couples. If the temperature found in this way is T_1 and the reading of the millivoltmeter is R_1, then assuming the fourth power law

$$R_1 \propto E_1 \propto (T_1^4 - T_o^4)$$

If now the pyrometer is sighted on some other body at some unknown temperature T_2 and a reading R_2 is recorded, then

$$\frac{R_2}{R_1} = \frac{E_2}{E_1} = \frac{T_2^4 - T_o^4}{T_1^4 - T_o^4}$$

---

* For details of this and a very complete account of radiation pyrometers see the article by Griffiths in the *Dictionary of Applied Physics*, Vol. I (Macmillan).

from which T_2 can be found. If T_1 and T_2 are very much higher than T_0 (which is usually the case), T_2 may be found from

$$T_2 - T_1 \sqrt[4]{\frac{R_2}{R_1}}$$

For various reasons* the readings of the instrument are not quite proportional to the fourth power law. Thus, to mention one or two causes producing discrepancies; the e.m.f. of the thermo-couple is not quite proportional to the temperature difference between its junctions, and the conduction of heat along the thermo-couple wires causes a slight temperature rise in the cold junction. The instrument must then be calibrated by comparing its readings with those of a standard thermo-couple at a series of points over a wide range of temperature, the scale of the millivoltmeter being usually graduated to give a direct reading of the temperature. Analysis of the calibration data shows that the deviation from the fourth power law varies with different instruments, the index of T lying within the limits 3·8 to 4·2.

Temperatures above 1400°C. are measured by interposing a rapidly rotating sector between the hot body and the pyrometer. The object of this is to cut down the radiation by some known fraction so as to obtain the same reading as when the instrument is sighted on a "black body" at some known temperature. This may be called the *apparent temperature* of the hotter source. Let these temperatures be T_1 and T respectively, then if the sector transmits a fraction n of the incident radiation, and R is the reading of the instrument in the two cases, we have, assuming the fourth power law:

$$R = \text{const. } T_1^4 = \text{const. } nT^4$$

from which
$$T = \frac{T_1}{n^{\frac{1}{4}}}$$

n being given by $\dfrac{360 - \theta}{360}$, where θ is the angle of the sector.

In some instruments the thermo-couple is replaced by a bimetallic spiral which uncoils due to the differential expansion of the strips when the radiation is focused on it. An aluminium pointer is attached to the extremity of the spiral and moves over a scale which is graduated in °C. A disadvantage of this instrument is the necessity for frequent adjustments of the scale zero which is affected by changes in the surrounding temperature, and also after the instrument has been sighted on a high temperature source.

Another modification is the *Forster fixed-focus pyrometer*. This has a diaphragm at the open end of the tube, the relative positions of this diaphragm and the thermo-couple being such that they constitute conjugate foci with respect to the concave mirror. This arrangement obviates the need for focusing the instrument.

* See Griffiths, *loc. cit.*

12.16 Optical Pyrometers

In optical pyrometry the temperature of a hot body is measured by comparing the intensity of the light from the body with that from a standard source—usually a filament lamp. To overcome the colour matching difficulty the light from the two sources is passed through a coloured filter which transmits only wavelengths between certain narrow limits, the instrument then being adjusted until the intensities of the light of the selected wavelengths from the two sources are equal. The temperature of the hot body must then be equal to that of the calibrated standard source. A good red glass filter is generally used as the red part of the spectrum is the first to appear when the temperature of a body is raised, thereby extending the lower limit of the instrument, and red glasses can be manufactured more nearly monochromatic than glasses of other colours (such as green or blue).

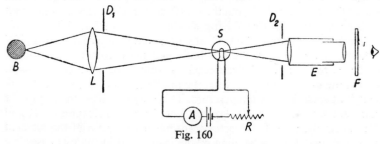

Fig. 160

The simplest form of optical pyrometer, which finds very general use, is the disappearing filament pyrometer. This was originally introduced by Morse in 1902, but has subsequently been modified by Holborn and Kurlbaum whose instrument is shown diagrammatically in Fig. 160. An image of the hot body B whose temperature is required, is projected by the objective L on to the filament of the standard source S which is viewed through a red glass filter F by an eyepiece E. D_1 and D_2 are diaphragms which limit the cones of light. The temperature of S is carefully adjusted by means of the rheostat R until the filament is invisible against the image of B. If the current through S is too small, the filament appears dark against the bright background of the image, whilst if it is too great the filament shines out brightly against a relatively duller background. When the brightness of S and the image of B have been exactly matched in this way, the reading of the ammeter A is taken. It should be noted that the brightness of the image formed by the lens L is independent of its distance from the source B, and consequently the readings of the instrument do not depend on its distance from the hot body.

The connection between the current and the temperature of the filament is of the form:

$$i = a + bT + cT^2$$

where a, b, and c are constants. These may be determined by sighting the instrument on a muffle furnace in which is placed a standard thermo-couple and taking observations at three different temperatures. Having thus obtained the necessary calibration data, the ammeter can then be graduated to read directly in °C.

The upper limit for empirical calibration using standard thermo-couples is about 1400°C. This temperature is also approximately the highest temperature at which the filament can be run without rapid deterioration. For higher temperatures an absorbing device must be used with the instrument to cut down the intensity by some known fraction. The usual practice is to place a piece of neutral tinted glass before the objective or to employ a rotating sector as with the total radiation pyrometer (page 325). To avoid any sensation of flicker the sector must have a minimum rotation speed of 30 to 40 revolutions per second. Calibration for the higher temperature range is effected by using Planck's radiation formula which may be applied with close approximation in the form of Wien's formula for values of λT less than 3000 μ degrees. Taking λ for red light as approximately 0·7 μ, this gives an upper limit with Wien's formula of just over 4000°C., which adequately covers the range of temperatures to be measured in practical work.

If the absorbing device transmits a known fraction n of the light of wavelength λ from the source at a temperature T, and the filament disappears when sighted directly on a "black body" at a temperature T_1, then

$$n (E_\lambda)_T = (E_\lambda)_{T_1}$$

where $(E_\lambda)_T$, $(E_\lambda)_{T_1}$ are the respective energies of the radiation with wavelengths between λ and $\lambda + d\lambda$ at the temperatures T and T_1. Using Wien's formula we therefore have

$$nc_1\lambda^{-5}e^{-c_2/\lambda T} = c_1\lambda^{-5}e^{-c_2/\lambda T_1}$$

or, taking logs

$$\log_e n = \frac{c_2}{\lambda}\left[\frac{1}{T} - \frac{1}{T_1}\right]$$

In this equation λ is the effective wavelength of the red light transmitted by the filter (i.e., the wavelength in best accord with the application of Wien's formula), and c_2 is a constant whose value is 14,350 micron-degrees. Thus, since T_1 is known, the value of T can be determined.

In another type of optical pyrometer due to Wanner the comparison between the radiations from the hot body and the standard source is effected by using a polarising spectrophotometer. The filament lamp is run at constant intensity, being checked periodically against a standard amyl acetate lamp, and the light from the hot body and the electric lamp is passed through a direct vision spectroscope which separates out two identical monochromatic beams. These beams are then passed through a polarising Nicol prism, plane polarised components in mutually perpendicular planes being selected from each beam by a biprism for transmission through a diaphragm to an analysing Nicol. The two beams are then brought to equal intensity by rotation of this Nicol prism. If the direct vision spectroscope is not incorporated in the instrument, the beams are viewed through a piece of red glass as in the disappearing filament pyrometer.

It is shown in textbooks of light that if the analysing Nicol has to be rotated through an angle ϕ to match the two beams of light of intensities I_1 and I_2, then

$$\frac{I_1}{I_2} = \tan^2 \phi$$

Using this equation with Wien's formula a calibration formula of the form

$$\log \tan \phi = a + \frac{b}{T}$$

where a and b are constants can be obtained.* In using this type of optical pyrometer when taking the temperature of a metallic surface it is necessary to sight the instrument normally on the surface as otherwise the light received will be largely polarised. Further the Wanner photometer does not give as well defined an image of the source as the disappearing filament pyrometer which is essentially a telescope and produces a very sharp focus of the distant body. On the other hand the polarising pyrometer does not require the use of absorbing devices for the higher temperature ranges, and gives readings which are less dependent on fluctuations of the standard filament lamp.

12.17 Comparison of Total Radiation and Optical Pyrometers

Both forms of pyrometer can be made up as robust, easily-handled instruments, and as such they are widely used in the practical determination of temperature. The radiation pyrometer has a somewhat lower temperature limit than the optical pyrometer and has the special advantage that after focusing, the temperature is automatically recorded without further manipulation. On the other hand it is more likely to be affected by any departure from ideal "black body" conditions of the hot source than the optical pyrometer with its restricted wavelength range, and the effects of atmospheric absorption are more pronounced in the radiation pyrometer. As we have seen both forms of pyrometer enable temperature scales to be obtained by extrapolation using the laws of radiation. Tests made at high temperatures using a "black body" furnace show that the readings recorded by both pyrometers are in very close agreement. Thus the difference between the temperatures

---

* See the article by Griffiths on Optical Pyrometers. *Op. Cit.*

recorded on the two scales is less than 0·5°C. at 1750°C., and about 4°C at 2800°C. The differences are within the error of experiment, and the results provide very satisfactory confirmation of the general principles of radiation and optical pyrometry.

It should be pointed out that both types of pyrometer give readings less than the true temperature if sighted on a body which is not radiating as a "black body." Under these conditions the most that can be said is that the temperature of the body is not lower than that recorded. The corrections to be applied which, as already indicated, are smaller for optical pyrometers than with the total radiation type, are difficult to estimate. The problem involves the determination of the emissive powers of surfaces at different temperatures, and although a large amount of experimental data has been accumulated, attempts to modify Planck's radiation law to represent the distribution of energy in the thermal spectrum of non-full radiators has not met with much success. The most satisfactory method for the determination of the melting points of refractory materials is to heat them in a crucible in a "black body" furnace whilst sighted by an optical pyrometer. The melting point is indicated by a break in the continuity of the temperature-time curve. In this way Kanolt determined the melting points of lime, and the oxides of magnesium, aluminium, and chromium.

12.18 The Temperature of the Sun

The estimation of the effective temperature of the sun affords an interesting application of the laws of black-body radiation. This has been done by finding the wavelength $\lambda_{max.}$ for which the energy is a maximum in the spectrum of the solar radiations, and applying Wien's displacement law,

$$\lambda_{max} \ T = 0.2892$$

Thus taking λ_{max} as 4900×10^{-8}cm. (as given in Kaye and Laby's tables), the temperature of the sun works out to be 5902°K.

The sun's temperature has also been deduced by applying Stéfan's law to the total radiation emitted. This method involves the determination of an important quantity in connection with solar radiation known as the *solar constant* (S). This is the amount of energy received from the sun by the earth (at its mean distance) in unit time per cm.² of surface placed normally to the sun's rays when absorption losses in the earth's atmosphere have been corrected for. It is usually expressed in calories per cm.² per min. or in watts per cm.² The measurement of this constant is carried out using a pyrheliometer such as that due to Ångström described in section 12.05. Allowances for atmospheric absorption are difficult to make as this depends on the amount of water vapour in the air and varies appreciably for different wavelengths. Various methods have been used to estimate these losses, the best being by taking simultaneous readings at high and low altitudes. The mean value of S

from recent observations may be taken as 1·94 calories per cm.² per minute, or ·135 watts per cm.²

Now if the mean distance of the earth from the sun is R, each square cm. of the surface of a sphere of this radius will receive S calories per minute from the sun, and hence the total amount of energy radiated per minute by the sun will be $4\pi R^2 S$. Thus, if r is the sun's radius, the amount of energy E radiated per cm.² of the sun's surface per minute is

$$E = \frac{4\pi R^2 S}{4\pi r^2} = \left(\frac{R}{r}\right)^2 \cdot S$$

Taking $R = 9\cdot28 \times 10^7$ miles, and $r = 4\cdot33 \times 10^5$ miles, and the value of S as given above,

$$E = \left[\frac{9\cdot28 \times 10^7}{4\cdot33 \times 10^5}\right]^2 \times 1\cdot94 \text{ cal. per min.}$$

or
$$= \left[\frac{9\cdot28 \times 10^7}{4\cdot33 \times 10^5}\right]^2 \times \cdot135 \times 10^7 \text{ ergs per sec.}$$
$$\text{(since 1 watt} = 10^7 \text{ ergs per sec.)}$$
$$= 6\cdot198 \times 10^{10} \text{ ergs per sec.}$$

Now by Stéfan's law

$$E = \sigma\,(T^4 - T_0^4)$$

where $\sigma = 5\cdot75 \times 10^{-5}$ ergs per cm.² per sec., and T, T_0 represent the temperatures of the sun and earth respectively. If T_0^4 is ignored compared with T^4 we have, to a sufficient degree of accuracy

$$E = 5\cdot75 \times 10^{-5} T^4$$

from which, using the value of E calculated above

$$T = 5730°K.$$

a result in fair agreement with that derived on the basis of Wien's law.

These estimates give the effective temperature of the sun's surface on the assumption that it radiates as a "black body." Actually its mean surface temperature will be somewhat higher than this although a value of about 6000°K. is generally accepted. The *internal* temperature of the sun is of course considerably higher than this. The sun is assumed to derive its vast source of energy from nuclear transformations taking place within it, and on this basis Jeans has calculated that the temperature of the sun's interior must be of the order of $10^7°K.$

QUESTIONS. CHAPTER 12

1. A hot metal ball is suspended in the atmosphere by a fine wire. Give an account of the process by which it loses heat.

Explain what is meant by *black body radiation*. How is this radiation obtained in practice? Show how the energy is distributed in the thermal spectrum and indicate how the distribution depends upon temperature.   (W.)

2. Describe and explain ONE method, based on an application of Stefan's radiation law, for the measurement of temperatures of the order of 2,000°C.; indicate the main disadvantage of the method.

Explain (a) why glass is suitable for use in the construction of greenhouses, (b) how heat losses are minimized in the Dewar flask. (W.)

3. Describe the disappearing filament pyrometer and explain carefully how it is used and how you would calibrate it.

A black-body radiator at 27°C. radiates energy at the rate of 4.7×10^5 erg. cm.$^{-2}$ sec.$^{-1}$ Calculate:

(a) the value of Stefan's constant, (b) the power in watts radiated by a filament 10cm. long and 2mm. diameter when it is at a temperature of 2,000°C., assuming that it behaves as a black-body radiator. (A.)

4. What is Stefan's law? Describe experiments to demonstrate Prevost's theory of exchanges.

A calorimeter, when cooling, obeys Newton's law of cooling. If it takes 5min. to cool from 60°C. to 50°C., how long will it take to cool from 40°C. to 30°C., the room temperature being 15°C.? (Camb. Schol.)

5. Explain the term *black body temperature*. Describe how it is found for an incandescent body, outlining the experimental work on which the method is based. (S.)

6. What is meant by a black body? What are the principal features of the radiation from a black body and how were they discovered?

The tungsten filament of a 60 watt lamp has a length of 20cm. and a diameter of 0·05mm. If the energy it radiates is only half that radiated by a black body at the same temperature, what is the temperature assuming that all the heat is lost by radiation.

(Stefan's constant $= 5.67 \times 10^{-5}$ c.g.s. units.) (S.)

7. Explain the terms " black body ", " thermal emissivity ", and " Stefan's law ". Why does a highly polished body radiate heat more slowly than a similar body with a blackened surface?

Describe how the energy radiated by a black body is distributed among the various wavelengths of the radiation and how this distribution depends on the temperature. How can the radiation from black bodies be studied in the laboratory?

(Camb. Schol.)

8. Describe ONE method of detecting thermal radiation. How would you demonstrate the presence of infra-red radiations in the spectrum of the sun?

Explain what is meant by a perfectly black body. Show how, on the assumption that the photosphere of the sun radiates as a black body, it is possible to estimate its temperature. (O.)

9. What are the factors which determine the net rate of loss of heat from a body by radiation?

Explain what is meant by a *black body* and state *Stefan's Law*.

A 60 watt electric lamp consists of a tungsten filament 15·0cm. long and 0·0032cm. in diameter enclosed in an evacuated bulb. Assuming that the filament obeys Stefan's law, and neglecting both radiation from the envelope to the filament and conduction through the leads, find a value for the working temperature of the filament.

(Stefan's constant $= 5.7 \times 10^{-5}$ erg. cm.$^{-2}$ sec.$^{-1}$ degC.$^{-4}$) (N.)

10. Describe some form of optical pyrometer and show how it is used to measure the temperature of a furnace (about 1,000°C.). How is the instrument calibrated?

A blackened sphere of iron, diameter 2·00cm., is suspended by a non-conducting thread at the centre of an evacuated spherical shell, whose inner surface is blackened and maintained at a temperature of 20°C. The sphere is at a temperature of 250°C. which is falling at the rate of 18·8 deg. min.$^{-1}$ Calculate a value for Stefan's constant.

(Specific heat of iron at 250°C. = 0·116 cal. gm.$^{-1}$ degC.$^{-1}$ Density of iron at 250°C. = 7·60 gm. cm.$^{-3}$)   (N.)

11. What is black body radiation? State two important laws relating to such radiation.

Assuming that the solar radiation falling normally on a surface at a distance equal to the radius of the earth's orbit is 1·90 cal. cm.$^{-2}$ min.$^{-1}$, that the distance of the earth from the sun is approximately 150 × 10^{6}km. and that the diameter of the sun is 1·39 × 10^{6}km., deduce a value for the surface temperature of the sun supposing that it radiates as a black body.

(Stefan's constant: 1·37 × 10^{-12} cal. cm.$^{-2}$ sec.$^{-1}$ degC.$^{-4}$)   [N.]

12. Discuss the methods employed for measuring temperature higher than 600°C. and describe ONE in detail.

Assuming that the earth as a whole absorbs and radiates like an ideal black body in space at 0° abs., obtain a value for the average temperature of its surface. Assume that the intensity of solar radiation at the surface of the earth is 2 calories per sq. cm. per minute and that the value of Stefan's constant is 5·7 × 10^{-5} erg. cm.$^{-2}$ sec.$^{-1}$ degC.$^{-4}$ (N.)

13. Give an account of how you would attempt to measure Stefan's constant. What accuracy would you expect to achieve?

By making the simplifying assumptions (a) that both the earth and the sun behave like black bodies, and (b) that the earth is at a uniform temperature throughout, calculate the equilibrium temperature of the earth. Comment on your result.

(Radius of sun = 7·0 × 10^{8}m.; radius of earth's orbit = 1·5 × 10^{11}m.; surface temperature of sun = 6,000°K.)   (Camb. Schol.)

14. Discuss the mechanism by which thermal equilibrium is attained when a hot body is placed in an evacuated space with cooler walls maintained at constant temperature. The body is not in physical contact with the walls.

A diode valve consists of two long coaxial cylinders. The inner cathode cylinder, of radius 0·05cm., radiates heat like a black body. The radius of the anode cylinder is large compared with that of the cathode. The cathode heater element dissipates one watt per centimetre length. If the steady anode temperature is 227°C., estimate the temperature of the cathode. Neglect end effects.

(Stefan's constant = 5·74 × 10^{-12} watt cm.$^{-2}$ °K.$^{-4}$.)   (Camb. Schol.)

15. State Stefan's law of radiation and the conditions for which it applies. Describe any experimental arrangement which could be used to verify the law.

Given that the solar constant (the solar energy incident normally at the earth's mean distance) is 2·0 cal. cm.$^{-2}$ min.$^{-1}$, calculate the surface energy of the sun. State any assumptions which are involved in your calculation.

Also estimate the energy in watts falling on the whole earth from all the stars in the galaxy. Assume that the galaxy contains 10^{11} stars identical to the sun and distributed uniformally in a sphere of radius 10^{22}cm. with the sun as centre.

(Stefan's constant = 5·7 × 10^{-5} erg. cm.^{2} sec.$^{-1}$ C.$^{-4}$; solar radius = 6·9 × 10^{10}cm.; solar distance = 1·5 × 10^{13}cm.; radius of earth = 6·4 × 10^{8}cm.; 1cal. = 4·2 joules.)   (Oxford Schol.)

CHAPTER XIII

THERMODYNAMICS

13.01 Transformation of Heat into Work—Heat Engines

The study of thermodynamics deals with the relation of heat to mechanical work and other forms of energy. One aspect of this relationship, namely, the conversion of mechanical work into heat, has already been studied in chapter VI, and the results, which are of fundamental importance and have general application, may be expressed in the following statement which constitutes the First Law of Thermodynamics (see also section 6.12). *When work is converted into heat a definite amount of heat is produced for each unit of work so converted; and, conversely, when heat is expended to produce work, this same amount of heat is used for every unit of work produced.* It will be seen that this is merely a particular statement of the Law of Conservation of Energy establishing the equivalence of heat and work. It deals only with the quantities of heat and work involved and has nothing to say about the direction of change, both being equally favoured. A detailed study of the conversion of heat into work, however, reveals that the process has certain limitations. The important principles governing the convertibility of heat into useful work, and the consequences resulting from them provide the material of the present chapter.

Any device by which heat is converted into work is called a *heat engine*. We shall not be concerned with the practical details of the mechanical arrangements of any particular form of engine, but to fix our ideas we shall imagine our engine to be a simple piston and cylinder device and concentrate on the thermal operations taking place. All heat engines function on account of the thermal properties of the *working substance* used. It forms the vehicle by which heat passes through the engine and in our case we shall consider this to be a perfect gas. Heat is supplied from a *source* at high temperature, the subsequent expansion of the gas forcing up the piston head. On absorbing heat from the gas by putting it in contact with a cold body called the *sink** the piston can be brought back to its original position and a new sequence of expansion and contraction started. This constitutes a *cycle of operations,* external work being performed by letting the heat down from a high to a low temperature during the cycle. The process is analogous to the production of work by a water

---

* The terms condenser and refrigerator are also used.

wheel by letting down water from a high to a lower level, but with the important difference that some heat disappears in each cycle of the heat engine, this heat being converted into an equivalent amount of work.

The fraction of the heat taken in which is converted into useful work determines the efficiency of the heat engine. It is clearly a matter of the utmost practical importance that this fraction shall be as large as possible. How may the greatest possible amount of work be obtained from an engine using a given source of heat? Is the efficiency of such an engine limited, and if so, how? Under what conditions will the efficiency be a maximum? The answers to these fundamental questions were provided by Sadi Carnot in a remarkable essay published in 1824 entitled "Réflections sur la puissance motrice du feu et sur les machines propres à développer cette puissance." In developing his theory of the ideal heat engine, Carnot based his work on the law of conservation of energy and the impossibility of perpetual motion. Carnot pointed out that the conditions for maximum efficiency were that the engine should be continuously performing a series of cyclic operations letting down the heat from a source at a fixed high temperature to a sink at a fixed low temperature. Further, all changes taking place during the operation should be perfectly reversible, so that by an expenditure of an equal amount of work on the engine, the amount of heat rejected to the sink in the forward cycle may be exactly restored to the source. Reversibility is an essential condition of the ideal heat engine, and before dealing with Carnot's cycle of operations in detail, some examples of reversible and irreversible processes will be discussed.

13.02   Reversible and Irreversible Processes

A reversible process from the thermodynamical point of view is one in which an indefinitely small change in the external conditions will result in all changes taking place in the direct process being exactly repeated in the inverse order and opposite sense. A reversible cycle is one which consists of a succession of reversible changes so that the cycle can be traversed in the reverse order to re-establish the initial state of the working substance.

A good example of a reversible change is provided by the infinitely slow expansion or compression of a gas at constant temperature. Imagine the gas to be contained in a vertical cylinder closed by a frictionless piston which is loaded so that the pressure exerted by the piston on the gas exactly balances the pressure of the gas on the piston when the cylinder is placed in contact with a constant temperature source. If the load on the piston is now decreased by an infinitely small amount, the gas will expand extremely slowly. The heat required by the gas to do work in pushing up the piston is taken from the source, and since the process is carried out very slowly the absorption of heat keeps pace with its expenditure and the temperature of the system remains constant. If at any stage of the expansion the load on the piston is increased so that the pressure exerted by the piston exceeds that of the gas by an infinitely small amount, the gas will now contract very slowly at constant temperature with every stage in the forward process being exactly reversed. An adiabatic expansion (or compression) can also be reversed provided again that the changes take place sufficiently slowly.

It is important to realise that throughout the process the gas is in mechanical and thermal equilibrium with its surroundings. These conditions will not be realised if there is any frictional restraint between the cylinder walls and the piston, or if the expansion takes place rapidly. With friction present it would require a finite difference between the pressures on the two sides of the piston to produce expansion or compression. Only part of the work done by the gas during expansion is available for external work as some work is used up in overcoming the frictional forces. Further, this work is not recovered during the compression process when indeed a greater amount of work is done on the piston as a result of the friction than was originally done by the gas on the piston during the expansion. A similar loss of energy takes place due to internal friction when eddies are formed due to rapid expansion. We thus see that dissipative forces disturb the conditions of equilibrium and produce irreversible effects.

A finite temperature difference between the working substance and the hot source also produces irreversibility. For although heat can be absorbed from the source during the expansion process, a small increase in the load will not cause the substance to yield heat to the hotter source during compression. The transfer of heat by radiation is another example of an irreversible process.

We may thus summarise the conditions for the reversible working of our simple heat engine as follows:

(i) the pressure and temperature of the working substance must never differ appreciably from its surroundings at any stage of the cycle of operations,

(ii) all the processes involved in the cycle must take place indefinitely slowly, and

(iii) the working parts of the machine must be free from friction.

It is clear that such reversible changes are ideal and can never be realised in practice. There must always be finite differences in temperature and pressure during any actual change as the effects of conduction and friction are always present in a practical engine. Nevertheless, the study of the ideal, perfectly reversible heat engine is of fundamental theoretical importance; such an engine constitutes a limiting case against which the performance of actual engines can be judged.

13.03 The Ideal Heat Engine—Carnot's Cycle

We shall now discuss the simple cycle of operations originally propounded by Carnot. The importance of Carnot's work was not immediately recognised, and it was not until Clapeyron (1834) reformulated much of the original work by representing the various changes geometrically with the aid of indicator diagrams that it became generally known. The working substance used, to which the cycle

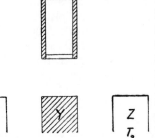

Fig. 161

is applied, can be anything, but again for simplicity we shall consider it to be a perfect gas contained in a cylinder closed by a perfectly fitting yet frictionless piston. We shall assume that both the piston and the cylinder walls are made of perfectly heat insulating materials, whilst the base of the cylinder is a perfect conductor. We shall also require a source X (see Fig. 161) which can supply any quantity of heat at a constant temperature T_1, and a sink Z which can absorb

any quantity of heat at a constant temperature T_2. The tops of X and Z are perfect conductors, and Y is a perfect heat insulating stand. All changes are to take place reversibly by infinitely small changes to the load on the piston, and we shall start with the gas corresponding to the point A of the indicator diagram (Fig. 162) when the base of the

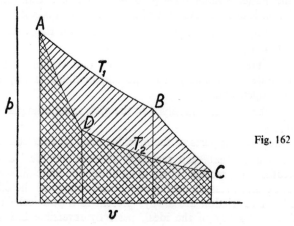

Fig. 162

cylinder is placed on X. The following is the sequence of operations performed during the cycle.

(i) With the cylinder placed on X, the gas is allowed to expand isothermally until in the state represented by the point B. During this stage heat is absorbed from the source and work is done *by* the engine.

(ii) The cylinder is now placed on Y, and the expansion continued under adiabatic conditions (since no heat is gained or lost by the gas) until the gas is in the state represented by C at a temperature T_2. The engine continues to do work during this stage.

(iii) The gas is now compressed isothermally with the cylinder placed on Z until the point D is reached. Work is done *on* the gas during this stage and a quantity of heat is given up to the sink.

(iv) Finally the cylinder is again placed on Y and the gas compressed adiabatically until it is again brought back to the initial condition represented by the point A. A further amount of work is done on the gas during this last stage.

The work done *by* the gas during the cycle is given by the areas down-shaded from right to left on the diagram, whilst the work done *on* the gas is given by the areas down-shaded from left to right. It is evident, therefore, that in each cycle the gas does an amount of external work equal to the area of the figure $ABCD$. Now since the gas has been brought back exactly to its original state at A, it follows that this work cannot have been obtained on account of any change in the intrinsic energy of the gas. The performance of an engine undergoing the cycle described is in fact independent of the nature of the working substance used, the amount of external work done by the engine being mechanically equivalent to the difference between the quantity of heat absorbed from the source at the temperature T_1 and the quantity of heat rejected to the sink at the lower temperature T_2. If these quantities of heat are respectively Q_1 and Q_2, it follows from the First Law of Thermodynamics that the external work done during the cycle is $J(Q_1 - Q_2)$. The engine thus converts heat into work and the *efficiency* of the process is given by

$$\frac{\text{external work done}}{\text{mechanical equivalent of the heat absorbed}}$$

$$= \frac{J(Q_1 - Q_2)}{JQ_1} \quad \text{or} \quad \frac{Q_1 - Q_2}{Q_1}$$

13.04 The Reverse Cycle—Refrigerators

Since every operation in Carnot's cycle is completely reversible, it follows that the cycle can be traversed in the reverse order (*i.e.*, $A D C B A$). In this case an amount of heat Q_2 will be taken from the sink and an amount of heat Q_1 put into the source; the amount of work done during the cycle is again given by the area $A B C D$ only this time the work is done *on* the working substance. Working in the reverse sense the engine thus acts as a "heat pump" or refrigerating machine (see also section 9.12), and the effectiveness of the process is measured by the ratio of the heat abstracted from the cold body to the work done on the machine. This quantity is called the *coefficient of performance* and is equal to

$$\frac{Q_2}{Q_1 - Q_2}$$

13.05 Carnot's Theorem

From the discussion in the foregoing sections it will be clear that the criterion for maximum efficiency is reversibility. We shall now discuss an important theorem enunciated by Carnot, namely, that *all reversible engines working between the same two temperatures possess the same efficiency.*

Let us consider two heat engines, A and B, both of which, when working directly in a continuous cyclic process, absorb heat from a source at a constant temperature T_1 and reject heat to a sink maintained at some lower constant temperature T_2. The mechanical details of the two engines and the nature of the working substances used are irrelevant details in the discussion; we shall assume only that the engines are working reversibly between the two temperatures specified. If the two efficiencies are not equal, let us suppose that of A to be greater than that of B, and let us consider that the quantities of working substance used in the two engines are such that both perform equal quantities of work when operating in the direct cycle. Now imagine the two engines to be so coupled that when A works *directly*, it drives B *reversely*. A will then absorb heat at the temperature T_1 and reject heat at the temperature T_2, whereas B will absorb heat at the temperature T_2 and reject heat at the temperature T_1. Let the quantities of heat absorbed and rejected by A be Q_1 and Q_2 respectively, and let the corresponding quantities for B be Q_1' and Q_2'. Then, since equal amounts of work are done in both cases, we have, by the First Law of Thermodynamics,

$$Q_1 - Q_2 = Q_1' - Q_2' \quad \ldots \ldots \quad [139]$$

But the efficiency of A is greater than that of B, and therefore

$$\frac{Q_1 - Q_2}{Q_1} > \frac{Q_1' - Q_2'}{Q_1'} \quad \ldots \ldots \quad [140]$$

Hence, from equations 139 and 140 it must follow that

$$Q_1 < Q_1'$$

and thus it further follows that

$$Q_2 < Q_2'$$

We thus see that a net amount of heat equal to $Q_1' - Q_1$ is *rejected* to the source, whilst an amount $Q_2' - Q_2 (= Q_1' - Q_1)$ is *absorbed* from the sink during each cycle. Hence, without any external agency, heat has passed from the cold to the hotter body. This is contrary to all experience, and we therefore conclude that

$$Q_1 = Q_1' \text{ and } Q_2 = Q_2'$$

i.e., $$\frac{Q_1 - Q_2}{Q_1} = \frac{Q_1' - Q_2'}{Q_1'}$$

or the two engines are equally efficient.*

As a corollary it follows that the efficiency of a reversible engine working between two temperatures must be a maximum for that temperature range. The efficiency is thus independent of the particular working substance used and of the details of design of the engine— being a function solely of the two working temperatures concerned.

13.06 The Second Law of Thermodynamics

The validity of Carnot's theorem rests upon the result of universal experience of heat transference when dealing with heat engines performing cyclical operations. This is embodied in an important principle known as the *Second Law of Thermodynamics* which has been stated by Clausius as follows:

"It is impossible for a self-acting machine, unaided by any external agency, to convey heat from one body to another at a higher temperature, or heat cannot of itself (that is, without compensation) pass from a colder to a warmer body."

The law has also been expressed in an equivalent statement by Kelvin, thus :

"It is impossible by means of inanimate material agency to derive mechanical effort from any portion of matter by cooling it below the temperature of the coldest of surrounding objects."

It is important to remember that this law applies only to reversible cyclic processes and does not hold when the operations are not continuous. Thus it is possible to conceive of work being obtained from the vapour of an evaporating liquid which is cooled below the temperature of its surroundings during the process. However, since the change as far as described is not a cyclic one, there is no violation of the law. Again, a highly compressed gas is capable of doing external work by expanding and forcing up the piston of the gas cylinder, the temperature of the gas falling as it expands. It must be realised, however, that the gas must have originally been compressed by external agency, the subsequent expansion of the gas releasing the work previously done on it. If after expansion the gas is brought back to its original condition, it will be found that the cyclic process is in

---

* It also follows that the efficiency is independent of the quantity of heat absorbed from the source, for if Q_1 is increased n times, the quantity of heat Q_2 rejected to the sink will increase in the same proportion giving the efficiency $\frac{Q_1 - Q_2}{Q}$ the same as before.

accordance with the law. The action of a refrigerating machine in conveying heat from a cold to a warmer body is not in contradiction to the law since the machine is not self-acting—it functions only when driven by another machine.

The Second Law of Thermodynamics is not capable of direct experimental proof, nor can it be deduced from any existing laws. It represents rather the summary of our experience in dealing with the process of transforming heat into work. The main evidence for its truth is that no machine has yet been devised which works in contradiction to the law, and further, deductions made from the law have been found to be in accord with experience. Our confidence in the law grows in proportion as these deductions and predictions from the law survive the test of experiment. Indeed, the Second Law of Thermodynamics is regarded as a universal law of nature which is more firmly established and of wider applicability than any other law.

The limitations of the First Law of Thermodynamics are immediately obvious from our considerations of the theory of heat engines. According to this law there is nothing to prevent the whole of the quantity Q_1 of heat absorbed at the source by the heat engine being converted into work, nor indeed does it exclude the performance of work by a substance when working from a lower to a higher temperature. Not only does the second law define the direction of the change when heat is converted into work, but emphasizes the limits of such a change. It draws attention to the uniqueness of heat energy which, unlike other forms of energy, is only available under certain specified conditions and then only partly so. (See section 13.17).

13.07 Kelvin's Absolute Scale of Temperature

We have seen in chapter I that the establishment of a practical scale of temperature depends upon the property of some particular substance. Scales so defined are arbitrary in character and frequently show a remarkable measure of disagreement. The desirability of defining a scale of temperature which is independent of the properties of some specific substance is thus clearly evident. Now the efficiency of a perfectly reversible engine working in a Carnot cycle depends only on the temperatures of the source and the sink between which it operates, and Kelvin realised that a scale of temperature could be defined in terms of the efficiency of such an engine which would indeed be quite independent of the properties of any particular substance. The scale of temperature based on the ideal heat engine is known as the *thermodynamical temperature scale*.

The efficiency of a perfectly reversible engine is given by

$$\frac{Q_1 - Q_2}{Q_1} = 1 - \frac{Q_2}{Q_1}$$

and this expression is a function only of the temperatures θ_1 of the source and θ_2 of the sink. Thus we may write

$$\frac{Q_2}{Q_1} = f(\theta_1, \theta_2)$$

Different temperature scales can be defined from this equation according to the form the function $f(\theta_1, \theta_2)$ takes, and Kelvin based his scale on the relation

$$\frac{Q_2}{Q_1} = \frac{\theta_2}{\theta_1} \quad\ldots\ldots\ldots\ldots \quad [141]$$

Thus on Kelvin's scale the temperatures θ_1 and θ_2 of two bodies are defined as being proportional to the quantities of heat transferred from and to them when acting as source and sink for a reversible engine executing a Carnot cycle.

On Kelvin's scale equal steps of temperature will be obtained if, starting off from some arbitrary temperature, a series of perfectly reversible engines perform equal amounts of work when undergoing a cycle of operations between these intervals, each absorbing the heat rejected by the engine immediately preceding it. Thus, if $A_1 B_1$, $A_2 B_2$, $A_3 B_3$, $A_4 B_4$, . . . (Fig. 163) represent a series of isothermals at temperatures θ_1, θ_2, θ_3, θ_4, . . . which intersect the adiabatics PQ and RS, and if we imagine a series of heat engines working between θ_1 and θ_2, θ_2 and θ_3, &c., the amount of heat absorbed by the first engine at θ_1, being Q_1

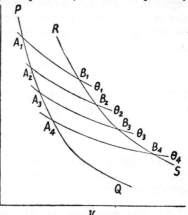

Fig. 163

and the amount rejected at θ_2 being Q_2, this latter amount being absorbed by the second engine at θ_2 an amount Q_3 being rejected at θ_3, and so on, then

$$Q_1 - Q_2 = Q_2 - Q_3 = Q_3 - Q_4 = \ldots$$

since the work done per cycle by each engine is the same. Now using equation 141 in the form

$$\frac{Q_1}{\theta_1} = \frac{Q_2}{\theta_2} = \frac{Q_3}{\theta_3} = \frac{Q_4}{\theta_4} = \ldots$$

it follows that

$$\theta_1 - \theta_2 = \theta_2 - \theta_3 = \theta_3 - \theta_4 = \ldots$$

thereby establishing the equality of the temperature intervals.

It is clear that as we proceed down the succession of temperature steps, the quantity of heat available for use at any given stage becomes progressively smaller, until ultimately a temperature is attained at which no heat is rejected. This temperature at which a substance will be entirely deprived of heat defines the absolute zero of the scale. No temperature lower than this is possible. Using equation 141 and writing the efficiency of a perfectly reversible engine in the form

$$\frac{\theta_1 - \theta_2}{\theta_1}$$

it is clear that when $\theta_2 = 0$ the efficiency of the engine is unity. We may thus define the absolute zero on the Kelvin scale as the temperature of the sink for a reversible engine which has an efficiency of unity; such an engine will convert *all* the heat into work.

Having thus defined the equality of temperature intervals and established the zero of the scale, all that remains is to fix the size of the degree of temperature. This is chosen so that the number of degrees between the ice and steam points is 100. If then a reversible engine functions with a source maintained at the steam point, and a sink at the ice point (temperature θ_0), and if Q_{100} and Q_0 represent the amounts of heat absorbed and rejected respectively during the cycle, then

$$\frac{\theta_0 + 100}{\theta_0} = \frac{Q_{100}}{Q_0}$$

This enables us to evaluate θ_0 which then makes it possible to obtain the value of any other temperature θ since

$$\frac{\theta}{\theta_0} = \frac{Q}{Q_0}$$

Q and Q_0 being the amounts of heat absorbed and rejected respectively by an engine working between the unknown temperature θ (as source) and the temperature of the ice point θ_0 (as sink).

13.08   Relation between Thermodynamic and Gas Scales of Temperature

We shall now consider how Kelvin's scale of temperature is related to that defined by the perfect gas relation $pv = RT$. For this purpose we shall consider one gram-molecule of a perfect gas used as the working substance for a reversible engine performing a Carnot's cycle of operations, the indicator diagram for which is shown in Fig. 164. AD and BC are adiabatics and AB and CD are isothermals at temperatures, *measured on the gas scale,* of T_1 and T_2 respectively. Let the

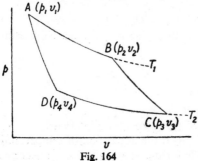

Fig. 164

corresponding temperatures on the thermodynamic scale be θ_1 and θ_2, and let the co-ordinates of the points A, B, C, D be as shown on the diagram.

Since the substance is a perfect gas, there will be no change in its intrinsic energy during an isothermal change. Hence in expanding from A to B, the quantity of heat Q_1 absorbed by the gas will be expended in doing external work, *i.e.*,

$$JQ_1 = \int_{v_1}^{v_2} pdv$$

$$= RT_1 \int_{v_1}^{v_2} \frac{dv}{v} \qquad \text{since } p = \frac{RT_1}{v}$$

$$= RT_1 \log_e \frac{v_2}{v_1} \qquad \ldots \ldots \ldots \ldots [142]$$

Similarly, for the isothermal change CD, when work is done *on* the gas to reject an amount of heat Q_2,

$$JQ_2 = \int_{v_4}^{v_3} pdv$$

$$= RT_2 \int_{v_4}^{v_3} \frac{dv}{v}$$

$$= RT_2 \log_e \frac{v_3}{v_4} \qquad \ldots \ldots \ldots \ldots [143]$$

Thus from 142 and 143

$$\frac{Q_1}{Q_2} = \frac{T_1 \log_e \frac{v_2}{v_1}}{T_2 \log_e \frac{v_3}{v_4}} \qquad \ldots \ldots \ldots \ldots [144]$$

To find the relation between v_1, v_2, v_3, v_4, we have from the isothermals

$$p_1 v_1 = p_2 v_2 \text{ and } p_3 v_3 = p_4 v_4$$

or

$$\frac{p_2 v_2}{p_1 v_1} = \frac{p_3 v_3}{p_4 v_4} = 1 \qquad \ldots \ldots \ldots \ldots [145]$$

and from the adiabatics

$$p_1 v_1^{\gamma} = p_4 v_4^{\gamma} \text{ and } p_2 v_2^{\gamma} = p_3 v_3^{\gamma}$$

or

$$\frac{p_2 v_2^{\gamma}}{p_1 v_1^{\gamma}} = \frac{p_3 v_3^{\gamma}}{p_4 v_4^{\gamma}} \qquad \ldots \ldots \ldots \ldots [146]$$

Hence, from 145 and 146 by division

$$\left(\frac{v_2}{v_1}\right)^{\gamma-1} = \left(\frac{v_3}{v_4}\right)^{\gamma-1}$$

and therefore

$$\frac{v_2}{v_1} = \frac{v_3}{v_4}$$

Thus

$$\log_e \frac{v_2}{v_1} = \log_e \frac{v_3}{v_4}$$

and equation 144 becomes

$$\frac{Q_1}{Q_2} = \frac{T_1}{T_2}$$

But θ_1 and θ_2 are defined by the relation

$$\frac{Q_1}{Q_2} = \frac{\theta_1}{\theta_2}$$

and thus it follows that

$$\frac{\theta_1}{\theta_2} = \frac{T_1}{T_2}$$

If $\theta_2 = 0$, then T_2 must also equal 0, and thus the two scales have the same zero. Further, since there are 100 degrees between the ice and steam points on both scales, then

$$\frac{\theta_0 + 100}{\theta_0} = \frac{T_0 + 100}{T_0}$$

from which it follows that the ice point temperatures θ_0 and T_0 on the two scales are identical. Finally, if θ and T represent the temperatures of a particular body on the two scales, then

$$\frac{\theta}{\theta_0} = \frac{T}{T_0}$$

and since $\theta_0 = T_0$, the numerical values of θ and T are the same, and thus the two scales are identical.

Actual gases must of course be used in the practical realisation of the thermodynamical scale of temperature. The gas scale readings require correction for any departure from the ideal conditions, and this can be done using experimental data for the deviations from the gas laws as obtained by Holborn (see section 5.09) and others.

Some Applications of Carnot's Cycle.

13.09 The Clausius-Clapeyron Equation. The curves $ABCD$ and $EFGH$ (Fig. 165) represent the isothermal curves at temperatures T and $T - \delta T$ respectively of 1gm. of a substance below its critical point. Along the sections BC and FG the liquid and vapour states of the substance co-exist in equilibrium. Hence T is the boiling point of the substance under the pressure p on the first isothermal, and $T - \delta T$ is the boiling point of the substance under the pressure $p - \delta p$ on the second isothermal. In order to transform the 1gm.

of the liquid substance at B completely into vapour at the point C, a quantity equal to the latent heat L of the substance must be supplied to it. In so doing let the volume increase from v_1 at B to v_2 at C.

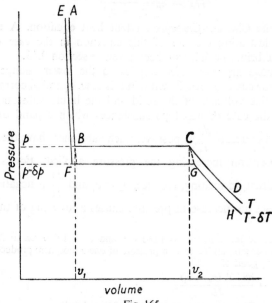

volume

Fig. 165

Let us now draw two adiabatic lines through B and C. These will intersect the lower isothermal at points very close to F and G respectively, and the figure $BCGF$ represents the indicator diagram when the substance is enclosed in the cylinder of an ideal heat engine and is taken through a Carnot cycle of operations commencing at B. A little consideration will show that the process is a reversible one since any slight change in the conditions will reverse the change from evaporation to condensation or *vice versa*. Now the efficiency of the heat engine working under these conditions will be

$$\frac{T - (T - \delta T)}{T} = \frac{\delta T}{T}$$

and this is equal to

$$\frac{\text{work done by the substance during the cycle}}{\text{work equivalent of heat absorbed at temperature } T}$$

$$= \frac{\text{area of figure } BCGF}{JL}$$

$$= \frac{\delta p \, (v_2 - v_1)}{JL}$$

Hence we have

$$\frac{\delta T}{T} = \frac{\delta p \ (v_2 - v_1)}{JL}$$

or

$$\frac{dp}{dT} = \frac{JL}{T \ (v_2 - v_1)}$$

which is the Clausius-Clapeyron latent heat equation. A numerical example illustrating the use of this equation in the case of water-steam equilibrium has already been given in section 9.13.

The above argument also applies to the lower change of state and the same equation results with, in this case, v_1 and v_2 corresponding to the specific volumes of the solid and the liquid substance respectively. In the case of wax-type substances which expand on melting $v_2 > v_1$ and therefore $\frac{dp}{dT}$ is positive. Such substances have their melting points raised on increasing the pressure applied. For substances which contract on melting, e.g., ice, $v_2 < v_1$ and $\frac{dp}{dT}$ is negative. Hence in such cases an increase in pressure causes a lowering of the melting point.

Thus for ice for which $v_1 = 1\cdot091$ c.c. and $v_2 = 1\cdot000$ c.c. at 273°K., and $L = 80$ cal. per gm., an increase in pressure of one atmosphere produces a change in the melting point of

$$dT = \frac{dp \cdot T \cdot (v_2 - v_1)}{JL}$$

$$= \frac{76 \times 13\cdot6 \times 981 \times 273 \times (1\cdot000 - 1\cdot091)}{4\cdot185 \times 10^7 \times 80}$$

$$= -0\cdot0075°C.$$

A result which is in satisfactory agreement with Dewar's experimentally determined value of $-0\cdot0072°C$. (See section 9.05).

13.10 Surface Tension and Surface Energy

The force, in dynes, acting at right angles to a line 1cm. long drawn in the surface of a liquid and tending to produce rupture is known as the *surface tension* (S) of the liquid. The value of S is independent of the area of the liquid surface, but for all liquids decreases as the temperature rises. Imagine now a soap film produced in a wire frame *ABCD* (Fig. 166) where *AB* is a movable wire. In order to keep *AB* at rest a force $F = 2lS$ dynes (two surfaces) must be

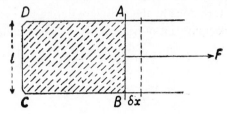

Fig. 166

applied, and the work done in stretching the film by moving AB a distance $\delta x = F\delta x = 2lS\delta x = S\delta A$ ergs where δA is the increase in area of the film. Let us assume that the stretching takes place under isothermal conditions and that in order to maintain the temperature constant an amount q of heat (in work units) is absorbed from the surroundings by each sq. cm. of the surface of the film. The total energy required to increase the area of the film as above is thus $(S + q)\ \delta A$ ergs and this energy (E) per unit area of surface is called the *surface energy*.

The quantity q is a function of the temperature and the form of this function can be obtained by an application of Carnot's principle when the film is taken through a reversible cycle. Starting with the film at A (Fig. 167) at a temperature of $T°$K. the area is increased isothermally by an amount δA to reach the point B. The work done

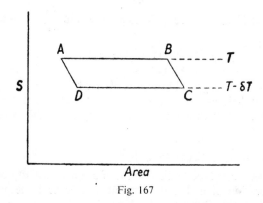

Fig. 167

on the film during this stage is $S\ \delta A$ ergs. The film is now adiabatically stretched along BC when the temperature falls to $T - \delta T$ and the surface tension becomes $S - \dfrac{dS}{dT} \cdot \delta T$ where $\dfrac{dS}{dT}$ is the rate of change of surface tension with temperature. The film is now allowed to contract isothermally along CD when the work done by the film is $\left[S - \dfrac{dS}{dT} \cdot \delta T \right]\ \delta A$ ergs. The cycle is then completed by an adiabatic contraction from D to A. The net external work done *by* the film in the cycle (also given by area of figure $ABCDA$) is thus

$$\left[S - \frac{dS}{dT} \cdot \delta T \right]\ \delta A\ -S\ \delta A$$

$$= -\frac{dS}{dT} \cdot \delta T \cdot \delta A \text{ ergs}$$

The heat absorbed to maintain the temperature constant at T is $q \, \delta A$, and hence the efficiency of the process is

$$\frac{-\dfrac{dS}{dT}.\delta T.\delta A}{q \, \delta A} = \frac{\delta T}{T}$$

i.e., $$q = -T \frac{dS}{dT}$$

Since $\dfrac{dS}{dT}$ is negative, q is a positive quantity, and thus it follows that a film is cooled if stretched under adiabatic conditions.

Assuming the above expression for q, we may now write the surface energy as

$$E = S - T \frac{dS}{dT} \text{ ergs per cm.}^2$$

For water at 0°C., $S = 75 \cdot 6$ dynes per cm. and $\dfrac{dS}{dT} = -0 \cdot 141$ dynes per cm. per °C.

$$\therefore \quad q = -273 \, (-0 \cdot 141)$$
$$= 38 \cdot 49 \text{ ergs per cm.}^2$$

and hence $E = 114 \cdot 09$ ergs per cm.^{2}

Thus, about half as much energy must be absorbed by the water surface when extended isothermally as is spent by the external forces in stretching it.

13.11 Reversible Electric Cells

In an electric cell a certain amount of chemical activity corresponds to the liberation of a definite amount of electrical energy. In some cells, such as the Daniell cell where there is no polarisation, this action is reversible—the chemical changes taking place in the reverse order when a current is passed through the cell in the opposite direction by some external means. Thus the deposit of copper from the copper sulphate solution which takes place in the direct action of the cell is consumed in the reversed electrolytic action, and an equivalent amount of zinc deposited on the other electrode from the zinc sulphate solution thereby restoring the cell to its original condition.

The chemical activity that takes place in the cell releases a quantity of heat ("heat of reaction") which is the source of the electrical energy when the cell is delivering a current. When so employed the cell may or may not absorb heat from its surroundings. The results of experiment show that some cells take in heat whilst others give out heat in this way when supplying electrical energy to an external circuit. In the former case the cell tends to become cooled when in normal use, and warmed when the action is reversed—these effects occuring the opposite way round in the latter case.

Let us now consider an electric cell with an e.m.f. of E volts when placed in an isothermal enclosure at a temperature of $T°K$. Let a quantity of charge q be delivered by the cell at this temperature (section AB of Fig. 168), when the work done by the cell will be Eq ergs. Let us suppose that a quantity of heat h (work units) is absorbed from the surroundings to maintain the temperature of the cell constant at T during this change. Now consider the cell to be thermally isolated

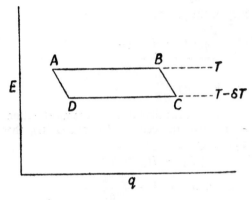

Fig. 168

and to deliver a very small amount of charge. This adiabatic change will cause the temperature to fall slightly as now the cell is itself the only source of energy. Let the cell now be represented by the point C of the diagram, the temperature having been reduced to $T - \delta T$ and the corresponding value of the e.m.f. being $E - \dfrac{dE}{dT} \cdot \delta T$ where $\dfrac{dE}{dT}$ is the rate of change of e.m.f. with temperature. Maintaining the cell at this lower temperature, a quantity of charge q is passed through the cell in an opposite direction to cause the chemical changes of the first operation to be exactly reversed. This operation is represented by the line CD on the diagram, and during this stage an amount of work $\left[E - \dfrac{dE}{dT} \cdot \delta T \right] q$ ergs is done on the cell. To complete the cycle of operations a very small charge is now passed through the cell to effect the adiabatic change DA when the cell is again in its initial condition at the temperature T. To ensure complete reversibility of the operations it is assumed that the resistances of the cell and associated circuit are negligibly small so that the irreversible effect of heat production in the conductors, &c., may be ignored. Irreversibility is also introduced by the inter-diffusion of the two liquids in the cell, and it is further assumed that this takes place to such a small extent during the cycle of operations that the effect can be disregarded.

The net work done *by* the cell in the cycle is

$$Eq - \left[E - \frac{dE}{dT}.\delta T \right] q$$

$$= \frac{dE}{dT}.\delta T.q \text{ ergs}$$

and the efficiency of the process is

$$\frac{\frac{dE}{dT} \cdot \delta T \cdot q}{h} = \frac{\delta T}{T}$$

i.e., $$h = qT \frac{dE}{dT}$$

If now the amount of heat liberated by chemical action when unit quantity passes through the cell is H (work units), then the total energy involved during the stage AB is

$$Eq = Hq + h$$

and $$\therefore E = H + T \frac{dE}{dT}$$

This is known as the *Gibbs-Helmholtz equation,* and from it we see that $E = H$ only when $\frac{dE}{dT} = 0$. This is very nearly the case for the Daniell cell which thus obtains all its electrical energy from the chemical activity taking place in it. Now the heat of reaction in this case is 775 calories for each gram of zinc dissolved. Hence, taking the electro-chemical equivalent of zinc as 0·00034gm. per coulomb,

$$H = 775 \times 0.00034 \times 4.185$$
$$= 1.10 \text{ joules per coulomb.}$$

Thus since the energy in joules required to discharge one coulomb gives the e.m.f. of the cell in volts, $E = 1.10$ volts. The observed value of E for a Daniell cell is 1·09 volts.

For the lead accumulator $\frac{dE}{dT}$ is positive, and in this case the energy released by the chemical activity in the cell is insufficient to maintain the current delivered and heat is taken from the cell itself with consequent lowering of the temperature. In the case of the Clark cell on, the other hand, $\frac{dE}{dT}$ is negative, and thus the heat liberated by chemical action exceeds the requirements for maintenance of the current. The excess heat energy is dissipated to the surroundings, the temperature of the cell rising when in use.

It should be noted that the Gibbs-Helmholtz equation can be used in a converse manner to obtain the heat of reaction of the chemical changes taking place in the cell if the values of E and its temperature coefficient have been determined experimentally.

13.12   Full Radiation—The Stefan-Boltzmann Law

As a consequence of the electro-magnetic theory of light Maxwell showed that when electro-magnetic waves are incident normally on a perfectly reflecting surface a pressure is exerted equal to the amount of radiation per unit volume of the space near the surface. This quantity E (in ergs per c.c.) is known as the *energy density* of the radiation. In the case of radiation incident in all directions, the pressure in dynes per sq. cm. exerted on the surface is given by

$$p = \frac{E}{3}$$

If then a reflecting surface is moved against incident radiation, work must be done.

Let us now consider a cylinder containing block body radiation and apply a Carnot cycle of operations using the radiation as the working substance. In order to avoid heat exchanges between the enclosure and the walls we shall assume that both the walls of the cylinder and the piston head are perfect reflectors. We shall further assume that the end of the cylinder—also a perfect reflector—can be removed to permit the enclosure being placed against a perfectly black surface maintained at a constant temperature. The following cycle of operations is then performed.

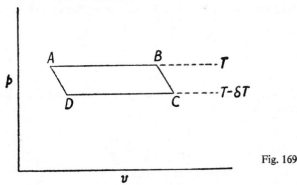

Fig. 169

The end of the cylinder is first closed by a perfectly black surface maintained at a constant temperature T. The cylinder will then be filled with diffused radiation of density E and pressure p corresponding to this temperature (point A of Fig. 169). The piston head is then slowly moved outwards increasing the volume of the enclosure by an amount v to the point B of the diagram. During this process an amount of work pv is done on the piston and the energy to do this is drawn from the black body surface which also supplies the additional radiation to maintain the energy density at E. Thus, if Q ergs is the total amount of energy supplied to the enclosure during this stage, we have

$$Q = pv + Ev$$

The black body surface is now removed, and the piston closed by a perfectly reflecting surface and a further slight increase in volume of the radiation is allowed to take place. No energy is received from any external source during this operation, the work done in forcing back the piston being obtained at the expense of the intrinsic energy of the radiation the temperature of which consequently falls slightly. Let the new temperature be $T - \delta T$ and the corresponding radiation pressure $p - \dfrac{dp}{dT} \delta T$, where $\dfrac{dp}{dT}$ is the rate of change of the pressure with temperature.

On completing the adiabatic expansion BC, the cylinder is closed by a perfectly black surface maintained at the temperature $T - \delta T$ and the piston is slowly moved inwards until the volume is reduced by an amount v to D on the diagram. The work done on the radiation during this stage is

$$\left[p - \frac{dp}{dT} \cdot \delta T \right] v$$

Finally the cylinder is again closed by a perfectly reflecting surface and a further small decrease in volume under adiabatic conditions restores the radiation from D to its original state at A.

The work done *by* the radiation during the cycle is

$$pv - \left[p - \frac{dp}{dT} \cdot \delta T \right] v$$
$$= v \frac{dp}{dT} \cdot \delta T \text{ ergs.}$$

Each stage in the cycle is perfectly reversible and the efficiency of the process is thus given by

$$\frac{v \dfrac{dp}{dT} \delta T}{(E + p)v} = \frac{\delta T}{T}$$

or

$$\frac{1}{E + p} \cdot \frac{dp}{dT} = \frac{1}{T}$$

But

$$p = \frac{E}{3} \text{ and } \frac{dp}{dT} = \frac{1}{3} \cdot \frac{dE}{dT}$$

$$\therefore \frac{dE}{E} = 4 \frac{dT}{T}$$

On integrating this equation we have

$$\log_e E = 4 \log_e T + \log_e k$$

where k is a constant of integration.

Hence $E = kT^4$.

Since the energy density determines the rate at which energy is emitted per sq. cm. through a small hole in the enclosure (cavity radiator), this must also be proportional to the fourth power of the absolute temperature—which is the Stefan-Boltzmann law. (See section 12.13).

13.13 Entropy

We shall now discuss a conception which has important applications in the development of thermodynamical theory and is also of great value in engineering practice when dealing with the actual behaviour of heat engines. This is the conception of *entropy*. It will be shown later (section 13.15) that the entropy of a substance is a function of its state and a change in the entropy occasioned by a reversible change in the state of a substance involving the addition of an amount of heat Q at a constant temperature T is defined by the ratio $\dfrac{Q}{T}$. This represents the entropy *gained* by the substance. If the substance loses a quantity of heat Q at a constant temperature T, then $\dfrac{Q}{T}$ represents the loss in entropy of the substance. If the temperature of the substance does not remain constant during the process, we may consider the heat to be communicated in successive elements of δQ such that the temperature remains sensibly constant for each element. The change of entropy involved will then be $\Sigma \dfrac{\delta Q}{T}$ or, in the notation of the calculus $\int \dfrac{dQ}{T}$.

Let us now consider the entropy changes which take place during a Carnot cycle of operations. Starting with the working substance in the state A (see Fig. 162) there will be a gain of entropy of $\dfrac{Q_1}{T_1}$ during the isothermal expansion AB as Q_1 units of heat are drawn from the source (which thus loses an amount $\dfrac{Q_1}{T_1}$ of entropy). During the adiabatic expansion BC no heat is supplied to or withdrawn from the working substance, and accordingly there is no change of entropy along the adiabatic. The working substance is now isothermally compressed at a temperature T_2 from C to D when it rejects a quantity of heat Q_2 to the sink. Accordingly, it loses an amount

12*

of entropy $\dfrac{Q_2}{T_2}$ (which is also the amount of entropy gained by the sink). The cycle is completed by the adiabatic compression DA (where again no change of entropy is involved), and thus for the whole cycle the *net* gain of entropy by the working substance is

$$\frac{Q_1}{T_1} - \frac{Q_2}{T_2}$$

Now the temperatures T_1 and T_2 on Kelvin's absolute scale are defined by the relationship
$$\frac{Q_1}{T_1} = \frac{Q_2}{T_2}$$

and thus we see that there is no change of entropy involved during a Carnot cycle of operations.

As indicated above, when a substance undergoes a (reversible) adiabatic change there is no gain or loss of entropy. Thus all points on an adiabatic are characterised by the same entropy, and adiabatic lines are referred to as "isentropics", *i.e.*, lines of equal entropy. In proceeding from one adiabatic to another by a reversible isothermal process a change of entropy will be involved. This characteristic property of adiabatics may be used to supply a definition of entropy as being that quantity which remains constant during (reversible) adiabatic operations.

13.14 Entropy Changes for a Reversible Cycle in which the Temperature is Continually Changing

Fig. 170 shows the indicator diagram for a working substance taken through a reversible cycle of operations in which the temperature is continually changing. In order to obtain the entropy

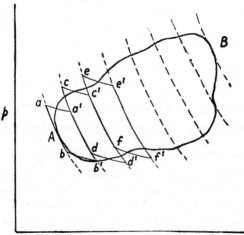

Fig. 170

changes taking place during the cycle we may imagine it to be made up of a large number of Carnot cycles obtained by drawing a series of adiabatics *ab, cd, ef,* .... and isothermals *aa', bb', cc',* ... &c. It will be seen that in following the elementary cycles the portions *a'd, c'f,* are traversed twice in the reverse order, and so cancel each other out. The effect is thus to represent the original figure by the zig-zag line *aa' cc' ee'* ... *ff' dd' bb'* .... and this can be made to approximate as closely as we like to the actual curve by progressively reducing the length of the isothermal sections *aa', cc',* ....

If now δQ represents the amount of heat transferred on any elementary isothermal at a temperature T, we have already seen that the sum of the $\dfrac{\delta Q}{T}$ terms for the corresponding Carnot cycle is zero. Hence for the set of Carnot cycles equivalent to the original cycle

$$\Sigma \frac{\delta Q}{T} = 0$$

or in the limit, for an infinite number of elementary Carnot cycles

$$\oint \frac{dQ}{T} = 0$$

where $\oint$ indicates the integration taken round the whole cycle. Thus we see that for *any* complete cycle of operations there is no change in the entropy of the working substance provided the operations are carried out reversibly.

Entropy is usually expressed by the symbol ϕ and thus the entropy change involved when a quantity of heat dQ is transferred to a body at a temperature T may be stated as

$$d\phi = \frac{dQ}{T}$$

and the above integral becomes $\oint d\phi = 0$

13.15 Entropy as a Condition of State

Let the points A and B (Fig. 171) represent any two states of a substance, *e.g.*, a gas, which are specified by particular values of the pressure, volume, and temperature. Let it be possible to pass reversibly from state A to state B either by the path through C or by that through C'. If now the substance is taken through the complete cycle of operations $ACBC'A$ we have for the cycle as a whole,

$$\oint \frac{dQ}{T} = 0$$

or, dealing with the entropy changes on the two parts of the cycle,

$$\int_{\substack{C \\ A}}^{B} \frac{dQ}{T} + \int_{\substack{C' \\ B}}^{A} \frac{dQ}{T} = 0$$

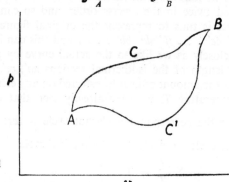

Fig. 171

Now, considering the section $AC'B$ (say) of the cycle, it is clear that if the direction of operations is reversed, the entropy changes will be reversed in sign, since the heat changes taking place in the forward process will now take place in the opposite direction. Hence

$$\int_{\substack{C' \\ B}}^{A} \frac{dQ}{T} = -\int_{\substack{C' \\ A}}^{B} \frac{dQ}{T}$$

Thus substituting in the previous equation we have

$$\int_{\substack{C \\ A}}^{B} \frac{dQ}{T} - \int_{\substack{C' \\ A}}^{B} \frac{dQ}{T} = 0$$

or

$$\int_{\substack{C \\ A}}^{B} \frac{dQ}{T} = \int_{\substack{C' \\ A}}^{B} \frac{dQ}{T}$$

We therefore see that the change of entropy in passing from state A to state B is independent of the path taken, provided only it is a reversible one. Accordingly, the entropy of a substance is a definite function of its state and is not in any way affected by the manner in which the particular condition of state has been attained.

It is important to realise that we do not know the absolute value of the entropy of a substance in any given condition. The equations above only define the *changes* of entropy that take place as a result of a change in the state of a body. Of course, it is possible to fix some arbitrary zero in reference to which the entropy changes may be stated. Thus, in engineering practice it is customary to assess entropy changes relative to a zero fixed at 0°C.

13.16 Entropy Changes in Irreversible Processes—Principle of Increase of Entropy

We have seen in section 13.13 that a reversible engine performing a Carnot cycle of operations transfers entropy from the source to the sink without altering its amount. Thus during the cycle the entropy changes in the working substance and the source-sink system are both zero. The Carnot cycle represents an ideal arrangement which gives the maximum efficiency for the operating temperatures. In practice the presence of irreversible effects will result in a lowering of the efficiency of the engine. Thus, if Q_1 units of heat is absorbed from the source at T_1 and a quantity Q_2 rejected to the sink at T_2, then

$$\frac{Q_1 - Q_2}{Q_1} < \frac{T_1 - T_2}{T_1}$$

or

$$\frac{Q_2}{Q_1} > \frac{T_2}{T_1}$$

from which

$$\frac{Q_2}{T_2} > \frac{Q_1}{T_1}$$

Thus the entropy gained by the sink $\dfrac{Q_2}{T_2}$ is greater than the entropy lost by the source $\dfrac{Q_1}{T_1}$. Now, since during the cycle the working substance is restored to its original state, its change of entropy must be zero. Hence for the system as a whole the cycle of operations has resulted in an increase in entropy. That is, for an irreversible cycle we have

$$\oint d\phi > 0$$

and the amount by which the entropy has increased gives a measure of the inefficiency of the cycle as compared with the performance under ideally reversible conditions.

Particular cases of irreversible processes are conduction and friction. Thus, if a quantity of heat Q passes from a body at temperature T_1 to a cooler body at temperature T_2, the entropy lost by the hot body $\dfrac{Q}{T_1}$ is less than that gained by the cooler body $\dfrac{Q}{T_2}$. Thus the entropy of the system increases during the process. Similarly the work done in overcoming friction produces heat which is absorbed by the surrounding bodies with a consequent gain of entropy.

Thus we see that in the ideal reversible changes of Carnot's cycle the entropy is conserved, but in all other cases it undergoes an increase. No real action is strictly reversible and accordingly for any isolated system the changes that occur within it will cause an increase in its aggregate entropy. This important principle is known as the *principle of increase of entropy*. It is a consequence of the Second Law of Thermodynamics, and indeed the Second Law has been stated in this form by Clausius to whom the name entropy is due.

13.17　Available and Unavailable Heat—the Degradation and Dissipation of Energy

Thermal energy is only available for conversion into work when it is let down from a high to a lower temperature. Thus for a perfectly reversible engine performing a Carnot cycle of operations during which a quantity of heat Q_1 is absorbed at a temperature T_1 and a quantity Q_2 rejected at a temperature T_2 the available energy is

$$J (Q_1 - Q_2)$$

$$= J Q_1 \left[1 - \frac{T_2}{T_1} \right] \qquad \text{since} \quad \frac{Q_1}{T_1} = \frac{Q_2}{T_2}$$

and the unavailable energy is

$$JQ_2$$

$$= JQ_1 \frac{T_2}{T_1}$$

We thus see that if T_2 represents the lowest possible working temperature under the given conditions, a fraction $\dfrac{T_2}{T_1}$ of the heat originally taken in is rendered unavailable for further use—or is "degraded." It is also evident that, whatever may be the heat content of a body, the heat will remain unavailable for conversion into work if other bodies at lower temperatures are not to hand.

Now through the processes of radiation and conduction there is a continual tendency to establish a uniformity of temperature throughout the universe, and accordingly there is a steady decrease in the proportion of the thermal energy which can be used for mechanical purposes. The quantities of other forms of energy available for conversion are also steadily decreasing as the effects of friction, eddies, &c., produce a constant dissipation of energy into heat—which, as we have seen, is never fully available for re-transformation.

Thus, although the total energy in the universe remains constant, it is progressively being dissipated and degraded into unavailable heat, until finally the universe will experience a "heat death" with all temperatures at the same uniform level. In this condition the entropy of the universe will be a maximum.

13.18　Temperature—Entropy Diagrams

An interesting alternative to the pressure-volume indicator diagram method of representing the performance of heat engines is provided by diagrams in which the changes of entropy of the working substance are plotted against the absolute temperature. These diagrams are referred to as $T\phi$ diagrams and provide much interesting information to the engineer. A special property of these diagrams is that the area subtended between any part of the T-ϕ curve and the entropy axis gives the heat absorbed (or rejected) during the change of state represented by the section of the curve considered. Thus for a body changing in a reversible manner from A (temperature T_1, entropy ϕ_1) to B (temperature T_2, entropy ϕ_2) by the curve shown in Fig. 172 we have

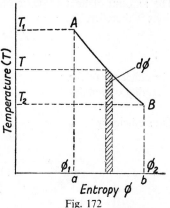

$$\text{Area } ABba = \int_{\phi_1}^{\phi_2} T\,d\phi = T(\phi_2 - \phi_1)$$

$$= \int T\frac{dQ}{T} \text{ since } d\phi = \frac{dQ}{T}$$

$$= \int dQ = Q$$

Q being the quantity of heat absorbed during the change from A to B.

For a complete cycle of changes with the substance returning to its original state the temperature-entropy curve forms a closed figure, and as with the indicator diagram, the area of the figure gives a measure of the

Fig. 172

work done (W) during the cycle. Thus if Q_1 is the heat absorbed and Q_2 the heat rejected, then

$$\oint T\,d\phi = Q_1 - Q_2$$

or, for the work done,

$$W = J(Q_1 - Q_2)$$

$$= J\oint T\,d\phi$$

In the case of the Carnot cycle the $T\phi$ diagram (Fig. 173) is rectangular in shape. During the isothermal expansion AB the entropy of the working substance increases from ϕ_1 to ϕ_2. In the subsequent adiabatic expansion BC the temperature falls to T_2 without change of entropy (see section 13.13). On being compressed isothermally at T_2 there is a decrease of entropy from ϕ_2, to ϕ_1, the cycle being completed along the constant entropy line ϕ_2 by the adiabatic compression DA.

The quantity Q_1 of heat absorbed during the isothermal expansion AB = area $ABEF$, i.e.

$$Q_1 = T_1(\phi_2 - \phi_1)$$

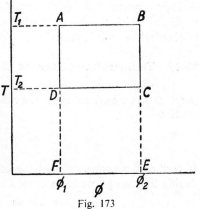

Fig. 173

and the quantity Q_2 of heat rejected (unavailable heat) during the isothermal compression CD = area $DCEF$, i.e.,

$$Q_2 = T_2 (\phi_1 - \phi_1)$$

The amount of heat converted into work during each cycle of operations is therefore $Q_1 - Q_2$

$$= (T_1 - T_2)(\phi_2 - \phi_1)$$
$$= \text{area } ABCD$$

The efficiency of the cycle in terms of the sink and source temperatures is readily obtainable from the above.

Thus efficiency
$$= \frac{Q_1 - Q_2}{Q_1}$$

$$= \frac{(T_1 - T_2)(\phi_2 - \phi_1)}{T_1 (\phi_2 - \phi_1)}$$

$$= \frac{T_1 - T_2}{T_1}$$

All the above quantities are very readily shown in the simple geometry of the figure from which it is obvious that for more efficient working the line CD should be made to approach more closely the horizontal axis. That is, the efficiency increases as T_2 tends to the absolute zero of temperature.

In this type of diagram the attention is focused on the heat changes involved, and since isothermal and adiabatic lines intersect at right angles, the area of the closed figure is easily assessed for computing the work done per cycle. In actual practice the presence of irreversible effects will destroy the extreme simplicity of the diagram as given above, and for many purposes a chart giving total heat H against entropy is commonly used. This is called a *Mollier chart*, for further details of which the student should consult any of the more practical books dealing with heat engines.

13.19 Temperature—Entropy Curves for Water and Steam

If a small amount of heat dQ is given to a substance at a temperature T, we have seen that the change of entropy $d\phi$ is given by the relation

$$d\phi = \frac{dQ}{T}$$

If further the temperature of the substance rises by an amount dT during the change, we have, considering unit mass

$$dQ = \text{specific heat} \times dT$$

For water the specific heat is unity and will be assumed to be constant at all temperatures. Hence in this case

$$dQ = dT$$

and therefore

$$d\phi = \frac{dT}{T}$$

Accordingly

$$\phi = \int \frac{dT}{T}$$

$$= \log_e T + C$$

where C is a constant of integration depending on the origin. If now we take the zero of entropy at 0°C. ($T = 273$°K.), $C = -\log_e 273$, and the entropy function for unit mass of water becomes

$$\phi_{water} = \log_e \frac{T}{273}$$

$$= 2\cdot3026 \,(\log_{10} \overline{273 + t°C.} - \log_{10} 273)$$

From this equation a range of values of ϕ for different temperatures can be calculated as set out in column 3 of Table 29.

To convert unit mass of water into steam at any temperature T it is necessary to supply an amount of heat equal to the latent heat L

TABLE 29

Entropy of water and steam at various temperatures

Temperature °C.	Latent Heat of steam	ϕ_{water}	ϕ_{steam}
0	606·5	0	2·274
10	599·5	·036	2·154
20	592·6	·071	2·093
30	585·6	·104	2·037
40	578·7	·137	1·986
50	571·7	·168	1·938
60	564·8	·199	1·895
70	557·8	·228	1·854
80	550·9	·257	1·818
90	543·9	·285	1·783
100	537·0	·312	1·752
110	530·0	·339	1·723
120	523·1	·364	1·695
130	516·2	·389	1·669
140	509·2	·414	1·647
150	502·3	·438	1·626
160	495·3	·461	1·605
170	488·4	·484	1·587
180	481·4	·506	1·569
190	474·4	·528	1·552
200	467·5	·550	1·538

at that temperature. Since the change takes place at constant temperature the increase of entropy during the process will be $\dfrac{L}{T}$, and the total entropy of the steam at any temperature (assuming the same origin as before) will be

$$\phi_{steam} = \phi_{water} + \dfrac{L}{T}$$

$$= \log_e \dfrac{T}{273} + \dfrac{L}{T}$$

Now the value of the latent heat of steam may be obtained at any temperature using Regnault's formula, viz.,

$$L_{t°c} = 606·5 - 0·695 \ t°\text{C}.$$

and hence

$$\phi_{steam} = 2·3026 \ (\log_{10} \overline{273 + t°\text{C}}. - \log_{10} 273) + \dfrac{606·5 - 0·695 \ t°\text{C}.}{273 + t°\text{C}.}$$

A set of values of ϕ_{steam} using this relation are given in column 4 of Table 29. With these values temperature-entropy curves have been

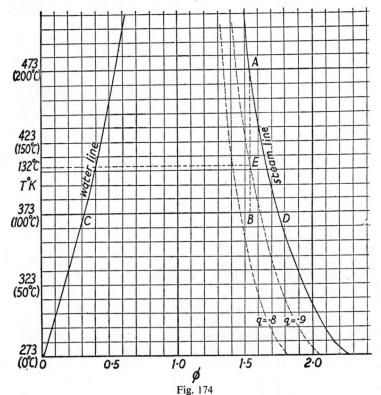

Fig. 174

constructed for water and steam as shown in Fig. 174 for the range of temperature indicated. Such a diagram is particularly useful in dealing with problems concerning water and steam when they both exist together. Thus, when dry steam at 200°C. (point A on the diagram) is adiabatically expanded to produce a reduction of temperature to 100°C. the fraction of the steam condensing may be found by following the line of constant entropy to the point B where it intersects the horizontal line through 373°K. The horizontal section CD of this line between the water and steam curves represents the increase of entropy for complete evaporation of each gram of water to steam at the given temperature. In proceeding from C to B, therefore, the entropy increase is that which is involved when a fraction $\dfrac{CB}{CD}$ of the water has been evaporated. Hence in the problem under discussion the fraction of the original mass of steam which has condensed as a result of the adiabatic expansion is $\dfrac{BD}{CD}$. From the graph this fraction is ·86.

By equal fractional divisions of such lines as CD a set of points can be obtained which represent the same steam-water ratio or *dryness fraction* (q) at the various temperatures. Lines joining such points are referred to as *constant dryness* or *constant quality* lines. Two such lines for q = ·8 and q = ·9 are given on the diagram. From this latter line we can see that to condense one-tenth (point E) of the mass of steam referred to above, an adiabatic temperature drop to 132°C. is needed.

13.20 Entropy of a Perfect Gas

In the general case when a quantity dQ of heat is given to a gas we have the relation

$$dQ = dU + dW$$

where dU is the increase in the intrinsic energy of the gas and dW represents the external work done (both in heat units). Considering now 1gm. of a perfect gas

$$dU = c_v\, dT \quad \text{and} \quad dW = \frac{pdv}{J} = \frac{R'T\,dv}{J\ v}$$

and therefore

$$dQ = c_v dT + \frac{R'T}{J} \cdot \frac{dv}{v}$$

The increase $d\phi$ of entropy corresponding to the gain of dQ units of heat at a temperature T is thus

$$d\phi = \frac{dQ}{T} = c_v \cdot \frac{dT}{T} + \frac{R'\,dv}{J\ v}$$

Integrating between the limits T_1 and T_2 we get for the change of entropy of the gas

$$\phi_2 - \phi_1 = \int_{T_1}^{T_2} \frac{dQ}{T}$$

$$= c_v \log_e \frac{T_2}{T_1} + \frac{R'}{J} \log_e \frac{v_2}{v_1} \quad . \quad . \quad . \quad . \quad [147]$$

This expression can be put into an alternative form by substituting $c_p - c_v$ for $\dfrac{R'}{J}$. Thus

$$\phi_2 - \phi_1 = c_v \log_e \frac{T_2}{T_1} + (c_p - c_v) \log_e \frac{v_2}{v_1}$$

and since

$$\frac{p_1 v_1}{T_1} = \frac{p_2 v_2}{T_2}$$

or

$$\frac{T_2}{T_1} = \frac{p_2 v_2}{p_1 v_1}$$

we have

$$\phi_2 - \phi_1 = c_v \left[\log_e \frac{p_2}{p_1} + \log_e \frac{v_2}{v_1} \right] + (c_p - c_v) \log_e \frac{v_2}{v_1}$$

$$= c_v \log_e \frac{p_2}{p_1} + c_p \log_e \frac{v_2}{v_1} \quad . \quad . \quad . \quad . \quad . \quad [148]$$

$$= 2 \cdot 3026 \left[c_v \log_{10} \frac{p_2}{p_1} + c_p \log_{10} \frac{v_2}{v_1} \right]$$

This is the most useful form for calculating entropy changes which are usually measured from an arbitrary zero when the gas is at 0°C. and exerts a pressure of 76cm. of mercury.

Cases of special interest occur when the gas is heated at *constant pressure* and *constant volume*. In the former case by putting $p_1 = p_2$ in equation 147 we get the change of entropy

$$\phi_2 - \phi_1 = c_p \log_e \frac{v_2}{v_1}$$

$$= c_p \log_e \frac{T_2}{T_1} \quad \text{since } \frac{v_1}{v_2} = \frac{T_1}{T_2}$$

and in the latter case with $v_1 = v_2$ in equation 148 we get

$$\phi_2 - \phi_1 = c_v \log_e \frac{p_2}{p_1}$$

$$= c_v \log_e \frac{T_2}{T_1} \quad \text{since } \frac{p_1}{p_2} = \frac{T_1}{T_2}$$

For an *isothermal change* the change of entropy involved may be obtained by putting $T_1 = T_2$ in equation 147 which then becomes

$$\phi_2 - \phi_1 = \frac{R'}{J} \log_e \frac{v_2}{v_1}$$

Power Cycles for Actual Heat Engines

13.21   Steam Engines—The Rankine Cycle

We have seen that the ideal limit of efficiency for a heat engine working between two given temperatures is obtained when the working substance is carried through a cycle of operations with all the heat being absorbed at the higher temperature, the rejected heat being given out at the lower temperature—all the operations being completely reversible In actual practice no engine works in a strictly reversible manner, and thus the ideal efficiency for reversible working is never realised. It does, however, provide a standard against which the performance of an actual engine may be judged, and this is measured by the "efficiency ratio" obtained by dividing the measured efficiency by the ideal efficiency.

Let us now consider the Carnot cycle for the simple heat engine described in section 13.03 when water is used as the working substance. At A (Fig. 175) the pressure in the cylinder is equal to the saturated vapour pressure of the water at the temperature T_1 of the source.

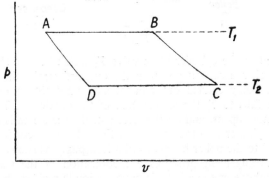

Fig. 175

The effect of supplying further heat at this temperature is to cause more water to vapourise at constant pressure—work being done as the piston moves outwards. At B the cylinder is thermally insulated and the expansion is continued adiabatically until the temperature falls to the sink temperature T_2. The cylinder is now placed in contact with the sink and the vapour compressed at constant pressure (the saturated vapour pressure at T_2) until the point D on the adiabatic through A is reached. The cycle is completed by again thermally insulating the cylinder and adiabatically compressing the vapour until the initial state at A is regained.

From a practical point of view the single cylinder heat engine undergoing the Carnot cycle of operations has two great objections. In the first place condensation within the cylinder would be an extremely slow process, and secondly, the thermal capacity of the cylinder itself would not be negligible and this would result in wasteful absorption of heat during the first stage of the cycle. In practice these difficulties are obviated by the use of a separate boiler (source) and condenser (sink) as shown in the simple arrangement of Fig. 176. Steam

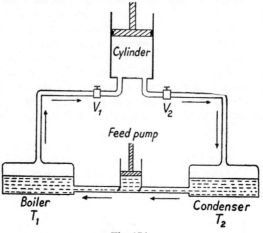

Fig. 176

is admitted to the cylinder via a valve V_1 which "cuts off" when the piston reaches a certain point on the outward stroke the remaining expansion being carried out adiabatically until the temperature of the steam falls to T_2 the temperature of the condenser. The exhaust valve V_2 now opens, and on the return stroke the steam is discharged from the cylinder into the condenser. From here the condensed water is returned to the boiler by means of a feed pump as shown.

This arrangement produces a modification of the Carnot cycle described above as the adiabatic compression is replaced by a process in which heat is absorbed in the boiler at T_1 to raise the temperature of the water from the feed pump from T_2 to T_1. Thus the liquid working

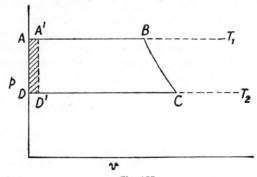

Fig. 177

substance does not absorb its heat at the highest temperature, and hence the cycle of operations is not as efficient as the Carnot cycle. This modified cycle is known as the *Rankine cycle* and is shown in Fig. 177. *AB* represents the admission of the high pressure steam from the boiler at a temperature T_1, *BC* is the adiabatic expansion after "cut-off", *CD* is the return stroke of the piston and the exhaust of the steam to the condenser. *DA* represents the transfer of the water from the condenser to the boiler by the feed pump, and it is in this stage, and the latter part of stage *CD*, that the divergence from the Carnot cycle occurs.

To obtain the net work done during the cycle the work done on the feed pump must be allowed for. If the volume of the condensed water returned in each cycle by the feed pump from the condenser at pressure p_2 to the boiler at pressure p_1 is v_w (*DD'* on the diagram), the amount of work involved is $(p_1 - p_2) v_w$. This is given by the shaded area *AA'D'D*, the net work done by the working substance being given by the area *A'BCD'*. If reversibility is assumed for all the operations, the Rankine cycle represents the maximum possible efficiency for the type of steam engine described. Accordingly it must replace the Carnot cycle when estimating the performance by means of the "efficiency ratio."

13.22   Efficiency of the Rankine Cycle

The efficiency of the Rankine cycle can be obtained from the $T\phi$ diagram shown in Fig. 178 and will be considered when saturated, wet, and superheated steam are used.

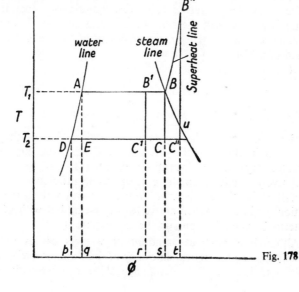

Fig. 178

(a) Saturated steam

For saturated steam the cycle is represented by the figure $ABCDA$.

$$\text{The efficiency} = \frac{\text{heat converted into work}}{\text{heat supplied}}$$

$$= \frac{\text{area } ABCD}{\text{area } pDABs}$$

Now area $ABCD$ = area DAE + area $ABCE$

$$= (\text{area } pDAq - \text{area } pDEq) + \text{area } ABCE$$

$$= \left\{ c_p (T_1 - T_2) - T_2 c_p \log_e \frac{T_1}{T_2} \right\} + (T_1 - T_2) \frac{L_1}{T_1}$$

where c_p is the average specific heat of water between T_1 and T_2, and L the latent heat of steam at T_1

Area $pDABs$ = area $pDAq$ + area $AB\,sq$

$$= c_p (T_1 - T_2) + T_1 \frac{L_1}{T_1}$$

$$= c_p (T_1 - T_2) + L_1$$

Taking $c_p = 1$, the efficiency is thus

$$\frac{(T_1 - T_2)\left[1 + \dfrac{L_1}{T_1} \right] - T_2 \log_e \dfrac{T_1}{T_2}}{T_1 - T_2 + L_1}$$

The corresponding Carnot cycle of operations is represented by the figure $ABCEA$ the efficiency of which is given by the ratio of the areas $ABCE$ and $AB\,sq$. The figure shows that this ratio is relatively greater than the ratio of the areas concerned in the Rankine cycle. This confirms the greater efficiency of the Carnot cycle as discussed in the previous section.

(b) Wet steam

For steam initially wet on admission to the cylinder the $T\phi$ diagram is the figure $AB'C'D$ where $\dfrac{AB'}{AB}$ represents the assumed dryness fraction q of the admitted steam. A consideration of the figure will show that in this case the efficiency will be lower than when using saturated (dry) steam as the ratio of the area $AB'C'D$ to area $AB'\,r\,p\,D$ is relatively smaller than the corresponding ratio for dry steam. The expression for the efficiency of the cycle with wet steam of dryness fraction q will be as given above for dry steam but with qL_1 replacing the term L_1 where it occurs in the expression.

Although a moderate amount of wetness causes only a slight reduction in the theoretical efficiency of the Rankine cycle, its practical effect is to increase the interchange of heat between the steam and the walls of the cylinder resulting in departure from the ideal conditions during adiabatic expansion with consequent further lowering of the efficiency.

(c) Superheated steam

In general engineering practice the steam is further heated after evaporation at the constant pressure of the boiler by passing it along pipes exposed to the furnace gases. In this way a considerable degree of *superheat* can be supplied, the temperature and entropy of the steam increasing along a line such as that indicated by BB'' in Fig. 178. An additional quantity of heat equal to the area $BB''ts$ is supplied to the steam during this process and the subsequent adiabatic expansion causes some of this heat to be given up. If carried far enough the steam will become dry at the point u where the adiabatic line through B'' cuts the dry steam line and may ultimately become wet as at C''. The efficiency of the cycle is now given by the ratio of the area $ABB''C''D$ to the area $ABB''tpD$ which is clearly somewhat greater than the efficiency of the dry steam cycle. The relative increase in the theoretical efficiency is however not very great even with a large degree of superheat, but superheating has the special advantage of reducing the extent of condensation with consequent reduction of heat interchanges with the cylinder walls. The ideal conditions for adiabatic expansion are thus more nearly attained and the practical efficiency considerably increased than with wet (or dry) steam working.

It should be realised that the diagrams given in Fig. 178 represent the theoretical efficiencies of the cycles dealt with. Apart from the question of heat leakage due to condensation, other factors—such as the necessity for a "clearance space" to prevent contact of cylinder head and piston during compression, pressure drop between boiler and cylinder, sluggishness in the response of the feed and exhaust valves, drop of temperature between the furnace gas and generated steam, &c.—are present which affect the performance and result in reduced practical efficiencies. From a thermodynamic point of view the large irreversible drop of temperature in the boiler is the most serious defect of the steam engine. This objection does not apply in the case of the internal combustion engine, the performance of which will now be discussed.

13.23 Internal Combustion Engines

The internal combustion engine employs air (to which has been added a small quantity of some combustible gas or vapour) as its working substance. It is thus a "hot air engine" although the name is usually reserved for the class of engines in which the heat supplied to the working substance through the walls of the cylinder by external combustion as in the case of the boiler of the steam engine. Although power cycles of high theoretical efficiency are possible with this class of engine, practical difficulties concerned with the slow rate of diffusion of heat through the gas, the limit to the maximum temperature necessary to avoid "softening" of the cylinder walls, and their large bulk with consequent slow speed, have resulted in their being superceded by the internal combustion engine where these objections are in large measure overcome. We shall consider here certain theoretical cycles of importance relating to the internal combustion engine only. The various cycles employed are categorised according as to whether the combustion takes place (i) isothermally, (ii) at constant volume, (iii) at constant pressure.

The Carnot cycle already described typifies the first class of cycle. The combustible mixture is compressed adiabatically to a temperature at which it ignites the heat released by the combustion process being supplied to the working substance at this temperature during the working stroke. The remainder of the cycle is followed as previously described, a fresh supply of mixture being used on each cycle. The efficiency of the Carnot cycle represents an ideal limit for the working temperatures concerned, but the practical difficulties involved make it unsuitable as an actual working cycle. Thus to obtain a suitable

$$\text{mean effective pressure} \left(= \frac{\text{work done per cycle}}{\text{volume of piston stroke}} \right)$$

it can be shown that a very high maximum pressure is required and this results in a bulky cumbersome engine. We shall now discuss theoretical cycles which are more readily adaptable to practical requirements.

13.24 The Otto Cycle

An important cycle which has wide general use with "gas" and "petrol" engines was first described by Beau de Rochas in 1862 and realised practically by Otto in 1876 and is typical of the second class of cycles in which the combustion takes place at constant volume. The Otto cycle is a four-stroke cycle, the sequence of operations being as follows:

(i) During the suction stroke EC (Fig. 179) the cylinder is filled with the combustible mixture at atmospheric pressure and at a temperature T_1 somewhat higher than that of the atmosphere (on account of the heating of the cylinder due to the previous working). (ii) The mixture is now compressed adiabatically to D the temperature rising from T_1 to T_2 during the process. (iii) At D the mixture is "sparked" when combustion takes place very rapidly at constant volume the pressure rising to that represented by the point A, and the temperature attaining some high value T_3 at this point. Adiabatic expansion follows to produce the working stroke AB during which the temperature falls to T_4. At B an exhaust valve opens the pressure inside the cylinder falling to atmospheric with the escape of some of the products of combustion (exhaust gases). (iv) In the final exhaust stroke *CE* the exhaust gases are completely discharged from the cylinder which is now ready for a further charge of combustible mixture for the next cycle of operations.

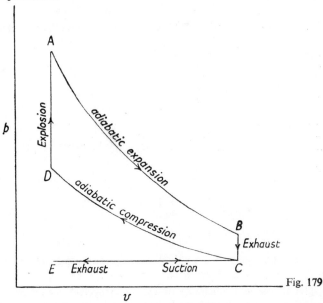

Fig. 179

It will be clear from the foregoing description that the efficiency of the cycle is less than that of the Carnot cycle since heat is supplied over the temperature range $T_3 - T_2$ and not at the higher temperature T_3. Nevertheless it can be shown that the cycle results in a much higher ratio of mean effective to maximum pressure which is a great practical improvement over the Carnot cycle.

From a thermodynamic point of view the action of the cycle may be conceived to take place with the same mass of air always in the cylinder as the cycle *ABCDA* is continuously traversed. If c_v is the specific heat of this air at constant volume, then the heat Q_1 absorbed during the stage *DA* is $c_v (T_3 - T_2)$, and the heat Q_2 rejected along *BC* is $c_v (T_4 - T_1)$. Hence the efficiency, which is the ratio of the work done (heat units) to the heat absorbed

$$= \frac{Q_1 - Q_2}{Q_1} = 1 - \frac{Q_2}{Q_1}$$

$$= 1 - \frac{c_v (T_4 - T_1)}{c_v (T_3 - T_2)}$$

$$= 1 - \frac{T_4 - T_1}{T_3 - T_2} \quad \cdot \quad \cdot \quad \cdot \quad \cdot \quad \cdot \quad \text{[149]}$$

If now the volumes of the mixture at C and D are v_1 and v_2 respectively, we have by equation 46 of chapter VII,

$$\frac{T_1}{T_2} = \left(\frac{v_2}{v_1}\right)^{\gamma-1} \text{ for the adiabatic line } CD.$$

and
$$\frac{T_4}{T_3} = \left(\frac{v_2}{v_1}\right)^{\gamma-1} \text{ for the adiabatic line } AB.$$
$$(AD \text{ and } BC \text{ being lines of constant volume}).$$

hence
$$\frac{T_1}{T_2} = \frac{T_4}{T_3} \text{ or } \frac{T_4}{T_1} = \frac{T_3}{T_2}$$

$$\therefore \quad \frac{T_4}{T_1} - 1 = \frac{T_3}{T_2} - 1$$

from which
$$\frac{T_4 - T_1}{T_3 - T_2} = \frac{T_1}{T_2} \quad \cdots \cdots \quad [150]$$

Thus from 149 and 150 the efficiency may be expressed as

$$1 - \frac{T_1}{T_2}$$

or as
$$1 - \left(\frac{v_2}{v_1}\right)^{\gamma-1}$$

$$= 1 - \left(\frac{1}{r}\right)^{\gamma-1}$$

where r is the ratio of the volumes of the mixture at D and C and is known as the *compression ratio*.

The theoretical efficiency of the Otto cycle thus depends on the compression ratio alone and not on the maximum temperature attained.* Increasing the value of r will result in increased efficiency although in practice a limit is imposed on the maximum attainable compression ratio by the nature of the fuel used and the design and strength of the engine.

13.25   The Diesel Cycle

As an example of the third class of cycle we shall consider that proposed by Diesel in 1900. In the Diesel engine the risk of spontaneous ignition of the combustible mixture which occurs under high adiabatic compression in the Otto cycle is obviated by injecting the fuel on the completion of the compression stroke using pure air only in the cylinder. Heavy liquid fuels are invariably used in this class of engine and the rate at which they are injected into the cylinder is carefully controlled by a subsidiary feed pump so that the combustion process takes place at constant pressure over a portion of the working stroke. The stages of the cycle can be followed by reference to Fig. 180.
(i)   During the suction stroke EC the cylinder is charged with air at atmospheric pressure.

---

* It should be realised, however, that *output* does depend on the maximum working temperature, and thus high efficiency with but low temperatures will result in a small amount of work being done per cycle.

(ii) The air is now compressed adiabatically to the point D the temperature rising from T_1 to T_2 during the process.

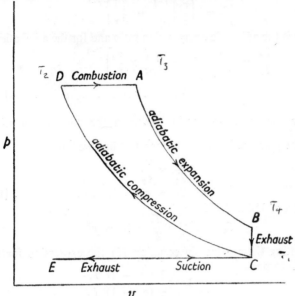

Fig. 180

(iii) The fuel is now injected at the point D at such a rate that during combustion the piston moves out for the portion DA of its working stroke under constant pressure conditions. At A the fuel supply is cut off and for the rest of the working stroke the expansion is adiabatic, the temperature falling from T_3 (at A) to T_4. At B an exhaust valve opens, the pressure immediately falling to atmospheric at the point C.
(iv) The cylinder is now completely emptied of the exhaust gases during the exhaust stroke CE, ready for the commencement of a fresh cycle of operations as above.

As with the Otto cycle we may again consider from a thermodynamic point of view that the same mass of air is always present in the cylinder in the continuous performance of the cycle. If now c_p is the specific heat of this air at constant pressure, and c_v its specific heat at constant volume, the amount of heat Q_1 received during state DA is $c_p(T_3 - T_2)$ and the amount Q_2 rejected along BC is $c_v(T_4 - T_1)$. The efficiency $1 - \dfrac{Q_2}{Q_1}$ is thus equal to

$$1 - \frac{c_v(T_4 - T_1)}{c_p(T_3 - T_2)}$$

$$= 1 - \frac{1}{\gamma}\left[\frac{T_4 - T_1}{T_3 - T_2}\right] \text{ since } \frac{c_p}{c_v} = \gamma \quad . \quad . \quad [151]$$

Now if v_1, v_2, v_3 are the volumes of the working substance at the points C (and B), D, and A we have, for the adiabatic line CD

$$\frac{T_1}{T_2} = \left[\frac{v_2}{v_1}\right]^{\gamma-1} = \left(\frac{1}{r}\right)^{\gamma-1}$$

where r is the adiabatic compression ratio and for the adiabatic line AB

$$\frac{T_4}{T_3} = \left[\frac{v_3}{v_1}\right]^{\gamma-1} = \left[\frac{v_3}{v_2} \times \frac{v_2}{v_1}\right]^{\gamma-1}$$

$$= c^{\gamma-1} \frac{T_1}{T_2} \quad . \quad . \quad . \quad . \quad . \quad . \quad . \quad . \quad . \quad . \quad [152]$$

c being the expansion ratio $\dfrac{v_3}{v_2}$ during the combustion stage DA.
Since the pressure is constant along this line we have

$$\frac{T_3}{T_2} = \frac{v_3}{v_2}$$

$$\text{or} \quad T_3 = cT_2 \quad . \quad . \quad . \quad . \quad . \quad . \quad . \quad . \quad [153]$$

Hence from 152 and 153

$$T_4 = c^{\gamma-1} \cdot \frac{T_1}{T_2} \cdot T_3 = c^{\gamma}T_1$$

Substituting these values in equation 151 the expression for the efficiency becomes

$$1 - \frac{1}{\gamma} \frac{(c^{\gamma} - 1)\, T_1}{(c - 1)\, T_2}$$

$$= 1 - \frac{1}{r^{\gamma-1}} \cdot \frac{c^{\gamma} - 1}{\gamma\,(c - 1)}$$

The efficiency of the Diesel cycle thus depends on the value of the constant expansion ratio c as well as the adiabatic compression ratio r. For a given value of r the efficiency increases as c is reduced. This however involves a corresponding reduction in the maximum temperature and the output per cycle is consequently decreased.

Diesel engines develop a somewhat higher mean effective to maximum pressure ratio than with an engine working on the Otto cycle and further there is considerable saving of fuel. On the other hand, on account of the high working pressures, a greater degree of precision is necessary in the design of the Diesel engine, and it must also be very robust in construction.

As with the Rankine steam engine practical considerations concerned with heat losses through the cylinder walls and in the exhaust stage, the fact that the working substance is not air but a varying mixture of air and combustible or exhaust gases the specific heat of which shows marked variation with temperature, and other factors of a technical nature, limit the performance of actual engines. Due to these various causes the efficiencies obtained in practice with engines performing the Otto and Diesel cycles are of the order of 70 per cent of the theoretical efficiencies.

QUESTIONS. CHAPTER 13

1. Describe in outline the Carnot cycle, explaining the terms *reversibility* and *efficiency*. (The derivation of the expression for the efficiency is **not** required.) What is the Kelvin scale of temperature and why is it useful? (S.)

2. Why is the word " ideal " used in the phrase " ideal heat engine " of the Carnot cycle? How does such an engine differ from a practical heat engine? Show that a soap film must cool when stretched if the temperature coefficient of surface tension is negative. (S.)

3. What is meant by *reversibility, efficiency, an ideal heat engine*? In what ways does an ordinary heat engine differ from an ideal heat engine, and how could its efficiency be increased? (S.)

4. Calculate the efficiency of a Carnot operation and show how it leads to the Kelvin scale of temperature. (S.)

5. State the First and Second laws of Thermodynamics and outline the arguments which show that no heat engine working between a source at $T_1°K$. and a sink at $T_2°K$. can have an efficiency greater than

$$\frac{T_1 - T_2}{T_1}$$

Discuss the operation of ONE practical heat engine or refrigerator. (Oxford Schol.)

6. The thermal efficiency of a Diesel engine is said to be 40 per cent. What is meant by this and how would you attempt to find this efficiency for a given engine? (Oxford Schol.)

7. Give an account of the Carnot cycle of operations and show clearly what you understand by a *reversible* operation. Describe a theoretical engine which would perform the Carnot cycle, and state the conditions which prevent an actual engine from being as efficient as the theoretical engine you describe. (Durham, B.Sc.)

8. Explain carefully what is meant by the terms " reversible " and " irreversible " applied to thermodynamic processes, and show that of all heat engines working between two given temperatures, the reversible Carnot engine is the most efficient. (Manch. Inter., B.Sc.)

9. What is meant by the " efficiency " of a heat engine? When is an engine said to be reversible? Discuss the conditions of reversibility. With a reversible engine, how would the work got by cooling 1lb. of water from 100°C. to 0°C. compare with that got by cooling 100lb. of water from 1°C. to 0°C.? (Camb. Schol.)

10. Describe Carnot's engine. Point out the conditions which are necessary to make it reversible, and discuss the effects produced when it is reversed. In an ice house, 10lb. of ice would melt per hour when the outside temperature is 20°C.; find the least horse-power which would be required to prevent this loss, the engine working between 20°C. and 0°C. ($J = 1,400ft.-lb.$ per 1lb. degC.; 1h.p. $= 33,000ft.-lb.$ per min.). (Oxford Schol.)

11. What is meant by the *efficiency* of a heat engine? Derive an expression for the efficiency of a Carnot engine, having a perfect gas as working substance, in terms of the temperatures of the source and sink. A Carnot engine, having a perfect gas as working substance, is driven backwards and is used for freezing water already at 0°C. If the engine is driven by a 500-watt

electric motor, having an efficiency of 60 per cent, how long will it take to freeze 9 kilograms of water?

(Take 15°C. and 0°C. as the working temperatures of the engine, and assume that there are no heat losses in the refrigerating system. Latent heat of fusion of ice = 80 calories per gm.; $J = 4\cdot2$ joules per calorie). (Camb. Schol.)

12. Assuming the properties of a Carnot engine, show how a scale of temperature may be devised which is independent of the properties of any particular substance. What fixes the zero of such a scale?

How may this absolute scale be realised approximately in practice?

(Manch., B.Sc.)

13. What do you understand by a perfect gas?

Show that the temperature measured by a perfect gas thermometer is the same as that measured on the Kelvin absolute scale. (Oxford Schol.)

14. Describe the Carnot cycle, explaining what is meant by reversibility.

Prove that all reversible engines working between the same temperature have the same efficiency.

Show how this leads to Kelvin's absolute scale of temperature.

Deduce the expression for the efficiency of a reversible engine in terms of the temperatures of its hot and cold reservoirs. (Birm., B.Sc.)

15. State and contrast the First and Second laws of thermodynamics. Discuss carefully the way in which each law is involved in the proof that no engine can be more efficient than a reversible engine, explaining what you mean by " reversible ".

(London, B.Sc.)

16. Discuss the effect of changes of pressure on the melting point of a solid and on the boiling point of a liquid. Deduce an expression for the change in the boiling point due to a small change in pressure.

The specific volume of steam at 100°C. and 76cm. of mercury pressure is 1,601c.c. per gram, and the latent heat of vaporisation of water 536 calories per gram. Find the change in the boiling point of water due to a change in pressure of 1cm. of mercury. (Camb. Schol.)

17. The density of water at its freezing point is $1\cdot00$ and the density of ice at its melting point is $0\cdot92$. Calculate the depression of the melting point by a pressure of one atmosphere (one million dynes per sq. cm.). Develop the thermodynamic argument you require. (Manch., 1st year)

18. Show from thermodynamic considerations that if a soap-film is stretched suddenly, its temperature falls.

Show also, that if a film is maintained at a temperature θ, while its surface is increased by an amount A, it absorbs an amount of heat $c\theta A$, where c is the temperature coefficient of surface-tension at any temperature. (Leeds, B.Sc.)

19. Show how thermodynamic considerations lead to the conclusion that bodies of a certain kind should emit radiation in proportion to the fourth power of their absolute temperature.

Differentiate between those bodies which obey this law, and others which do not. (Leeds, B.Sc.)

20. Define " entropy " and show that the change in entropy of a self-contained system in passing from one state to another depends only upon the initial and final states. Show also that the entropy of a system is increased by processes which tend to equalise the temperature of its parts. (Manch., B.Sc.)

21. State Carnot's theorem and describe the entropy changes in the working substance of a reversible heat-engine during a cycle.

Calculate the change in entropy occurring when 10gm. of water at 100°C. are mixed with 20gm. of water at 15°C. (Birm., B.Sc. Exams.)

22. Define entropy, and discuss the statement that entropy tends to a maximum. Draw pv and $\theta\phi$ diagrams for a reversible engine performing a Carnot cycle. Describe the operations involved and interpret the enclosed areas in each diagram. (Manch., B.Sc.)

23. On a p-v diagram, AB, BC represent changes at constant volume and constant pressure respectively, AC being an isothermal. D is a point on the line BC and is on the adiabatic through A. Show the relative positions of the points A, B, C and D on a $\theta\phi$ (temperature-entropy) diagram.

If the temperature difference between B and C is $\delta\theta$ and the entropy difference is $\delta\phi$ use your diagrams to show that

$$k_p \cdot \delta\theta = \theta \, \delta\phi + \tfrac{1}{2}\delta\phi \cdot \delta\theta$$

where k_p is the specific heat at constant pressure and θ is the absolute temperature. (Durham, B.Sc.)

24. Distinguish between a reversible and an irreversible cycle.

Explain carefully the physical conditions necessary for reversibility.

Illustrate by means of a p-v diagram the cycle of changes in any internal combustion engine with which you are familiar. (Oxford Schol.)

25. Show that no heat engine can be more efficient than Carnot's reversible engine, and compare the conditions with those of (i) the Rankine cycle, (ii) the Otto cycle, (iii) the Diesel cycle. (Manch., B.Sc., Eng.)

26. Explain the factors which govern the efficiency of a heat engine, illustrating your answer by reference to a Carnot engine and a steam engine.

To what extent may an internal combustion engine be regarded as a heat engine? Why does the efficiency of a petrol engine increase with increasing compression ratio? (Oxford Schol.)

27. Deduce an expression for the efficiency of a steam engine working on the Rankine cycle. Show how the quality of the steam at the end of the expansion may be found. (Manch., B.Sc., Eng.)

28. Deduce an expression for the efficiency of a perfect air engine operating on the constant volume cycle. What are the chief reasons why the efficiency of actual internal combustion engines falls short of this efficiency? (Manch., B.Sc., Eng.)

H.T.-13+

CHAPTER XIV

SOME ASPECTS OF METEOROLOGICAL PHYSICS

14.01 Introductory

The subject of meteorology is concerned with the study of climatic conditions and weather changes, but in its wider sense it also embraces the optical and electrical phenomena of the atmosphere. The science of physics forms the basis on which the entire subject rests, the co-ordination and interpretation of meteorological observations being effected by the application of well established physical principles. The main branch of physics concerned in both theoretical and practical meteorology is heat. Most of the problems of interest to the meteorologist arise from the distribution and exchange of heat from one part of the atmosphere to another, and it is these aspects of the subject which we shall discuss in outline in the present chapter.

14.02 The Structure of the Atmosphere

The layer of air which surrounds the earth and is carried along with it during its rotation is called the atmosphere. It is a mixture of gases, and in the lower layers of the atmosphere, the relative proportions of the main constituents show no appreciable variation. The percentage composition of dry air is given in table 30 from which it will be seen that the chief gases are nitrogen and oxygen which together comprise by weight nearly 99 per cent of the air, the molecular weight of which may be taken as approximately 29.

TABLE 30.   Percentage composition of dry air

Gas	By volume	By weight
Nitrogen	78·03	75·48
Oxygen	20·99	23·18
Argon	0·94	1·29
Carbon dioxide	0·03	0·045
Neon		
Helium	Slight traces, less than	
Krypton	0·001 per cent.	
Xenon		

In addition to these gases there are certain non-permanent constituents the chief of which is water vapour. The amount of water vapour present in the air varies greatly according to the conditions and may be as much as 4 per cent. On account of the large latent heat associated with it, the water vapour content is of the utmost importance to the meteorologist, and its presence is responsible for many of the diverse manifestations of our weather. This aspect of the subject will receive fuller consideration in later sections.

There is no definite upper limit to the atmosphere but it becomes increasingly attenuated with height so that although atmospheric phenomena have been observed at heights of 300 kilometres and beyond, most of the total air is contained in the lower levels up to about 20 or 30 kilometres. At this latter height the pressure of the air is about 10 millibars. Our knowledge of the pressure and temperature of the air at different heights is obtained chiefly from instruments carried by small free balloons. Earlier instruments gave an automatic written record, but the instrument had to be found before the information was available. Present day instruments transmit continuous radio signals to a receiving station on the ground and by this means direct recordings of temperature and pressure have been obtained up to heights of 25 km. and, in exceptional cases, slightly above 30 km.

The results of such observations show that there is a steady decrease of temperature with height of approximately 6°C. per km. to about 10 or 11km. after which the temperature remains practically constant at about − 55°C. This was first discovered by de Bort in 1900 who called the region above 11km. the isothermal zone. Further observations however revealed that there were slow variations in temperature in a horizontal plane and the name was later changed to the *stratosphere.* The lower region of the atmosphere is called the *troposphere,* and the surface separating the two regions the *tropopause.* This boundary is quite sharp although its actual height varies with latitude from about 8km. at the poles to 18km. at the equator. In addition there are diurnal variations the height being lower over depressions than in anticyclonic areas. The temperature is everywhere extremely low in the stratosphere although the temperature over the poles is not nearly so low as that over the equator.

The troposphere is marked by considerable vertical mixing and the presence of cloud, but in the stratosphere there is less interchange and turbulence, and the main control of temperature must be by radiation. There can be little water vapour in air at stratospheric temperatures, although the presence of higher humidities on occasions is shown by the rather rare irridescent "mother of pearl" clouds seen at dawn and sunset in middle latitudes at a height of about 25km.

Temperatures above about 30km. have not been directly measured as yet, but in 1922 Lindemann and Dobson from observations on meteors concluded that there is much more air at meteoric heights than was to be expected if the stratospheric temperature remained constant throughout. They inferred that the temperature must therefore increase in the upper regions of the stratosphere to make possible the greater spread of air and estimated temperatures for this zone which were actually in excess of those at the earth's surface! Whipple (1923) at once realised the significance of their conclusions from the point of view of the abnormal propagation of sound to great distances. This could be explained if at a level of 40 or 50km. there was an upper warm region with the temperature increasing upwards from which the sound waves could be reflected. Calculations based on the received sound suggested that at these heights the temperature rises at approximately 6°C. per km. *i.e.*, at about the same rate as it falls in the troposphere (some of Whipple's estimates for stratospheric temperatures are shown in Fig. 181). The high temperatures at these

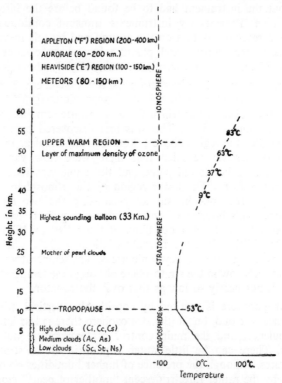

Fig. 181. The Structure of the Atmosphere

heights are ascribed to the presence there of a quantity of ozone. This is assumed to be formed by the partial dissociation of oxygen as a result of absorption of ultra-violet light from the sun's rays. The ozone thus formed absorbs the ultra-violet light even more strongly than the oxygen, the temperature of the ozone layer rising in consequence.

Above the upper warm region the air is extensively ionised as is evidenced by the phenomena of the aurora and from experiments on the reflection of radio waves (Heaviside and Appleton layers). This region is called the *ionosphere* and as yet our knowledge of it is somewhat scanty. Observations on persistent meteor trails reveal the existence of winds up to 100 miles per hour at levels of about 100 km., and spectra of the highest level auroral rays indicate the presence of nitrogen and atomic oxygen at these great heights (if the constituents of the atmosphere are assumed to be separated by the effects of diffusion above some level in the stratosphere, helium should be overwhelmingly predominant at very great heights). The auroral spectrum further indicates that at heights of about 100 km. the air has become much colder than in the higher stratospheric layers, but in the upper ionosphere the air may attain (by day) a temperature of 1000°C. or more.

14.03 Pressure of the Atmosphere

To the practical meteorologist the accurate determination of the atmospheric pressure is of profound importance. The preparation of the weather charts (see section 14.23) on which forecasts of future weather conditions are based demands precise knowledge of the pressure conditions at a given time at numerous places over a wide area of observation. For this purpose the pressure of the atmosphere is measured by mercurial barometers. Aneroid barometers are not accepted as giving readings of sufficient accuracy for the absolute measurement of pressure, although they are widely used in barographs to record *changes* in pressure conditions.

The standard mercury barometer used at meteorological stations in the British Isles is known as the Kew pattern. The mercury column is enclosed in a glass tube contained in a protective casing which is swung on gimbals to ensure an accurately vertical setting. The scale against which the mercury level is read is automatically corrected for changes in the level of the mercury in the cistern which is rigidly attached to the tube. The observed reading of the barometer is subject to various corrections before an accurate value of the atmospheric pressure can be obtained. These corrections are as follows:

(i) Correction for index error—caused by slight errors in the scale graduations.

(ii) Correction for temperature—this has been dealt with in section 4.11.

(iii) Correction for latitude—this is necessary on account of the variation of the acceleration of gravity with latitude, and the correction is made using the latitude of 45° as standard.

(iv) Reduction to mean sea level—the above three corrections give the true value of the pressure at station level. For use on weather charts a further correction for height above sea level is required.

All four corrections are usually reduced to a simple operation by the use of a correction card specially made for each instrument.

The unit of pressure adopted in meteorology is the *bar* which is equal to a pressure of one million dynes per square centimetre. This is equivalent to the pressure of a column of mercury of length 29·531 inches or 750·1mm. at a temperature of 0°C. in latitude 45°. As a convenient working unit the *millibar* (mb.), a thousandth part of the bar, is used. Expressed in these units the average pressure of the atmosphere is 1013·2 millibars.

14.04   Horizontal Distribution of Pressure

The pressure readings, corrected as above, from the various observing stations are collated and represented pictorially on a map covering the area of investigation. Lines are then drawn through places of equal pressure. These lines, which are called *isobars,* are usually drawn for intervals of two millibars, and it is often necessary to interpolate between the actual pressures observed to fix the position of the lines. The appearance of the lines on a complete map resembles that of the contour lines on a topographical map. The isobars form closed curves (if extended over a sufficiently wide area) much as shown on the hypothetical chart given in Fig. 182. The chart shows several distinct

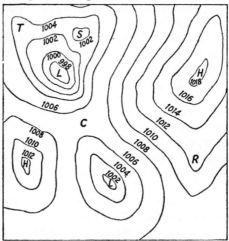

Fig. 182

types of pressure distribution. The areas marked *L* are regions of low pressure and are referred to as *depressions* or *cyclones.* They may contain within them smaller areas of low pressure (*S*) and these are known as *secondary depressions.* The one shown in the figure is a fully developed secondary with its own system of isobars; other forms of secondary depressions are often mere bulges or irregularities in the isobars from the centre of the main depression. Regions of high pressure *H* are called *anticyclones* and generally the isobars are somewhat further apart than in depressions—especially so near the centre. Of the remaining configurations shown, R represents a *ridge* or *wedge* of high pressure, T a *trough* of low pressure (formerly called a V-shaped depression), and C is a *col* which is a region between two high and two low pressure regions, being analogous to the term "saddle back" used in reference to topographical maps.

The rate of change of pressure horizontally across the isobars is called the *pressure gradient* and this is a factor of great importance in determining wind velocities (see section 14.13). Where the isobars are crowded together the pressure gradient is large and the area is one of strong winds. On the other hand if the isobars are well spaced the pressure gradient is small and the region calm.

14.05 Vertical Distribution of Pressure

It is well known that the pressure of the atmosphere decreases with increasing height above sea level, the average rate of decrease near the surface of the earth being 1 millibar per 30 feet approximately. The pressure at any given height is also affected by temperature, and of two places A and B at which the barometric readings are identical, the pressure at a given height above A will be greater than that at a corresponding height above B if the temperature at A is greater than that of B. This is because the density of the air, and hence the weight of the air column above A to the height concerned, is less than at B, and thus a smaller amount has to be subtracted from the surface pressure at A than at B.

The variation of pressure with height in a gas of molecular weight M and at a temperature $T°K$. has already been expressed in mathematical form in section 8.22, as

$$\log_e \frac{p}{p_o} = - \frac{Mg}{RT} \cdot h \quad \ldots \ldots \quad [154]$$

p_o and p being respectively the pressures of the gas at ground level and at a height h above it. This equation will represent the facts only in an *isothermal* atmosphere, but if we assume the air temperature to decrease uniformly with height according to the formula

$$T = T_o - ah$$

T_o being the ground temperature and a the rate of decrease of temperature with height, the equation to be integrated becomes

$$\frac{dp}{p} = - \frac{Mg}{R} \cdot \frac{dh}{(T_o - ah)}$$

giving $\log_e p = \frac{Mg}{aR} \log_e (T_o - ah) + \text{const.}$

or $\log_e \frac{p}{p_o} = \frac{Mg}{aR} \log_e \left[\frac{T_o - ah}{T_o} \right] \quad \ldots \ldots \quad [155]$

since $p = p_o$ when $T = T_o$.

14.06 Barometric Altimetry

Equations 154 and 155 can be adapted for the determination of heights from barometric readings if the value of the air temperature (and its rate of decrease in the latter equation) are known. Thus equation 154 may be written in the form

$$h = \frac{RT}{Mg \log_{10}e} \cdot (\log_{10} p_o - \log_{10} p)$$

which at sea level in latitude 45° becomes in metres

$$h = 67 \cdot 4 \; T \, (\log_{10} p_o - \log_{10} p)$$

and in feet

$$h = 221 \cdot 1 \; T \, (\log_{10} p_o - \log_{10} p)$$

The standard altimeter is an aneroid barometer with a scale graduated in inches of mercury or in millibars, and with a movable height scale round the outer edge of the dial. This scale can be rotated to read zero on the height scale at the ground, thus automatically correcting for variations in p_o. The scale is graduated on the assumption that the air is at a mean temperature of 283° absolute corresponding to 10°C. or 50°F. If the mean temperature from the ground to the height of observation differs from the assumed temperature of 10°C. the observed height is in/de -creased by $\frac{1}{283}$ of its value for every 1°C. above/below 10°C. (the corresponding fraction per °F. is $\frac{1}{509}$).

The International Commission for Air Navigation with a view to uniformity in the estimation of aircraft performances recommended the adoption of an altimeter based on the specification of an *international standard atmosphere*. This is defined by a mean sea level temperature of 15°C., a pressure of 760mm. (1013·2mb.), and a rate of decrease of temperature of 6·5°C. per kilometre from sea level to 11 kilometres above which the temperature is assumed constant at −56·5°C. Re-writing equation 155 as

$$h = \frac{T_o}{a} \left\{ 1 - \left[\frac{p}{p_o}\right]^{\frac{aR}{Mg}} \right\}$$

and using the data above, the formula for the height in metres becomes

$$h = 44309 \left\{ 1 - \left[\frac{p}{p_o}\right]^{\cdot 1904} \right\}$$

In practice there is no special advantage in using the I.C.A.N. altimeter, as it requires corrections for deviations from standard conditions just as the isothermal instrument does.

14.07 Temperature Distribution

A knowledge of the horizontal and vertical distribution of temperature in the atmosphere is of great importance in meteorology as it plays a very large part in determining weather conditions. The air immediately above the earth's surface tends to attain the temperature of the ground with which it is in contact, the lower layers acquiring heat by conduction, which is subsequently transferred to the higher levels by the process of convection. Before discussing the observed results it will be first necessary to deal with the source of the earth's heat. This is derived almost entirely from the sun's rays some of which are reflected, absorbed, or scattered in the atmosphere, the rest being absorbed by the earth's surface which gets warmed in consequence. The presence of small particles and droplets in the atmosphere is responsible for the diffuse reflection and scattering of some of the solar radiation, and this plays little or no further part in heat exchanges in the atmosphere. Of the remaining radiation we have already seen (section 14.02) that the presence of ozone in the upper stratospheric heights is responsible for strong absorption in the ultra-violet, but subsequently there is no appreciable absorption by the normal constituents of the atmosphere. On arriving at the earth's surface some of the incident radiation will be reflected back into space—particularly from water vapour—the effect increasing with the obliquity of the rays. The rest of the radiation is absorbed by the earth's surface, and the resulting rise in temperature will vary with the nature of the surface.

In the case of land the heating effect is confined to the surface layers, the actual temperature rise being determined by the thermal capacity and conductivity of the surface material. Low thermal capacity and low thermal conductivity (as with light soils and sandy areas) will result in a large increase of the surface temperature. The heating effect will also depend on the obliquity of the rays, being less as we move from the equatorial zones on account of the larger area covered by the sun's rays as a result of the increased angle of incidence.

Over the sea the effects are noticeably different, as here the absorbed heat is distributed to some depth and this, together with the much larger thermal capacity of water, results in only slight increases in the surface temperatures. Maps showing the distribution of temperature by the use of *isotherms* (lines of equal temperature) reveal this difference in behaviour of land and sea masses towards the incident radiation—the low temperatures over the oceans being in marked contrast with the much higher temperatures recorded over the continents. The depth of the surface heating also plays an important part in controlling the surface temperatures at night when the earth's surface is losing heat by radiation. There is a pronounced fall of temperature over land surfaces, whereas the decrease over the sea and large water surfaces is very much less as the loss of heat is here shared through a considerable depth.

Although the water vapour in the atmosphere does not absorb an appreciable amount of the sun's (short wave) radiation, the presence of moisture in the atmosphere plays an important part in controlling the cooling rate of the earth's surface. This is because water vapour absorbs strongly the longer wave infra-red radiation from the earth's surface

13*

(see section 12·08), much of which is re-radiated back to the earth. A cloud layer thus acts as a protecting blanket by night, although by day the top surfaces of clouds will reflect some of the incident solar radiation thereby reducing the surface heating at the earth.

14.08   Vertical Distribution of Temperature

It will be evident from the foregoing discussion that the atmosphere is heated from below, and its temperature in the lower layers is determined mainly by the effects of convection currents. When a mass of air rises vertically through the atmosphere it does so without appreciable loss of heat by conduction or radiation to the surrounding air, and its temperature at any height is dependent on the pressure there and the cooling produced by its expansion under closely approximate adiabatic conditions. For an atmosphere in convective equilibrium there is thus a vertical fall of temperature from the earth's surface. For dry air, the rate of fall of temperature, or the *dry adiabatic lapse rate* as it is called, may be calculated as follows:

For a column of gas in gravitational equilibrium we have (see section 8·22)

$$dp = - g \, \frac{M}{v} \cdot dh$$

where dp is the change in pressure for a change dh in height, and v is the volume occupied by the gram-molecular weight (M) of the gas. Using the perfect gas equation $pv = RT$ the above equation may be written:

$$\frac{dp}{p} = - \frac{gM}{RT} \cdot dh \quad . \quad . \quad . \quad . \quad . \quad . \quad . \quad . \quad [156]$$

Now by equation 45 of section 7.10 the relation between the temperature and pressure in an adiabatic change is given by

$$\frac{T^\gamma}{p^{\gamma-1}} = \text{constant} \, (k)$$

which on differentiation gives

$$\gamma T^{\gamma-1} \cdot dT = k \, (\gamma - 1) \, p^{\gamma-2} \cdot dp$$

and thus from these last two equations

$$\frac{dp}{p} = \frac{\gamma}{\gamma - 1} \cdot \frac{dT}{T} \quad . \quad . \quad . \quad . \quad . \quad . \quad . \quad . \quad [157]$$

Hence from equations 156 and 157

$$dT = - \left[\frac{\gamma - 1}{\gamma} \right] \frac{Mg}{R} \cdot dh$$

which on integration gives

$$T = - \left[\frac{\gamma - 1}{\gamma} \right] \frac{Mg}{R} \cdot h + C$$

where C is a constant of integration. Since $T = T_o$ (the surface temperature) when $h = 0$, then $C = T_o$, and hence

$$T_o - T = \left(\frac{\gamma - 1}{\gamma}\right) \frac{Mg}{R} \cdot h$$

Substitution of the values $M = 29$gm., and $g = 981$cm. per sec^2., $R = 8\cdot31 \times 10^7$ ergs per gm. mol. per °C., $\gamma = 1\cdot40$, gives a rate of fall of temperature of $9\cdot8 \times 10^{-5}$°C. per cm. or $9\cdot8$°C. per km. Thus the dry adiabatic lapse rate corresponds to practically 1°C. fall of temperature for every 100 metres of ascent or 5·4°F. for every 1,000ft. of ascent. Average conditions in the atmosphere give a lapse rate which is somewhat lower than this, and as we have already seen, for the purposes of defining the standard atmosphere a lapse rate of 6·5°C. per kilometre is assumed to represent these conditions.

Moist air, provided it is not saturated, will cool at approximately the dry adiabatic lapse rate with ascension, but when the air becomes saturated on reaching the dew point as a result of the adiabatic cooling, condensation begins and the released latent heat causes a pronounced lowering of the lapse rate. The rate of decrease of temperature with height for ascending saturated air is known as the *wet adiabatic lapse rate*. Its actual value depends on the temperature of the air ranging from about 2°F. per 1,000ft. for cold air where the amount of water vapour condensed is low to about 4°F. for hot (tropical) air from which a relatively larger amount of condensation takes place. For temperate regions an approximate average value of 2·7°F. per 1,000ft. is generally assumed.

Near the ground layers abnormally high lapse rates often occur over strongly heated areas as vertical currents do not always develop freely in the immediate vicinity of the ground. At night on the other hand when the earth is radiating to a cloudless sky the surface layers of the air may be cooled below those at a higher altitude. A negative lapse rate or an *inversion* is then set up. Inversions are also formed when warm air passes over cold land surfaces and are often found above cloud layers. This is due to the fact that the top surface of a cloud is radiating freely into space and thus tends to become cooler than the air layers immediately above it. Inversions represent conditions of great vertical stability, and when they occur at the surface of the earth fog is frequently produced.

14.09 Vertical Stability and Instability of the Atmosphere

A stable atmosphere is one in which the conditions do not favour the development of vertical motion or convection currents in the air, whereas if such vertical currents freely occur in the atmosphere the conditions are those of instability. The criterion for stability or instability is determined by whether the actual lapse rate in the air is less than or greater than the dry adiabatic lapse rate. Thus if a mass of air is forced upwards it will cool at the adiabatic lapse rate, and if the actual lapse rate is less than this the air will always be cooler than the surrounding air at the same level. Accordingly the mass of air will be relatively denser than the surrounding air and will thus tend to sink back to its original position, and the atmosphere is therefore in stable equilibrium. Conversely if the actual lapse is greater than the dry adiabatic lapse rate the ascending mass of air will find itself surrounded by colder and hence heavier air. It thus continues to rise with increasing speed, and violent vertical currents quickly develop under the unstable conditions. In this way the heat in the layers in contact with the earth's surface is conveyed to the upper layers, the downward currents of cooler air warming at the adiabatic lapse rate as they descend. If this process of convective mixing continues for some time the conditions tend to approach those of an atmosphere with a lapse rate equal to the dry adiabatic rate. In such an atmosphere a vertically moving mass of air will always find itself surrounded by air of the same temperature, and is thus not acted on by forces which tend to restore it to or further displace it from its original position. Under these conditions the air is in neutral equilibrium.

It may be noted here that under stable atmospheric conditions, determined by a low lapse rate (or an inversion), a cloud layer of little vertical thickness usually of the stratus family (see section 14.17) develops which may persist for days. Rain is almost unknown in these conditions, although in certain circumstances mist or fog may be present. Under conditions of instability however a vast vertical development of cumulus cloud results which is accompanied with heavy showers and sometimes thunderstorms.

The upper limit of convective equilibrium is defined by the tropopause which marks the greatest height at which ordinary clouds are formed. Above this limit, in the stratosphere, the lapse rate will be determined by radiative equilibrium, and is in consequence negligibly small.

14.10 Conditional Stability

We have seen in the foregoing section that if the lapse rate is greater than the dry adiabatic lapse rate vertical mixing takes place as a result of the unstable conditions. The lower layers of air may be lifted to such a height that the initial moisture content is sufficient to produce a state of saturation with subsequent condensation and heating of the air on account of the released latent heat. The criterion of stability is now determined by the wet adiabatic lapse rate, and if this is greater than the observed lapse rate the air will again become stable. A little consideration will show that the critical limit for *absolute stability* is that the actual lapse rate should always be less than the wet adiabatic lapse rate (region *A* in Fig. 183), and for *absolute instability* the actual lapse rate must be greater than the dry adiabatic rate (region *C*). These represent the extreme cases irrespective of the moisture content of the air. If the observed lapse rate lies in between these extreme limits

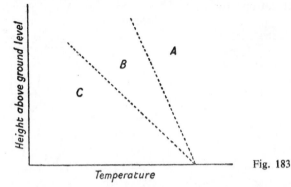

Fig. 183

(region *B*) the air is said to be *conditionally unstable*, the stability being determined by considerations of humidity. Thus for air of high relative humidity a small upward impulse would be sufficient to bring about condensation at a low level resulting in conditions of actual instability (the wet adiabatic line providing the stability criterion). On the other hand with air of low relative humidity the condensation level would be high, and the air would remain stable with respect to small impulses (the dry adiabatic line providing the stability criterion).

14.11 The Tephigram

The $T - \phi$ diagram discussed in section 13.18 has been adapted by Sir Napier Shaw for meteorological use as a means of evaluating the physical changes that take place in air masses moving vertically through the atmosphere, thereby providing a method of obtaining a short term forecast of local weather conditions. When prepared for this purpose the $T - \phi$ diagram is commonly referred to as a *tephigram*, and has a horizontal scale of temperature (°F.), and a vertical scale which gives the logarithm of the *potential temperature* (θ). This is the temperature that a specimen of air would have if it were brought to a standard pressure of 1000mb. under dry adiabatic conditions. Thus the potential temperature corresponding to an actual temperature $T°$K. when the pressure is $p$mb. is, using equation 45 of section 7.10,

$$\theta = T \left(\frac{1000}{p} \right)^{\frac{\gamma-1}{\gamma}}$$

$$= T \left(\frac{1000}{p} \right)^{\cdot 286} \quad \text{taking } \gamma = 1 \cdot 40 \text{ for air.}$$

Potential temperature is related to the entropy ϕ by the relation (see section 13.20), $\quad \phi = c_p \log_e \theta + \text{const.}$
where c_p is the specific heat at constant pressure.

A somewhat simplified, and reduced, form of the tephigram chart is shown in Fig. 184. As mentioned above the vertical lines are lines of equal temperature (°F.), and the horizontal lines represent lines of equal entropy (or potential temperature). A sample of air moving

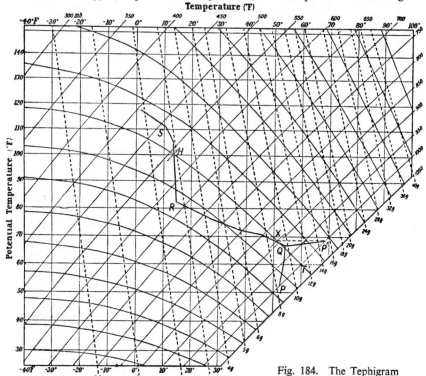

Fig. 184.   The Tephigram

vertically through the atmosphere under dry adiabatic conditions will follow these horizontal lines. Saturated, or wet, adiabatic conditions are represented by the curved lines sloping upwards towards the left. These lines tend to flatten out and become parallel to the dry adiabatic lines at low temperatures and pressures. The lines (almost straight) sloping upwards to the right indicate the atmospheric pressure and are marked off in millibars. Height values in feet (or other units) can be assigned on any particular occasion. The dashed lines rising almost vertically are the humidity lines and give the quantity of water vapour in grams required to saturate one kilogram of dry air under the appropriate conditions of temperature and pressure.

In using the tephigram the temperature (and humidity) readings of the upper air, obtained by means of a radiosonde (a hydrogen balloon fitted with radio-sounding instruments) are recorded against the pressure values, and the resulting plot makes it possible to see at a glance the state of the air as regards stability, and thus to define the local weather possibilities. Thus if the curve slopes downwards to the left, the lapse rate exceeds the dry adiabatic rate, and the air is in a state of absolute instability. A curve sloping upwards to the right indicates an inversion, and a vertical line represents isothermal conditions. Conditional instability is shown by the curve lying between the dry and wet adiabatic lines, and the type and extent of cloud (and associated weather) likely to develop in this case is determined from a consideration of the humidity data.

To illustrate the use of the tephigram we shall consider the temperature-pressure curve plotted in Fig. 184 from data obtained in an early morning with clear skies. The ground temperature is represented by the point P (ground level pressure 1000mb.), the temperature increasing with height to the point Q. PQ thus represents an inversion which is typical of clear mornings. Above Q the temperature falls off with height along the curve QR, the air in these layers being conditionally unstable as the slope of QR is intermediate between the slopes of the dry and wet adiabatic lines. Beyond R there is a sharp fall in the lapse rate below that of the wet adiabatic rate, and the air in the region above R will become absolutely stable. The future cloud and weather possibilities are dependent on the likelihood and extent of instability developing in the surface layers later in the day, and by a method due to Gold* it is possible to estimate the temperature to which the surface air is likely to rise during the day. Let this temperature be represented by the point P' (68°F.) with a corresponding wet bulb temperature of 60°F. (point T) provided by the humidity data of the air mass. The lapse rate in the surface layers is now somewhat greater than the dry adiabatic rate and in consequence any slight disturbance will upset the stratification of the air layers and air from the surface will rise, its temperature falling along the dry adiabatic. Now it has been shown by Sir Charles Normand† that the corresponding decrease in the wet bulb temperature occurs at the saturated adiabatic lapse rate, the condensation point being given by the intersection of the dry adiabatic through the dry bulb temperature (P') and the saturated adiabatic through the wet bulb temperature (T). (The line through the dew point at ground level drawn parallel to the humidity line also passes through this point).

On this occasion therefore the cloud base will be formed at a level through the point X at a pressure of 890mb. On account of the deep layer of conditionally unstable air above this level thick cumulus cloud (section 14.17) will develop up to the level indicated by the point H where the wet adiabatic line through T intersects the observed temperature curve. This gives a cloud top at 560mb. The stable lapse rate above this point prevents further vertical development of cloud, although the upward momentum of the rising air may cause the cloud to push its

* E. Gold, F.R.S. *Professional Notes of the Meteorological Office.* Vol. 5, No. 63.
† C. W. B. Normand. *Memoirs of the Indian Meteorological Department for* 1921.

head here and there into the clearer air above. It will be seen that the cloud top penetrates well above the freezing level and thus heavy rain (see section 14.18) with the likelihood of thunder will occur.

14.12   Wind

Wind is air in motion, the component of the motion parallel to the surface of the earth being that which is generally measured. The primary cause of wind is the irregularity of temperature set up in the air in consequence of the inequalities of solar and terrestrial radiation at the earth's surface. The resulting air movements occurring may take the form of general atmospheric circulations or they may be more local in character. To the former class belong the *trade winds*, the general structure of which are well described and illustrated in books on physical geography. The latter class of winds will receive fuller attention in later sections of this chapter.

In specifying winds it is necessary to give both the velocity and the direction from which they blow. The direction is given in degrees measured from true north in a clockwise sense. The velocity may be expressed in miles per hour or in kilometres per hour. The latter is the international unit, and for purposes of quick conversion 8km. per hour may be taken as approximately equal to 5 miles per hour. The measure-ment of the "surface wind" velocities is usually taken at a height of 33ft. (10 metres) by a cup anemometer* which records the average wind over a given period. To record gusts and lulls an instrument recording the variations in the pressure of the wind blowing into the open end of a tube and called a Dines anemograph is employed. For the details of wind velocities in the free air ("upper winds") observations are taken of pilot balloons through specially designed theodolites. This method gives the average wind over the range of height between consecutive obser-vations, and for wind velocities at a definite height observations may be taken of the movement of clouds (nephoscope observations) carried along by the wind.

The earliest systematic attempt at estimating wind velocities was made in 1805 by Admiral Beaufort. Originally devised as a numerical scale with corresponding descriptive terms assessing the wind force according to its effect on a "man-of-war" of the period, it has been adapted and modified for land use, and is now generally adopted as the standard wind scale. The Beaufort scale is given in table 31 and includes the arrows used to represent the wind strengths as given in the daily weather maps issued by the Meteorological Office. The direction of the wind is indicated by the orientation of the arrow (which flies with the wind), and the force of the wind by the number of feathers attached to the arrow; each full-length feather denoting two steps on the Beaufort scale, and a short feather one step.

---

* For a description of this and other forms of wind measurement the reader is referred 'o *Meteorological Observers' Handbook* published bv H.M.S.O

TABLE 31

The Beaufort Wind Scale

Beaufort No.	Wind	Arrow	Specification for use on land	Average speed in m.p.h.	Range of speed in m.p.h.
0	Calm		Calm; smoke rises vertically.	0	1
1	Light air		Direction of wind shown by smoke drift; but not by wind vanes.	2	1–3
2	Light breeze		Wind felt on face; leaves rustle; ordinary vane moved by wind.	5	4–7
3	Gentle breeze		Leaves and small twigs in constant motion; wind extends light flag.	10	8–12
4	Moderate breeze		Raises dust and loose paper; small branches moved.	15	13–18
5	Fresh breeze		Small trees in leaf begin to sway; crested wavelets form on inland waters.	21	19–24
6	Strong breeze		Large branches in motion ; whistling heard in telegraph wires; umbrellas used with difficulty.	28	25–31
7	Moderate gale		Whole trees in motion; inconvenience felt when walking against wind.	35	32–38
8	Fresh gale		Breaks twigs off trees; generally impedes progress.	42	39–46
9	Strong gale		Slight structural damage (chimney pots and slates removed.)	50	47–54
10	Whole gale		Seldom experienced inland; trees uprooted; considerable structural damage occurs.	59	55–63
11	Storm		Very rarely experienced; accompanied by widespread damage.	69	64–75
12	Hurricane		—	above 75	

From the analogy of the flow of water downhill it would appear that air should move from regions of high to regions of low pressure. This however is not the case (for reasons given in section 14.13), and observations show that winds tend to blow approximately *along* the isobars. The relation between wind direction and the isobars is regarded as one of the fundamental principles of dynamical meteorology and is expressed by the following law due to Buys Ballot:

If an observer stands with his back to the wind in the northern hemisphere the region of lower pressure will be on his left; and in the southern hemisphere on his right.

The general validity of the law may be confirmed on any weather map, although occasional variations, due to local peculiarities, may be observed. In practice it is found that the wind blows slightly across the isobars from the region of high to that of low pressure. The inclination is approximately 30° over land surfaces, being much less (about 10°) over sea areas. As will be explained later the effect is due to surface friction which is more pronounced over land than over the sea. In the higher layers (free air) of the atmosphere the wind blows along the isobars, a rough rule being to allow 10° veer for every 1,000ft. of ascent.

Observations reveal definite diurnal variations in the wind which are more pronounced in the summer months. During the day, due to turbulent mixing (convection currents) of the surface and upper layers of the air, the surface wind increases in speed and is found to veer somewhat, the upper wind backing slightly. At night however, with the cessation of thermal turbulence, the free upper wind resumes its original value and veers, the surface wind falling and backing. Over the sea there is ordinarily little change in the thermal lapse rate and the effects of turbulence are absent. Hence there is no observable diurnal variation in the wind over the open sea.

14.13 Geostrophic Wind

The effect of the rotation of the earth on a body travelling on its surface in the northern hemisphere is to cause an acceleration to the right (to the left in the southern hemisphere) which is proportional to its velocity and perpendicular to the direction of travel as will be demonstrated more fully below. Thus a small mass of air in a pressure system will be subject to two forces, one caused by the pressure gradient which tends to move the air perpendicular to the isobars towards the region of low pressure, and the other due to the earth's rotation tending to deflect it to the right of the direction in which it is moving. For steady motion these two forces must balance, and thus the force due to the pressure gradient must act perpendicular to the air movement but to the left; in other words the air must move *along* the isobars with the low pressure on the left. Under these conditions of dynamical equili-brium the wind is said to be *geostrophic.* If the pressure gradient is large, the effects of the earth's rotation must be correspondingly large, and thus the wind velocity will increase proportionately with the pressure gradient. The relationship between the geostrophic wind velocity and the pressure gradient may be derived as follows:

(i) *Effect of pressure gradient:* Fig. 185 represents a cylinder of air of length dl and with vertical faces A, B of unit area set parallel to the

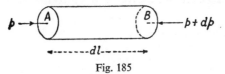

Fig. 185

isobars. The net force acting on the air is dp, and if ρ is the density of the air, the mass of the air in the cylinder is ρdl. Hence the acceleration of the air transverse to the isobars is

$$\frac{dp}{\rho dl} = \frac{\gamma}{\rho}$$

where γ is the pressure gradient $\dfrac{dp}{dl}$.

(ii) *Effect of the earth's rotation:* If the angular velocity of the earth about the polar axis NS (Fig. 186) is w, that about a vertical axis through a point P in latitude ϕ by resolution is $w \sin \phi$. Thus the tangent plane through P is rotating in an anti-clockwise sense with this velocity, the effect of which in a time interval t is to cause a point A to be displaced to A' to the left of A. Hence a mass of air originally

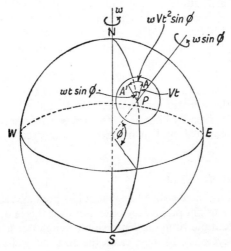

Fig. 186

moving towards A with a velocity V finds itself a distance AA' to the *right* of the new position of A (*i.e.*, A').

The displacement distance AA' $= AP \times$ angle $A'PA$
$$= Vt \times wt \sin \phi$$
$$= \tfrac{1}{2} (2Vw \sin \phi) \, t^2$$
$$= \tfrac{1}{2} f t^2$$

where f is the acceleration tranverse to AP to the right. Thus

$$f = 2Vw \sin \phi$$

On balancing the above accelerations we have

$$2Vw \sin \phi = \frac{\gamma}{\rho}$$

which gives the equation from which to determine the geostrophic wind velocity. It is usual to denote the value of this by G, and hence we have

$$G = \frac{\gamma}{2 \, w\rho \sin \phi} \quad \cdot \quad \cdot \quad \cdot \quad \cdot \quad \cdot \quad \cdot \quad \cdot \quad [158]$$

The effect of trees, buildings, &c. is such as to reduce the wind velocity at the surface of the earth below its geostrophical value—the surface "friction" resulting in a decrease of the geostrophic acceleration. The effect of pressure gradient then predominates, and the surface wind has a reduced value and moves in a direction somewhat inclined to the isobars towards the side of low pressure. Over the sea the effects of surface friction are not so pronounced and the deviation from geostrophic conditions is accordingly small. With increasing height the wind will approach its geostrophic value as given by equation 158 providing the isobars show only slight curvature (see section 14.12).

For a given latitude ϕ, G is directly proportional to the pressure gradient γ, and hence inversely proportional to the distance apart of the isobars. This fact is used in the construction of geostrophic wind scales* by means of which close estimates of the upper wind values may be obtained when conditions are not favourable for their direct determination. It should also be noted that G is inversely proportional to $\sin \phi$, and hence for a given gradient the geostrophic wind velocity increases as the equator is approached. At the equator ($\sin \phi = 0$) the value of G given by equation 158 becomes infinitely large. This is manifestly impossible, and in practice the wind scale is not used for latitudes less than $10°$.

14.14   Gradient Wind

In calculating the geostrophic wind it has been assumed that the air is following a straight path. Very often however this is by no means the case, as for example round a depression (cyclone) or a region of high pressure (anticyclone). In these cases the air is subject to a centripetal

---

* See *Meteorology for Aviators* R. C. Sutcliffe. H.M.S.O

acceleration directed at right angles to the isobars to the concave side. This acceleration, known as the *cyclostrophic component,* will be the resultant of the geostrophic acceleration and the acceleration due to pressure gradient. Thus for isobars with a radius of curvature r, for a wind velocity V, we have

$$\frac{V^2}{r} = \frac{\gamma}{\rho} - 2Vw \sin \phi \text{ (cyclonic motion)}$$

$$\text{or } \frac{V^2}{r} = 2Vw \sin \phi - \frac{\gamma}{\rho} \text{ (anticyclonic motion)}.$$

These equations may be combined as

$$2Vw \sin \phi \pm \frac{V^2}{r} = \frac{\gamma}{\rho}$$

taking the positive or negative sign according as to whether the motion is cyclonic or anticyclonic. This equation is known as the *gradient wind equation* the solution of which provides a value of V known as the *gradient wind.*

In middle and high latitudes it is generally assumed that the cyclostrophic component is negligible compared with the geostrophic component of the acceleration, in which case the geostrophic wind may be taken as a sufficiently close approximation of the true wind. In the tropics however violent circular storms of small diameter (hurricanes) are not infrequent, and in these circumstances the cyclostrophic component cannot be ignored, and indeed often far outweighs the geostrophic component. In such circumstances the cyclostrophic wind velocity will be given by

$$\frac{V^2}{r} = \frac{\gamma}{\rho}$$

$$\text{or } V = \sqrt{\frac{\gamma r}{\rho}}$$

14.15 Variation of Wind with Height—Thermal Wind

In the layers of the atmosphere immediately above the earth's surface the effect of surface friction produces a reduction in the wind force and causes it to back as already mentioned. These disturbing effects extend to a height of about 1,500ft. above which the "free air" wind velocity is given to a sufficiently close approximation by its geostrophic value corrected where necessary for the cyclostrophic component, and for this purpose an isobaric map prepared at sea level may be used.

For the determination of winds at great heights however it would be necessary to use maps giving the pressure distributions at those heights which show variations from the distribution at sea level. It will be seen from equation 155, section 14.05, that the decrease in pressure with height is greater over a region of low temperature than over a higher temperature region, and thus a depression tends to develop above a cold region and an anticyclone above a warm region. In consequence of this alteration in the pressure distribution due to vertical variation of temperature, an additional component must be added to the geostrophic wind to obtain the true wind at high levels. This component is called the *thermal wind.*

Under average conditions the temperatures in the troposphere decrease from the tropics to the poles resulting in a depression forming over the polar regions. There is thus a thermal wind in the northern hermisphere circulating from west to east (opposite direction for southern hemisphere), and this, superimposed on the geostrophic wind will cause all winds to become increasingly westerly with height. Thus in the upper levels surface westerlies become stronger whilst easterlies weaken (after an initial increase due to the reduction of surface friction effects) and above about 3,000ft. are replaced by westerlies. It should be realised that these are very general rules however which are often upset by the unequal heating of the surface areas (*e.g.*, over land and sea), and do not apply in the tropics where the gradient wind equation breaks down.

14.16   Winds of Local Origin

We shall now briefly consider certain winds which do not arise as a result of the general pressure distribution and thus do not obey Buys Ballot's law. These winds owe their origin to special circumstances related to local topography, and the main type of such winds are dealt with below:

(*a*) *Föhn wind.* This is a warm dry wind which blows down the leeward side of a mountain range. On meeting the mountain barrier the wind is forced up, condensation taking place as a result of the cooling produced. The continued development of cloud is often followed by the precipitation of orographic rain (see section 14.17). The ascending air on the windward side may be considered to cool at the wet adiabatic lapse rate of 2·7°F. per 1,000ft., and the air that descends on the leeward side, having lost most of its moisture will warm at the dry adiabatic rate of 5·4°F. per 1,000ft. and reaches the valleys as a warm dry wind. The name of the wind originated in the Alps where the föhn is prevalent, another important example being the Chinook of the Rocky Mountains.

(*b*) *Anabatic wind.* This is a wind which blows up a mountain slope or hillside as a result of surface heating. During the day the air on the hillside is warmed by contact with the surface and thus acquires a temperature in excess of the air at the same level over the valleys. The warmed air being lighter tends to rise and an upcurrent of cooler air from the valleys flows up the hillside to take its place producing an anabatic wind. These winds are of no great strength and are generally considerably masked by irregular convection effects.

(c) *Katabatic wind*. This is a reverse effect to that described under (b) above and occurs as a result of surface cooling on sloping ground by radiation to a clear sky at night. The air in contact with the ground eventually acquires a temperature lower than that over the valley at the same height, and as a result of the decrease in density the cold air flows down the slope into the valley giving rise to the katabatic wind. Since the effect develops under conditions of inversion, irregular convection is absent and the katabatic wind, although shallow, is often quite strong. The cold air accumulates in the low lying ground where frequently fog and mist develop.

(d) *Land and sea breezes*. These are due to the unequal heating of the land and sea by the sun's rays during the day, and to the different cooling rates during the night. In the daytime the air over the land becomes warmer than that over the sea, and as a result of the lighter air above it the pressure over the land is reduced below that over the sea. In consequence of this pressure difference there is a surface wind from sea to land. At night, the earth being colder than the sea, the effects are reversed and a land breeze blows off shore.

The sea breeze usually sets in during the forenoon and may continue until the evening. In the British Isles it rarely exceeds force 3, but its strength increases in lower latitudes. The land breeze, which may not set in until midnight or the early morning, is generally less developed than the sea breeze.

14.17 Water Vapour in the Atmosphere—Clouds

The presence of water vapour in the atmosphere results from the continued evaporation from free water surfaces provided by the sea, rivers, lakes, &c. The actual amount of water vapour present at a given level in the atmosphere at any given time is defined in terms of the relative humidity (see section 9.21) and the prevailing air temperature. If this temperature should fall below the dew point the excess moisture will condense out in the atmosphere in the form of minute water droplets (or even tiny ice particles if the temperature is low enough) thus producing cloud, and it may be precipitated back to the earth. Cases of precipitation will be discussed in a subsequent section, and we shall concern ourselves here with the details of cloud formation only.

In the absence of suitable nuclei to promote condensation (see section 9.27) a considerable degree of supersaturation would develop in the atmosphere. This however nevers occurs to any marked extent as the presence of minute crystals of water soluble salts in the atmosphere provides the means whereby the initial stages of drop formation are overcome. Condensation may be brought about by cooling (a) due to loss of heat by radiation, or (b) by adiabatic means due to upward movement. Strictly speaking the former method of cooling, as exemplified by the cooling of the surface layers by radiation from the earth on a clear night, results only in the formation of dew, or, with slight turbulence, fog (radiation fog). Accordingly, we may consider cloud formation as due solely to the process of vertical motion of the air. Unsaturated air as it ascends will cool at approximately the dry adiabatic lapse rate until the dew point is reached when condensation will begin. The difference in cloud types is accounted for by the difference in the manner of ascent which may be brought about as follows:

(i) By turbulent motion in the surface layers (dynamical turbulence),

(ii) by local orographical effects,

(iii) by large scale convection as a result of surface heating,

(iv) by the interaction of air masses (this will be dealt with in section 14.21).

The clouds thus formed are infinite in variety, but it is possible to classify them in ten main types belonging to four main groups as given in table 32.

TABLE 32
Cloud classification*

Group	High clouds†	Medium clouds	Low clouds	Clouds of great vertical development (Heap clouds)
Type	Cirrus (Ci) Cirro-cumulus (Cc) Cirro-stratus (Cs)	Alto-cumulus (Ac) Alto-stratus (As)	Strato-cumulus (Sc) Stratus (St) Nimbo-stratus (Ns)	Cumulus (Cu) Cumulo-nimbus (Cb)
Mean levels	Above 20,000ft.	7,000–20,000ft.	Ground level–7,000ft.	1,500ft. to cirrus level

† These clouds are formed of ice crystals as is evidenced by the production of halos when the sun or moon are viewed through them.

Clouds formed by dynamical turbulence ((i) above) are usually of the stratus type and their formation demands a high degree of humidity so that condensation is effected at a comparatively low level, together with marked turbulence (as on a boisterous day) in order to lift the air to the necessary height. In addition there should be a stable lapse rate above the cloud as otherwise heap clouds would result. There

---

* A fuller description of the various types of cloud together with photographic illustrations will be found in the H.M.S.O. publications listed in the bibliography at the end of this book.

is little or no rain associated with clouds originating as a result of dynamical turbulence since they do not usually possess the necessary thickness (see section 14.16).

The development of orographic cloud has already been discussed (section 14.16). The cloud so formed is of the stratus type of no great vertical thickness and with a flat base which clearly defines the condensation level of the uprising current of air. If rain falls on the windward side of the hill or mountain range the descending air becomes clear of cloud at a higher level on the leeward side as indicated in Fig. 187. It

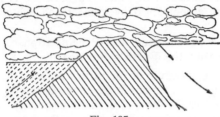

Fig. 187

sometimes happens that the rising air does not reach saturation at any stage of its ascent, but that condensation is brought about in a layer of air at some distance above the hill as a result of orographic lifting. The cloud thus produced appears stationary, being formed as the air rises, and melting away as the air descends on the leeward side of the hill. In shape it resembles a lens (see Fig. 188), and is thus called *lenticular* cloud.

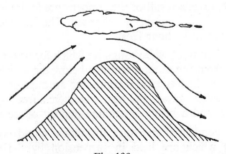

Fig. 188

Instability in the lower layers of the atmosphere formed by heating of the earth's surface (or by other means) will set up large scale convection currents with the formation of cumulus cloud the development of which will depend on the conditions prevailing in the higher layers of the atmosphere. Thus if the lapse rate for saturated air above the cloud base is stable there will be little or no vertical development and a layer of alto-cumulus cloud will result which may spread and cover the sky. If however the lapse rate is only just stable the initial momentum of the air may be sufficient to carry it 2,000 to 3,000ft. beyond the condensation level, and the cloud will have a more marked vertical development. Such clouds are often seen forming on a summer's day and are usually referred to as "fair weather cumulus," and apart from light local showers are not accompanied by rain. On the other hand if there is a steep lapse rate in the atmosphere (condition of instability), the ascending air will continue to rise at great speed with the formation of clouds of very large vertical development of the cumulus or cumulonimbus type. The marked fall in temperature through the cloud results in a copious condensation of water vapour which is precipitated as heavy rain often accompanied by thunder and lightning.

14.18   Direct Condensation and Precipitation of Atmospheric Moisture on the Earth's Surface

Dew. If the temperature of the earth's surface falls below that of the dew point the water vapour in the atmosphere is deposited directly on to the ground in the form of dew. As mentioned in the previous section this occurs on a clear still night when the earth is losing its heat by radiation to the sky. The most marked dew deposits occur on objects of relatively poor thermal conductivity, which are therefore not able to replace the heat lost by radiation by drawing upon the reserve of heat in the ground. For this reason grasses become heavily coated with dew, and loose stones more so than rocks, &c.

Hoar frost. This forms either when the temperature of the dew point is below freezing point, in which case the air's moisture condenses directly into ice, or as a result of the temperature falling below freezing point after dew has already been deposited. The ice layer resulting in either case is known as hoar frost.

Rime. Is a deposit of white rough ice which forms on the windward side of objects from fog at temperatures below the freezing point. The liquid droplets of which the fog is composed are supercooled, but when blown by the wind against railings, trees, &c., freezing takes place instantaneously, and a heavy crystalline deposit of ice results.

Rain. The condensed moisture forming a cloud consists of tiny water droplets which are held in suspension by the ascending air currents in the cloud. If the cloud droplets grow by coagulation these upward currents may not be sufficient to support them, and the moisture is precipitated to the earth as rain. Now it is known that under still air conditions water droplets will fall with a terminal velocity which increases with the diameter of the drop. For a drop of diameter 0·1mm. the terminal velocity is 0·32 metres per second, and this increases to a maximum terminal velocity of 8 metres per second for drops of about 4·5mm. in diameter. Thus no rain will fall if the upward currents of air have a velocity exceeding 8 metres per second. Droplets having diameters less than 0·1mm. quickly evaporate in falling from the condensation level through air layers of lower humidity, and this represents an approximate lower limit to the size of rain drops. The upper limit of size is about 5·5mm., as drops with diameters in excess of this become deformed when falling with the above maximum terminal velocity and finally break up.

It is interesting to note that in middle latitudes heavy rain falls only from clouds with tops above the freezing level, drizzle or very light rain falling from other clouds which do not reach this level. This has brought forward the suggestion that large raindrops may initially start as ice crystals which continue to grow as they fall through layers of supercooled droplets, and ultimately melt on passing through the warmer air layers below the cloud.

Snow and sleet. If the temperature is below the freezing point when condensation takes place the water vapour passes directly into the solid state forming a cloud of minute ice crystals. As a result of further condensation these grow into feathery crystalline structures which fall as snow if the temperature of the intervening air layers is at or below the freezing point. With higher temperatures the snow melts and is deposited as rain or, if only partially melted, as sleet.

Hail. Although snow is the normal form of frozen precipitation, conditions in certain clouds favour the formation of ice pellets often quite large in size known as hailstones. In most big clouds there exists above the lower layer of condensed water droplets a supercooled region in which the temperature may fall as far as − 20°C. without the droplets freezing. Above this there is a "snow region" which is comprised entirely of minute ice crystals. If a small droplet from the supercooled region is forced by an upward current of air into the snow region it will immediately freeze and grow by the accumulation of small particles of snow imprisoning a certain amount of air. In falling it thus commences as a ball of soft white ice, and when descending through the supercooled region it will increase further in size by gathering supercooled droplets on its surface. These freeze immediately on contact forming a coating of clear ice. The ball may now fall from the base of the cloud, or a sudden upward gust may carry it as far as the snow region again for the whole process to be repeated. It may be carried up and down the cloud in this way several times, and when finally deposited will show a record of its history by the number of alternate shells of white and clear ice. The formation of large hailstones demands ascending currents of considerable force, and the cloud must possess very great vertical development. Hail is thus typical of instability showers associated with heavy cumulus cloud, and is frequently an accompaniment of thunderstorms.

Glazed frost. A necessary condition for the formation of glazed frost is an inversion of temperature over the ground. It can be formed either as a result of warm moist air passing over very cold ground, in which case the moisture is deposited directly on the ground as ice, or by raindrops falling on ground which is below the freezing point. The result in either case is a hard layer of glassy ice which constitutes glazed frost.

Synoptic Meteorology

14.19 Air Masses—the Polar Front

The term "air mass" is given to air which has substantially the same characteristics over a wide area. The existence of such homogeneous masses of air is clearly revealed from a detailed consideration of synoptic charts, and they play an all important part in determining the weather sequence of the areas over which they pass. The horizontal dimensions of these masses are normally hundreds, or even thousands of miles, and although they are capable of wide variation in properties, it is usual to classify them for practical convenience into two main types according to their origin. Thus air originating in polar regions is referred to as a "polar air mass," whilst that proceeding from tropical regions is known as a "tropical air mass." Polar air is cold and dry near the surface, and as it travels southwards it becomes heated from below. This gives rise to unstable conditions producing gusts and squalls and broken cumulus cloud with some showers and good visibility. Tropical air on the other hand is warm and moist, and as it travels northwards over progressively cooler surfaces, stable conditions result. Atmospheric conditions depend on the amount of moisture present, and may vary from fine and clear skies to extensive low cloud with fog and drizzle. With both types of air mass it is further usual to consider whether the air was of maritime or continental origin. If of the former type there is in each case a higher humidity ratio which leads to more marked cloud development and showery conditions.

Generally speaking polar air travels from north-east towards south-west whilst tropical air travels in the opposite direction. Between the two air streams there will be a surface extending for hundreds of miles across which there will be marked discontinuities of humidity and temperature. This surface separating the two air masses is called the *polar front,* and it is on this front that the majority of depressions in middle latitudes develop. We shall now proceed to trace the origin and life history of such a polar front depression.

14.20 The Polar Front Depression

The modern theory of depressions was formulated by the Norwegian meteorologists V. and J. Bjerknes during the 1914–18 war as a result of accurate observations taken from a number of closely linked observing stations in their own country. While there is no doubt that the majority of depressions do form in the manner to be described it should be emphasised that the account put forward is more a detailed description than an actual explanation of the origin and subsequent history

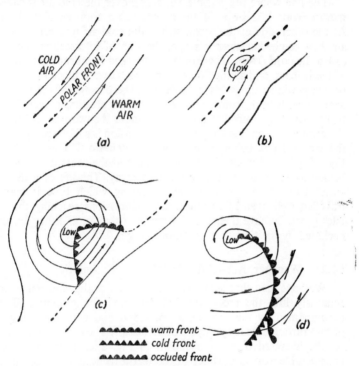

Fig. 189.   Life history of a depression

of the depression. This can be followed by reference to Fig. 189(*a*) which shows the existence of the polar front in the simplest case with the cold and warm currents flowing in opposite directions (approximately south-west and north-east respectively). This arrangement is evidently an unstable one and waves tend to develop in the front as shown in Fig. 189(*b*), where a bulge of warm air intrudes into the polar air producing a distortion in the front. The pressure falls at the top of the bulge, which thus becomes the centre of the depression and the isobars are distorted and assume a roughly oval form. As the bulge of warm air (known as the *warm sector*) increases a well marked depression develops as shown in Fig. 189(*c*), the pressure at the centre falling and the winds becoming stronger. The general distribution of the isobars is as indicated; they run almost parallel in the warm sector and are close together, the winds thus being strong here. It is found that the centre of the depression moves in a direction approximately parallel to the isobars in the warm sector with a variable speed which, in the vicinity of the British Isles, is usually between 20 and 30 miles per hour.

The line where the warm air is rising over the cold air is called the *warm front,* the slope of the ascent being approximately 1 in 150. At the other side of the warm sector the cold air undercuts the warm air at a somewhat steeper slope, the line of intersection there being called the *cold front.* The fronts are marked by abrupt discontinuities of the isobars, and as the depression passes over a place there will thus be sharp wind changes associated with the passage of the fronts. The changes with both fronts will be such as to produce a veer. The cold front ultimately overtakes the warm front as the warm air is lifted from the ground. When this happens the depression is said to be "occluded," the single front being known as an *occlusion.* This is shown in Fig. 189(*d*), and when this stage is reached the pressure at the centre begins to rise and the depression dies away. The life of a depression varies from 2 to 5 days or more during which time it may have travelled well over 1,000 miles. The depressions reaching the British Isles generally have their origin over the Atlantic, and they are usually occluded by the time they reach our shores.

14.21   Weather Associated with a Depression

We have seen that as a depression develops the cold polar air continuously lifts the moist tropical air from the ground and this is accompanied by the production of cloud and with the precipitation of rain through the lower layers of colder air. The rain areas are marked on the idealised depression shown in Fig. 190. They extend over a considerable area in advance of the warm front, and are confined to a more restricted area behind the cold front. The lower diagram gives a vertical section through the depression along the line *XX* and shows the cloud formation at various points. Thus as a depression approaches a place there will be a succession of layer clouds from very high cirrus down to nimbo stratus heralding the advance of the warm front. Precipitation of rain may occur from the higher layers of alto stratus cloud, but this often evaporates before reaching the ground. As the cloud base lowers the precipitation increases and becomes continuous until the front has passed. A sharp rise in temperature accompanies the passage of the warm front, and the cloud tends to lift and break in the warm sector. The weather here is characteristic of warm moist tropical air and there may be some light rain.

The advent of the cold front is recognised by a sharp wind veer and marked fall in temperature. The undercutting cold air forces the warm air up at a vigorous rate with the development of towering cumulus cloud accompanied by heavy showers of rain. The retardation of the

lower layers by surface friction often causes the cold air above to over-
hang the front. This results in conditions of great instability with strong
convectional currents giving intermittent heavy squalls of wind and

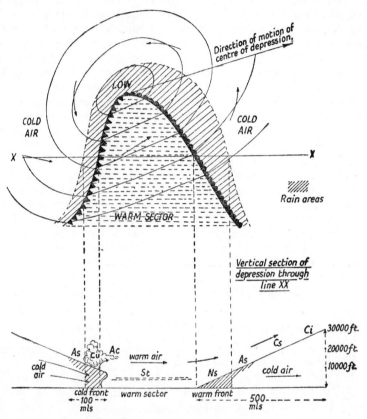

Fig. 190.   Depression showing rain areas

rain. When these effects accompany the cold front it is referred to as a
line squall. The rain at the cold front is confined to a relatively narrow
belt, and with the passage of the front the weather takes on the characteristics of polar air with good visibility, cumulus cloud, and instability
showers.

At the occlusion of a depression the warm sector disappears as
the cold front overtakes the warm front. The occluded front will have
the characteristics associated with a cold or warm front according as
to whether the air behind the advancing cold front is relatively colder
or warmer than the air in advance of the warm front.

O•

14.22 The Anticyclone

The anticyclone is a region of high pressure and is represented on the weather map by a series of closed isobars with the pressure decreasing outwards from the centre. The winds blow clockwise round an anticyclone in the northern hemisphere and anticlockwise in the southern hemisphere (in conformity with Buys-Ballot's law), and near the surface they tend to blow outwards across the isobars from the central high pressure region on account of surface friction. Anticyclones usually cover a very large area and the isobars are widely spaced. Due to the small pressure gradient the accompanying winds are accordingly light, and the central region is often one of calm or of very light winds. In contrast with the depression, the anticyclone is a region of *descending* air; high and medium cloud are dissolved in consequence of the adiabatic heating, and thus there is no precipitation and the weather is fair with clear skies. This weather is typical of the summer anticyclone, but with conditions favourable for radiation, the nights are often cool, resulting in the formation of fog or mist which may persist well into the morning. In winter the weather associated with an anticyclone is very varied, often being dull and gloomy ("anticyclonic gloom"). These conditions arise from a layer of strato-cumulus cloud covering the sky, which develops in consequence of an inversion of temperature at heights of about 2,000 to 3,000ft.

The movements of anticyclones are usually slow and irregular. After formation they often remain almost stationary for periods of several days.

14.23 The Weather Map

The accumulated meteorological observations taken simultaneously at a number of observing stations can best be appraised and interpreted when plotted on a map of the area under investigation. Such a weather map, or synoptic chart as it is frequently called, forms the basis of modern weather forecasting. The map shows the weather characteristics over the area at the time of compilation, and a study of a sequence of such maps yields information about the movement of the pressure systems, and a skilled forecaster can give details of the impending weather. In the British Isles data for the preparation of synoptic charts is available every three hours starting at 01·00 hours G.M.T. The most important feature of the chart is the visual representation of the pressure distribution by a system of isobars which are drawn by a process of interpolation from the data supplied from the observing stations. The meteorological information from each station is shown on the chart by a set of code numbers and symbols indicating the pressure (reduced to mean sea level), barometric change in the past three hours, temperature (°F.), humidity (or dew point), visibility, cloud (amount and type), wind force and direction, and present and past weather. Details of the scheme of arrangement of this information around the position of the station and of the code systems used can be obtained from Air Ministry Meteorological Office Form 2459.* A typical weather map (simplified) showing the isobaric distribution, surface temperatures (°F.), and wind data is given in Fig. 191.

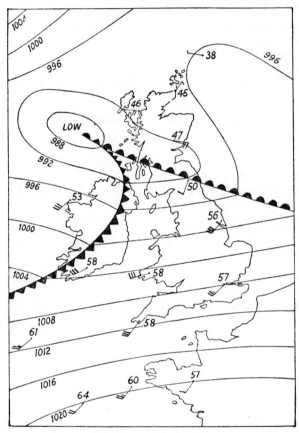

Fig. 191. The depression of November 12, 1947

QUESTIONS. CHAPTER 14

1. Write a short account of the constitution of the earth's atmosphere.

2. Explain the method of measuring altitude by means of a barometer.

Show that the difference in altitude (z) between two places on a mountain may be given by an expression of the form

$$z = RT \,(\log_e h_2 - \log_e h_1)$$

where h_2 and h_1 are barometric readings of the places, T is the temperature of the air between them, and R is a constant. Upon what factors does the value of R depend? (Camb. Schol.)

---

* Instructions for the Preparation of Weather Maps with Tables of Specifications and Symbols H.M.S.O.

H.T.–14+

3   Prove that, provided the force of gravity does not vary with height, the ratio of the atmospheric pressure at height h to that at the earth's surface is $e^{-\frac{gh}{RT_o}}$ when the temperature is assumed to have the constant value T_o. In actual fact the ratio is more nearly given by the expression

$$\left(1 - \frac{gh}{RT_o} \cdot \frac{\gamma - 1}{\gamma}\right)^{\frac{\gamma}{\gamma - 1}}$$

where T_o is the temperature at the surface of the earth and γ the ratio of the two specific heats of air. Can you suggest any explanation of this fact?
    (R is the gas constant per unit mass).   (Oxford Schol.)

4.   Define *dry abiabatic lapse rate.*
    Give a critical account of the vertical stability of the atmosphere indicating the general weather characteristics associated with (a) stable, (b) unstable conditions.

5.   What is meant by "potential temperature"?
    Describe the tephigram and explain its use in weather forecasting.

6.   Write explanatory accounts of the following:—(a) conditional instability, (b) sea breeze, (c) Foehn principle.   (Birm., B.Sc.)

7.   What is meant by (a) geostrophic wind, (b) thermal wind?
    Discuss the variation of wind with height.

8.   Draw up a classification of clouds. Explain the importance of clouds in forecasting weather phenomena.   (Birm., B.Sc.)

9.   What is meant by the dew point? How can it be measured? Under what conditions do hoar frosts occur?
    Discuss the effects produced by the presence of water in the atmosphere pointing out the physical causes that produce clouds, rain and hail-stones, and the part played by latent heat in these phenomena. What is the cause of a "cloud-burst"?
                    (Manch. Inter.)

10.   Give an account of the polar front theory of depressions and describe the weather changes associated with the passage of a depression over a place.

SOME MATHEMATICAL AND PHYSICAL CONSTANTS

$\pi = 3.1416$ $\qquad\qquad$ $e = 2.71828$

$\pi^2 = 9.8696$ $\qquad\qquad$ $\log_{10} e = 0.4243$

$\dfrac{1}{\pi} = 0.3183$ $\qquad\qquad$ $\log_e 10 = 2.3026$

$\sqrt{\pi} = 1.7725$ $\qquad\qquad$ $\log_e N = 2.3026 \log_{10} N$

$\log_{10} \pi = 0.4971$

Density of mercury (at 0°C. and 76cm. pressure) $= 13.595$gm. per c.c.

Acceleration of gravity (at sea level in latitude 45°) $= 980.616$cm. per sec.^{2}
$\qquad\qquad\qquad\qquad\qquad\qquad\qquad\qquad\quad = 32.1725$ft. per sec.^{2}

Gas constant $(R) = 8.314 \times 10^7$ ergs. per gm. molecule per °C.

Mechanical Equivalent of Heat $(J) = 4.1852$ joules per (15°) calorie.

Some useful conversion factors:

$\qquad$ 1 inch $\quad = \quad$ 2.540cm.

$\qquad$ 1lb. $\qquad = \quad$ 453.6gm.

$\qquad$ 1ft.-lb. $\quad = \quad$ 1.356 joules

$\qquad$ 1 H.P. $\quad = \quad$ 746 watts (1 watt $=$ 1 joule per sec.)

$\qquad$ 1 B.Th.U. $=\quad$ 252 cal.

$\qquad$ 1 pint $\quad = \quad$ 568 c.c.

$\qquad$ 1 radian $\quad = \quad$ 57.3°

$\qquad$ 1° $\qquad = \quad$ 0.00175 radian

ANSWERS TO NUMERICAL QUESTIONS

CHAPTER I (p. 16)

(4) 830mm. of mercury.　(5) 39°C.　(7) 21·7°C.　(9) 80·1cm. of mercury.

(12) (i) 46·8°C. (ii) 25°C.　(13) $t_c = 100 \left(\dfrac{\log \dfrac{X_c}{X_i}}{\log \dfrac{X_s}{X_i}} \right)$ where X_i, X_s are the values of X

at the ice, steam points respectively.　(17) 151·7°.

CHAPTER II (p. 52)

(1) 10·9°C.　(2) 188°C.　(3) 0·033.　(4) 0·031, 31·31 cal. per °C.　(5) 14·00 cal. per gm.　(6) 0·5415.　(7) 15·5°K.　(8) Exothermic reaction with 5,000 cals. of heat released. Specific heat of liquid = 0·5 cal. per gm. per °C.　(9) 86,000 cals.　(10) 12gm.　(11) 1·685cm.　(12) 0·917 gm. per c.c.　(13) 13·6cm.　(15) $V + \dfrac{Wt\alpha}{J\varrho\sigma}$.
(16) 0·0727.　(17) 202·3sec.　(18) 6·26 cal. per gm.　(19) 0·425 cal. per gm. per °C.; 31·3 cal. per °C.　(20) 9·03km.　(21) 68cm.　(22) 167 joules; 40cals.　(23) 2hrs.
(24) 0·514 cal. per gm. per °C.　(25) 43·6secs.　(26) 0·37 cal. per gm. per °C.; 29·8 cal. per gm.　(27) 26°C.　(34) 15·18min.　(29) 840 watts.　(30) 40·5 cal. per gm.
(31) (a) 88 cal. per min., (b) 1·36°C. per min.　(32) 79·3mins.

CHAPTER III (p. 71)

(1) 29·1°C.　(2) 139·2°C.　(3) Brass rod 2ft., iron rod 3ft.　(4) 1¼mm.　(5) 0·0424 sq. cm.　(6) 40secs.　(8) 11·7 × 10⁻⁶ per °C.; 13·8 × 10⁻⁶ per °C.　(9) 2·592 secs. lost per day.　(11) 0·0093 per cent.　(12) $\frac{7}{83}$ or approx. $\frac{1}{12}$.　(13) 3·202 Kgm. wt.　(14) 3·016 × 10⁷ dynes.　(15) 4·8 × 10⁶ dynes; 2·88 × 10⁴ ergs.　(16) 193·9cm.

(17) Clock loses at rate of 0·0049 per cent.　(18) $\dfrac{\alpha a L^2}{3}$.

CHAPTER IV (p. 89)

(2) 0·000237 sq. cm., 0·044c.c.　(3) $\dfrac{r_1 \text{ (mercury thermometer)}}{r_2 \text{ (alcohol thermometer)}} = \dfrac{1}{2}$.　(4) (a) 0·9 per cent. increase; (b) 0·0016 per cent. decrease.　(5) 0·0217 sq. mm.　(6) 12·4cm.
(7) 140·4°C.　(8) 116·0 cm. of mercury; 0·775c.c.　(9) 5·22gm.; 0·0006 per °C.
(10) (a) 1·0546:1, (b) 0·00104 per °C.　(11) 0·000196 per °C.　(13) 1·14:1.　(14) 82·5 gm. wt.; decreases by 0·135 gm. wt.　(15) 29·3°C.　(16) 82·4 × 10⁻⁵ per °C.
(17) 0·00028cm.　(18) 74·606cm.　(19) $l_0(1 + \alpha t)$; $l_0[1 + (\alpha - 3\lambda)t]$;
$l_0 \dfrac{1 + (\alpha - 3\lambda)t}{1 + \lambda t}$.　(20) 762·43mm. of mercury.

CHAPTER V (p. 109)

(1) 25·41cm.　(2) 1·366 × 10⁸ ergs.　(3) 758·8mm.　(4) 76·75cm.　(5) 26 complete strokes.　(6) 26·7cm. of mercury.　(8) 14 litres.　(10) Top of tube 131·64ft.

below surface.   (11) 95·85 cu. ft.   (12) 0·006°C.   (13) 0·248.   (14) 0·718 atmos.
(15) 0·003658 per °C.   (16) 0·204 joules per gm. per °C.   (17) 2·62 × 10⁻³.
2·67 × 10⁻⁷cm. of mercury.

CHAPTER VI (p. 129)

(2) 150·6°F.   (3) 780·7.   (5) 15 per cent.   (6) 13min. 56sec.   (7) 0·0335 B.Th.U.
per min.   (8) 11·8lb.   (9) 93·4gm.   (10) 20°C.; 20·23°C.   (11) (a) 6·378 × 10¹⁰ergs;
(b) 3·384 × 10⁶ dyne-cm.   (12) 3·312 × 10⁴ ft.-lb. wt. per min.   (13) 10·4 H.P.
(effective).   (14) 1·453 gm. per sec.   (15) 85·6 metres.   (16) 0·5425 cal. per gm.
per °C.; 0·57°C. per min.   (17) 4·15 × 10⁷ ergs. per cal.   (18) 3·922 × 10⁷ ergs. per
cal.   (19) 0·423 cal. per gm. per °C.   (20) 4·16 joules per cal.   (21) 238 gm. per sec.
(22) 50·6gm.   (24) 17·95 gm. per sec.   (25) 4·88°C.   (26) 4·168 × 10⁷ ergs. per cal.

CHAPTER VII (p. 156)

(1) 23·27 atmos.   (3) 8·31 × 10⁷ ergs. per mole per °C.   (4) (a) 0·508 atmos.;
−43°C., (b) 393·7c.c.   (5) 457·2c.c.; 379·2c.c.   (6) 0·6391$mR'T$ ergs;

$$\frac{mR'T}{\gamma - 1}\left(1 - \frac{1}{2^{\gamma-1}}\right) \text{ ergs,}$$

where m is the mass of the gas in gm., R' the gas constant per gm. of the gas, $T°$(abs.)
the initial temperature of the gas.   (7) −52·5°C.; 38cm. of mercury.   (8) 166°C.;
69·82 lb.-wt. per sq. in.   (9) (a) 548·3°K.; 9·519 atmos.; (b) 6·866 × 10⁹ ergs.
(10) 7·23 × 10³mm. of mercury; 247°C.; 2·29 × 10⁸ ergs.   (11) 0·246 cal. per gm.
per °C.   (12) 1·41.   (13) 99·9cm. of mercury, 105·5°C.; 3·64 litres, 76·5°C.   (14)

14·5:1.   (15) 413°C., 9,700 joules, 5,241 joules.   (16) $\dfrac{2R}{J}$.   (17) $T\left(\dfrac{p_0}{p}\right)^{\frac{\gamma-1}{\gamma}}$;

$$\left[\frac{6V}{\pi}\left\{\left(\frac{p}{p_0}\right)^{\frac{1}{\gamma}} - 1\right\}\right]^{1/3}; \ \left(\frac{p}{p_0}\right)^{\frac{1}{\gamma}} - 1. \quad (18) \ RT_2\left\{\left(\frac{Cv}{R} + 1\right)\log_e\frac{T_2}{T_1} + \log_e\frac{P_1}{P_2}\right\}$$

(21) 1·32atmos.   (22) 670 metres.   (23) 1·44.

CHAPTER VIII (p. 188)

(1) (i) 3·553 × 10¹⁰; (ii) 3·109 × 10¹⁰.   (6) 83·7 cal. per gm.   (8) 4·732 × 10⁴ cm.
per sec.   (9) 2495°C.   (10) (a) 18·5 × 10⁴ cm. per sec.; (b) 1,270 metres per sec.;
(c) 8·4 × 10⁷ ergs per mole per °C.; (d) 1·39 × 10⁻¹⁶ ergs per °C.   (12) $\dfrac{p_1}{p_2} = \sqrt{\dfrac{T_1}{T_2}}$;

$\dfrac{p_1}{p_2} = 1$.   (13) 2·34 × 10⁴secs.   (14) 3·85 × 10⁶ dynes.   (15) 1·305 × 10⁵ cm. per
sec.; 20,070°K.

CHAPTER IX (p. 230)

(2) (a) 150·8cm.; (b) 38·0cm.   (3) 33·7cm.   (4) (a) Becomes saturated before
volume is halved; (b) 0·9cm. (S.V.P. 1·4cm.).   (5) Volume must be reduced to 20c.c.
(6) 15·83cm.   (7) (a) 430mm.; (b) 55mm.   (8) 165·62cm.   (10) 1698mm. of mercury.
(12) 37·8c.c.   (13) 75·37cm., 0·37cm. of mercury.   (14) 29·24, 50·85cm.   (15) 326mm.
(16) 1·1787gm.   (17) 9·095gm.   (18) $\dfrac{76}{(B - p)\varrho}$ litres; 0·01gm.   (19) 9,630gm.
(20) 168cm. of mercury.   (21) 1·93cm.   (22) 60 per cent.   (23) 35·23 per cent.
(24) 65c.c.   (25) 4·085gm.   (26) 1,190gm.   (27) 0·00146cm.   (29) 68 per cent.

CHAPTER X (p. 265)

(5) 87·1cm. of mercury.   (11) 1·98 cal. per mole per °C.   (12) 1·272 × 10¹¹ ergs; 1·199 × 10¹¹ ergs.

CHAPTER XI (p. 296)

(2) 166⅔:1.   (3) 15·4cm.   (4) 0·001625 cal. per sq. cm. per sec.   (5) 48·9°C.; 135·8°C.   (6) 1·124 × 10⁻⁴ c.g.s. units.   (7) 366·3°C.   (8) 117·6cals.   (9) 12·6 per cent.   (10) 2,776 cals per sec.   (11) Temp. of junction, 40°C.   (12) 3·6 × 10⁶cals.   (13) 125·7°C.   (14) 17 cals. per sec.   (15) 100·5°C.   (16) 10⁻⁴ cal. per sq. cm. per sec. per °C.   (17) 190 cals. per sec.; 12·07; 11·93°C.   (18) 19·5°C., 7·0°C.; 31·25 cals. per sq. metre per sec.   (19) 57·25 gm. per min.   (20) (a) 50·58°C.; (b) 8·157 × 10⁷cals.   (21) 2·61cm.   (22) 0·0060 cal. per sq. cm. per sec. per °C.   (23) 4hr. 23min. 30sec.

(24) $T^2 = 16 - \left(\dfrac{250H}{\pi}\right) x$ where x is the distance along the bar measured from the hot end, and H is the rate at which heat is conducted along the bar; 1·885 × 10⁻² cal.

per sec.   (25) Rate of heat flow $= \dfrac{aA}{l} \log_e \dfrac{T_1}{T_2}$ where a is the constant in the relation

$k = \dfrac{a}{T}$; $\sqrt{T_1 T_2}$.   (26) 8·63 × 10⁷ years.   (27) $T_2 = \dfrac{T_1 kr + T_2 dE(r + d)}{dE(r + d) + kr}$.

CHAPTER XII (p. 330)

(3) 5·8 × 10⁻⁵ ergs per sq. cm. per sec. per °C.⁴; 972·5watts.   (4) 10·16mins.   (6) 2590°C.   (9) 2891°K.   (10) 1·366 × 10⁻¹¹ cals. per sq. cm. per. sec. per degC.⁴   (11) 5728°K.   (12) 333°K.   (13) 136·8°C.   (14) 490·5°C.   (15) 5837°K.; 1·55 × 10¹¹ watts.

CHAPTER XIII (p. 375)

(9) 83:1.   (10) 0·041 H.P.   (11) 9·23mins.   (16) 0·36°C.   (17) 0·0071°C. taking latent heat of ice as 80 cals. per gm. and $J = 4·2 × 10⁷$ ergs per cal.   (21) 0·2256 cal. per °C. gain.

INDEX TO TABLES OF PHYSICAL CONSTANTS AND DATA

INDEX